Animal Stress and Welfare During Transport and Slaughtering

Animal Stress and Welfare During Transport and Slaughtering

Guest Editor

Elbert Lambooij

MDPI

Basel • Beijing • Wuhan • Barcelona • Belgrade • Novi Sad • Cluj • Manchester

Guest Editor
Elbert Lambooij
Wageningen University & Research
Wageningen
The Netherlands
(retired)

Editorial Office
MDPI AG
Grosspeteranlage 5
4052 Basel, Switzerland

This is a reprint of the Special Issue, published open access by the journal *Animals* (ISSN 2076-2615), freely accessible at: https://www.mdpi.com/journal/animals/special_issues/98W48VWSP9.

For citation purposes, cite each article independently as indicated on the article page online and as indicated below:

Lastname, A.A.; Lastname, B.B. Article Title. *Journal Name* **Year**, *Volume Number*, Page Range.

ISBN 978-3-7258-2657-5 (Hbk)
ISBN 978-3-7258-2658-2 (PDF)
https://doi.org/10.3390/books978-3-7258-2658-2

Cover image courtesy of Elize van den Brink

Contents

animals

Editorial

Animal Stress and Welfare During Transport and Slaughtering: An Outline for Future Policies

Elbert Lambooij

Retired from Wageningen University & Research, Livestock Research, PO Box 9101, 6700 HB Wageningen, The Netherlands; bertlambooij@kpnmail.nl

Citation: Lambooij, E. Animal Stress and Welfare During Transport and Slaughtering: An Outline for Future Policies . *Animals* **2024**, *14*, 3064. https://doi.org/10.3390/ani14213064

Received: 3 October 2024
Revised: 17 October 2024
Accepted: 22 October 2024
Published: 24 October 2024

1. Introduction

There is a lot of discussion about the transport of farm animals. The main issues include welfare, stress, legislation, consumer concern, and economic aspects. The main hazards identified for transported animals are common across all species. Hazards characterized as serious are inadequate ventilation, insufficient space allowance, transport duration, lack of appropriate food and water during transport, incorrect handling during (un)loading, poor fitness prior to transport, introduction of pathogens before and during transport, and the inappropriate application of resting periods during transport (Figure 1). Loading at the farm and unloading at the slaughterhouse have been considered one of the most stressful preslaughter events [1]. It is stated in the Treaty of Amsterdam [2] that all animals should be protected from avoidable sources of disturbance, pain, or suffering during transport, lairage, restraint, stunning, slaughter, or killing. The public pressure towards the welfare of animals during transport concerns the industry, and in response, they demand the introduction of new and more strict regulations. It is also stated that the management has to set up the welfare standards and monitor them via non-invasive tools and audits [3]. Here, the present progress in handling of farmed animals during transit and slaughterhouse is described.

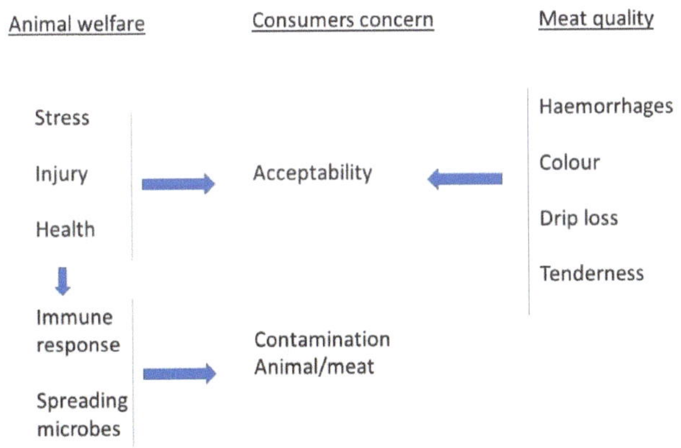

Figure 1. Welfare, concerns, and quality factors.

2. Welfare and Stress

According to the Terrestrial Animal Health Code [4], animal welfare means the physical and mental state of an animal in relation to the conditions in which it lives and dies. The five components of animal welfare are freedom from hunger and thirst, discomfort, pain and injury or disease, fear and distress, and expression of normal behavior [5].

Until now, it is thought that a lot of behavior and cognition were exclusive to humans and some primates. Recently, it is considered that animals have knowledge of their own state, deal with their own knowledge, and evaluate the physiological state of their conspecifics [6]. Moreover, individual animals may have their own conscious profile since the possible presence of different profiles varies across species [3].

Death itself is not related to welfare, i.e., an animal's welfare is not compromised by its dying. Events can influence welfare as long as the animal is alive, either at a farm or at an abattoir. Slaughter can be regarded as premature killing of healthy animals.

The adverse effects of animal welfare include several aspects, which can be assessed separately and possibly on different scales. Such components include the following: pain, fear, anxiety, frustration, behavioral disruption, malaise, discomfort [7]. These are all different aspects of welfare, which can be present at slaughter and killing, and which together, in some way, contribute to the total level of welfare [3]. Comparing values for the assessment and measurement of welfare is complicated as they are scored on different scales. The severity of these effects could be critical, severe, moderately limited, or negligible [8]. In the context of a relatively short slaughter process, some of the components seem more relevant to consider than others. The most relevant seem to be negative emotional states, like pain, distress, fear, and frustration, as well as physical injuries [9,10].

When an animal's coping strategy cannot cooperate with changes in circumstances, we call this stress. In that case, the pituitary–adrenal system (glucocorticoids) and the sympathetic–adrenal medullary system (catecholamines) are activated, which results in behavioral and clinical deviations from normal functions; the control system overloads and fitness declines [11]. Tools to assess acute and chronic stress are used to observe behavior and measure heart rate, body temperature, and blood stress metabolites. Other measurements include skin damage score, meat pH, rigor mortis, temperature, color, water-binding capacity, and contamination [7].

3. Transport

Slaughter animals are transported at least once, but sometimes more, before they arrive at a slaughterhouse. A lot of animals are transported from the breeding farm to the fattening farm, and, when finished, to the slaughterhouse, sometimes via assembly places, which means one or more extra transports. The global animal slaughter was in 2022: 140 billion chickens, 1.5 billion pigs, 625 million sheep, and 300 million cows. Moreover, 90 million tons of live-weight fish are slaughtered [12].

According to an investigation by The Guardian, more than 20 million farm animals die during road transport from the farm to the slaughterhouse in the United States. Analysis of publicly available data shows that about 20 million chickens, 330,000 pigs, and 166,000 cattle were dead on arrival or died in lairage. The main causes are heat stress, especially in summer, freezing, or trauma [13]. In the Czech Republic, an investigation on the mortality rate after road transport showed pigs 0.065%, cattle 0.027%, and sheep 0.015%. The rate in laying hens was 0.507%, for broiler chickens it was 0.425%, for geese it was 0.003%, and for rabbits it was 0.199% [14].

An alternative for the transport of animals to a slaughterhouse is a mobile abattoir for partial or full slaughter. Because the slaughterhouse is transported to the animals, this is one method to avoid death in transport. Both stress and mortality are significantly lower. Problems related to this option include legislation and economic consequences [15,16]. In short, mobile slaughter would be too expensive for large scale use.

4. Lairage

At the arrival of the slaughterhouse, animals need to be checked for signs of poor welfare at farm level during loading and transportation. For instance, questions like space allowance or animals unfit to be transported are points to be assessed in the slaughterhouse that are related to the farm of origin; others, like some injuries in the animals, can be linked with the transportation conditions; and finally, during unloading, the condition of the ramp (slippering or not) and the animals (fearful, exhausted, and panting) must be assessed. When moved to lairage, the state of animals and handling of animals by the slaughterhouse personnel must also be considered. The lairage is the period between the entry of the animals into the resting pens (after being unloaded off the truck) until they are taken out of the pen to move to the stunning point. During lairage, animals are exposed simultaneously to a variety of stressors that may result in high levels of psychological and physical stress, thus compromising their welfare. These potential stressors can include fasting, mixing of unfamiliar individuals, handling by humans, exposure to a novel environment, noise, forced physical exercise, and in some cases extremes of temperature and humidity and water deprivation [17,18].

5. Stunning

It is necessary that slaughter animals need to be well restrained for an optimal stunning. Before restraining, animals are separated individually or in a group. In both situations, stress can occur. Individual restraining includes fixation by rope at the head of the animal or positioning in a gondola, V-type/rail conveyor, or in a (rotating) box. A group of animals can be driven automatically in a gondola. Control points for restraining are the size of the restrainer in relation to the species, throughput, duration, injuries, effect on bleedout, and worker safety [19].

In many countries, laws were introduced to avoid pain and suffering in animals during the slaughter process. Stunning is applied to induce unconsciousness and insensibility for the duration of death by exsanguination [20]. In general, unconsciousness means that the brain structures do not function. Stunning and killing methods for mammal, bird, and fish species include physical methods, electronarcosis, gas induction, and chemical methods. Physical methods include head percussion, brain penetration, spiking, decapitation, live chilling, and asphyxia. When electronarcosis is used, the electrodes can be placed on the head, or on the head and body, or include the entire body. The gasses used are carbon dioxide or nitrogen, and for chemical stunning, clove oil and Aqui-S™ are used [21].

Under laboratory conditions, the brain activity during stunning and thereafter can be measured with an EEG (electroencephalogram), VERs (visual evoked responses), VEPs (visual evoked potentials), or SERs (somatosensory evoked responses). Power spectral parameters can be used to quantify the depth of anesthesia. The EEG signals can be classified in delta (<4 Hz), theta (4 to 7 Hz), alpha (8 to 13 Hz), beta (13 to 32 Hz), and gamma (32 to 45 Hz) rhythms. During the delta and theta rhythms animals are considered to be unconscious [22]. In the slaughter line, important indicators for unconsciousness and insensibility are a physical collapse, rhythmic breathing, vocalization, eye reflex, wide open and relaxed eye and pupil, and blinking. Other indicators include the cognitive threat test, righting, nose pinch, and tongue hanging out [23,24].

Key factors that should be considered during bleeding are maintenance of devices, severed vessels, wound conditions, cardiac arrest, tonic and clonic muscle activity, and orientation of the carcass. When checks indicate that the animal shows signs of consciousness, intervention needs to be applied [23,24].

6. Recommendations for Developments

The state of consciousness of farmed animals continues to be a point of discussion. Further research regarding this state may result in improvements in legislation and management tools related to farms, transport devices, and slaughterhouses.

The environment of animals during transport can be measured using electrotechnical devices. However, the control, regulation, and adaptation during transport need improvement. Software and management tools can be developed to resolve welfare issues.

Indicators for unconsciousness and insensibility in the slaughter line are just clinical signs. A combination of electronic components may result in a device that measures the state of the brain, which refers to the subjective or inner qualitive experience of an animal. When the brain's ability to integrate information is blocked or disrupted, the animal is unconscious [25].

Funding: This research received no external funding.

Conflicts of Interest: The authors declare no conflict of interest.

References

1. van Reenen, K.; Reimert, H.; Gerritzen, M.; Leenstra, F.; Lambooij, B. *Hazard Identification and Characterization of Welfare Aspects During Transport of Farm Animals*; Report 152; Animal Science Group: Wageningen, The Netherlands, 2008; Available online: http://edepot.wur.nl/121065 (accessed on 16 October 2024).
2. EFSA (European Food and Safety Authority). Welfare aspects of the main stunning and killing the main commercial species of animals. *EFSA J.* **2004**, *45*, 1–29.
3. Terlouw, C.; Bourguet, C. Quantifying animal welfare preslaughter using behavioural, physiological and carcass and meat quality measures. In *Preslaughter Handling and Slaughter of Meat Animals*; Faucitano, L., Ed.; Wageningen Academic Publication: Wageningen, The Netherlands, 2022; pp. 13–61.
4. World Organisation for Animal Health (OIE). *Terrestrial Animal Health Code*; WOAH: Paris, France, 2019; pp. 333–336.
5. Brambell, R. *Report of the Technical Committee to Enquire into the Welfare of Animals Kept Under Intensive Livestock Husbandry Systems*; H.M. Stationery Office: London, UK, 1965; 84p.
6. EFSA (European Food and Safety Authority). *Animal Consciousness*; EFSA Supporting Publications: Parma, Italy, 2017; Volume 14, pp. 1–165.
7. Faucitano, L.; Ribas, J.; Lambooij, E. Transport of pigs. In *Livestock Transport and Handling*, 6th ed.; Grandin, T.A., Ed.; CABI Publishing: Wallingford, UK, 2024; in press.
8. Blokhuis, H.J.; Miele, M.; Veissier, I.; Jones, B. *Improving Farm Animal Welfare*; Wageningen Academic Publication: Wageningen, The Netherlands, 2013; 232p.
9. Algers, B.; Anil, H.; Blokhuis, H.; Fuchs, K.; Hultgren, J.; Lambooij, B.; Nunes, T.; Paulsen, P.; Smulders, F. *Project to Develop Animal Welfare Risk Assessment Guidelines on Stunning and Killing*; (Question No EFSA-Q-2007-168) project ID CFP/EFSA/AHAW/2007/01; EFSA Supporting Publications: Parma, Italy, 2009.
10. Atkinson, S. Assessment of Cattle and Pig Welfare at Stunning in Commercial Abattoirs. Ph.D Thesis, Swedish University of Agricultural Sciences: Uppsala, Switzerland, 2016.
11. Broom, D.M.; Johnson, K.G. *Stress and Animal Welfare*; Chapman and Hall: London, UK, 1993.
12. Orzechowski, K. *Global Animal Slaughter Statistics and Charts*; Faunalytics Organization: San Diego, LA, USA, 2024.
13. Kevany, S. *Animals Farmed, Animal Welfare*; The Gardian.org: Manchester, UK, 15 June 2022.
14. Valkova, L.; Vecerek, V.; Voslarova, E.; Kaluza, M.; Takacova, D.; Brscic, M. Animal welfare during transport: Comparison of mortality during transport from farm to slaughter of different animal species and categories in the Czech Republic. *Ital. J. Anim. Sci.* **2022**, *212*, 914–923. [CrossRef]
15. Lambooij, E.; Puister-Jansen, L.; Graven, W.; Bemelman, Y. *Haalbaarheid Mobiel Slachthuis en Uitsnijderij*; Livestock Research report 460; Wageningen UR Livestock Research: Lelystad, The Nederland, 2011.
16. Hultgren, J. Avoiding live-animal transport to slaughter: Mobile abattoirs. In *Preslaughter Handling and Slaughter of Meat Animals*; Faucitano, L., Ed.; Wageningen Academic Publication: Wageningen, The Netherlands, 2022; pp. 390–434.
17. EFSA (European Food and Safety Authority). Welfare of pigs at slaughter. *EFSA J.* **2020**, *18*, 6148.
18. EFSA (European Food and Safety Authority). Welfare of cattle at slaughter. *EFSA J.* **2020**, *18*, 6275.
19. Temple, G. Practical methods to improve animal handling and restraint. In *Animal Welfare and Slaughter*; Velarde, A., Raj, M., Eds.; 5M Publishing Ltd.: Sheffield, UK, 2016; pp. 71–90.
20. Blackmore, D.K.; Delaney, M.W. *Slaughter of Stock*; Publ. No. 118. Vet. Con. Ed.; Massey University: Palmerston North, New Zealand, 1988.
21. Villarroel, M.; Lambooij, E. Fish. In *Preslaughter Handling and Slaughter of Meat Animals*; Faucitano, L., Ed.; Wageningen Academic Publication: Wageningen, The Netherlands, 2022; pp. 118–150.
22. Schomer, D.L.; Lopes da Silva, F.H. *Niedermeyer's Electroencephalography Basic Principles, Clinical Applications and Related Fields*; Wolters Kluwer: Philadelphia, PA, USA, 2012; 1296p.
23. EFSA (European Food and Safety Authority). Scientific opinion on monitoring procedures at slaughterhouses for pigs. *EFSA J.* **2013**, *11*, 3523.
24. AVMA. *Guidelines for the Humane Slaughter of Animals: 2024 Edition*; American Veterinary Medical Association: Schaumburg, IL, USA, 2024.
25. AVMA. *Guidelines for the Euthanasia of Animals: 2020 Edition*; American Veterinary Medical Association: Schaumburg, IL, USA, 2020.

animals

MDPI

Review

Importance of Knife Sharpness during Slaughter: Shariah and Kosher Perspective and Scientific Validation

Pavan Kumar [1,2], Ahmed Abubakar Abubakar [1], Jurhamid Columbres Imlan [3], Muideen Adewale Ahmed [1], Yong-Meng Goh [4], Ubedullah Kaka [5,6], Zulkifli Idrus [1,7] and Awis Qurni Sazili [6,7,*]

1 Institute of Tropical Agriculture and Food Security, Universiti Putra Malaysia, UPM, Serdang 43400, Selangor, Malaysia; pavankumar@gadvasu.in (P.K.); ahmadsadeeq7@gmail.com (A.A.A.); muideenahmed13@gmail.com (M.A.A.); zulidrus@upm.edu.my (Z.I.)

2 Department of Livestock Products Technology, College of Veterinary Science, Guru Angad Dev Veterinary and Animal Sciences University, Ludhiana 141004, Punjab, India

3 Department of Animal Science, College of Agriculture, University of Southern Mindanao, Cotabato 9407, Philippines; jurhamidimlan@yahoo.com.ph

4 Department of Preclinical Sciences, Faculty of Veterinary Medicine, Universiti Putra Malaysia, UPM, Serdang 43400, Selangor, Malaysia; ymgoh@upm.edu.my

5 Department of Companion Animal Medicine and Surgery, Faculty of Veterinary Medicine, Universiti Putra Malaysia, UPM, Serdang 43400, Selangor, Malaysia; dr_ubedkaka@upm.edu.my

6 Halal Products Research Institute, Universiti Putra Malaysia, Putra Infoport, UPM, Serdang 43400, Selangor, Malaysia

7 Department of Animal Science, Faculty of Agriculture, Universiti Putra Malaysia, UPM, Serdang 43400, Selangor, Malaysia

* Correspondence: awis@upm.edu.my

Simple Summary: A sharp knife of appropriate dimension (blade length) is very important during halal and kosher slaughtering of animals without stunning for a rapid and clean neck severance. It improves bleeding and alleviates pain and stress in animals by early onset of unconsciousness. An efficient bleed-out improves meat quality and food safety. With the ever-increasing demand for halal and kosher meat due to its awareness, authenticity, nutritive value, and animal welfare compliance, there is an urgent need to emphasize the role of knife sharpness during slaughter as per the prescribed religious practices. Other issues such as neck cut positions, blade length of the knife, proper training of slaughterhouse workers, infrastructure, and constant monitoring of the slaughtering process also need to be addressed to improve animal welfare and meat quality.

Abstract: Halal and kosher slaughter have given the utmost importance to the sharpness of knives during the slaughter of animals. A sharp knife of appropriate dimension (blade length) makes slaughter less painful during neck severance and facilitates desirable bleeding. The role of knife sharpness has not been given due credit from an animal welfare perspective and is likely ignored by the people involved in slaughterhouses. A neat, clean, and efficient neck cut by an extremely sharp knife reduces the pain. It improves the bleeding out, thus making animals unconscious early without undergoing unnecessary pain and stress. It also helps in improving meat quality and food safety. A slight incremental improvement in knife sharpness could significantly improve the animal welfare, productivity, efficiency, and safety of meat plant workers. The present review critically analyzed the significance of knife sharpness in religious slaughter by reducing stress and pain and improving meat quality and food safety. The objective quantification of knife sharpness, proper regular training of slaughterers, and slow slaughter rate are the challenges faced by the meat industry.

Keywords: religious slaughter; knife sharpness; neck cut position; slaughter skill; restraint; animal welfare

Citation: Kumar, P.; Abubakar, A.A.; Imlan, J.C.; Ahmed, M.A.; Goh, Y.-M.; Kaka, U.; Idrus, Z.; Sazili, A.Q. Importance of Knife Sharpness during Slaughter: Shariah and Kosher Perspective and Scientific Validation. *Animals* **2023**, *13*, 1751. https://doi.org/10.3390/ani13111751

Academic Editors: Gareth Pearce and Elbert Lambooij

Received: 13 April 2023
Revised: 19 May 2023
Accepted: 22 May 2023
Published: 25 May 2023

1. Introduction

The World Organization for Animal Health (WOAH) defines animal welfare as *"the physical and mental state of an animal concerning the condition in which it lives and dies"* [1]. It consists of the responses of an animal to socio-physiological factors, escapes or avoidance behavior, animal physiology, and biochemistry [2]. Each society has its interpretation of animal welfare based on its moral and ethical values and welfare standards [3]. With increasing awareness, and education, consumers prefer food not only for good nutritive quality, sustainability, and processing but also for ethical and spiritual quality. Consumers are more concerned about handling animals and management practices involved during meat production, such as stunning prior to slaughter, and free-range chicken production. The animal welfare issue has taken center stage in global meat production and marketing [4–9]. Further, a well-established link between pain or stress during slaughter and its negative impact on meat quality warrants the immediate attention of researchers and policy-makers for proper animal welfare compliance during animal slaughtering [3].

The pre-slaughter handling and slaughter process is crucial from an animal welfare perspective as it comprises converting the live animal into edible pieces of meat for human consumption. Several pre-slaughter factors cause an animal stress, fear, pain, and distress, thus compromising their welfare [10,11]. Pain is *"an unpleasant sensory and emotional experience associated with or resembling real or potential tissue injury"* [12]. Stress is a complex physiological state that comprises a range of integrative and behavioral changes in response to a real or perceived threat to homeostasis. In contrast, fear denotes a condition of danger perception or potential harm that could compromise an animal's safety [13]. Distress denotes a negative and aversive state under which the ability of an animal to cope and adapt is impaired [14]. Pre-slaughter handling practices have a significant effect on animal stress, pain, distress, and fear, such as transportation conditions, loading and unloading, stocking density, water and feed availability, attitude and training of animal handlers, and slaughtering practices (such as stunning, restraints, knife sharpness, and training of butchers). Knife characteristics, especially knife sharpness, are very crucial among these factors.

During kosher and halal slaughter, animals should be restrained by using minimum stressful methods to hold the animals for neck cuts. Gentle handling and less stressful restraints lower the issue of delayed periods of consciousness after a neck cut and improve animal welfare [15]. These considerations also affect the reaction of the animal to a throat cut [16–18], the issue of prolonged consciousness after the neck cut [19], and the perfusion of blood in the respiratory tract [20]. Tight/robust restraints cause stress—especially in cases where animals are turned on their side or back in a rotatory casting pen—substantial tissue damage, and prolonged time to reach the stage of unconsciousness, thereby feeling pain, distress, anxiety, and suffering [21]. The issue of gentle handling and less stressful restraining methods are more critical in the slaughter of cattle without stunning as compared to sheep. This could be due to the larger body size of cattle [15] and anatomical differences in the blood supply to the brain; thereby taking longer time to reach the stage of unconsciousness in cattle as compared to sheep [22,23]. The World Organization for Animal Health (WOAH, previously known as OIE) recognized fully inverted, upright, and lateral/sideway restraints for animals, with suspending only allowed in poultry [24].

Even though religious authorities (both Islam and Judaism) and sacred texts emphasize the importance of using a sharp knife and maintaining its sharpness during slaughter, there is very little scientific evidence available to corroborate this element, especially in the context of ritual slaughter (halal and kosher slaughter). Religious slaughter implies the slaughter of permissible animals by severing the trachea, esophagus, and blood vessels using a sharp blade and following laws as prescribed by rituals (blessings/invocations) that characterize its purity/ethical value [25]. Islam and Judaism have given utmost importance and zero tolerance towards animal welfare compliance during slaughter and strongly advocate benevolent and compassionate treatment of animals.

A clean neck cut at the proper position using a very sharp knife facilitates rapid blood loss and fewer incidences of false aneurysm formation, thereby leading to the early onset of unconsciousness. Specific details about sharpness and its dimensions for the *Chalaf* are mentioned for kosher slaughter. However, this very critical factor is largely overlooked in scientific literature. In this context, this paper critically reviews the significance of knife sharpness in religious animal slaughter without stunning under the broad ambit of animal welfare principles by alleviating stress and pain, this ultimately improving meat quality. The manuscript also reviewed various factors that could render knife sharpness as an important determinant of animal welfare during slaughter such as the position of the neck cut, knife dimension, training/expertise of the slaughterman, restraints, and presentation of the animal during slaughter.

2. Sharia's Perspective of Knife Sharpness

A sharp knife is recommended for halal and kosher slaughter for efficient bleeding, alleviating pain, and producing quality meat [26,27]. Islam prescribes proper guidelines for ensuring the sharpness of knives and the slaughter of an animal by complying with animal welfare principles mentioned in the Holy Quran and Hadiths.

> "Certainly Allah has decreed proficiency in all things. Thus ... if you perform slaughter (zabh), perform it well (painlessly). Let each of you sharpen his knife/blade and let him minimize suffering to the animal he slaughters (zabiha die painlessly/peacefully)."
>
> "Allah calls for mercy in everything, so be merciful when you kill and when you slaughter; sharpen your blade to relieve its pain." (Al-Qaradawi, 1994).
>
> "Allah has commanded you to treat all creatures with kindness. When you slaughter an animal, do so kindly. Sharpen the knife well and give comfort to the animal being slaughtered."

The Islamic tradition strongly advocates the humane slaughter of animals to alleviate pain and suffering. For example, *Sahih Muslim (Book 21, Chapter 11, Number 4810)* records Prophet Mohammad (PBUH) saying:

> "Verily Allah has enjoined goodness to everything; so when you kill, kill in a good way, and when you slaughter, slaughter in a good way. So every one of you should sharpen his knife and let the slaughtered animal die comfortably."
>
> "When one of you slaughters, let him complete it."
>
> حَدَّثَنَا أَبُو بَكْرِ بْنُ أَبِي شَيْبَةَ، حَدَّثَنَا إِسْمَاعِيلُ ابْنُ عُلَيَّةَ، عَنْ خَالِدٍ الْحَذَّاءِ، عَنْ أَبِي قِلاَبَةَ، عَنْ أَبِي الأَشْعَثِ، عَنْ شَدَّادِ بْنِ أَوْسٍ، قَالَ ثِنْتَانِ حَفِظْتُهُمَا عَنْ رَسُولِ اللَّهِ صلى الله عليه وسلم قَالَ إِنَّ اللَّهَ كَتَبَ الإِحْسَانَ عَلَى كُلِّ شَىْءٍ فَإِذَا قَتَلْتُمْ فَأَحْسِنُوا الْقِتْلَةَ وَإِذَا ذَبَحْتُمْ فَأَحْسِنُوا الذَّبْحَ وَلْيُحِدَّ أَحَدُكُمْ شَفْرَتَهُ فَلْيُرِحْ ذَبِيحَتَهُ

In accordance with the Prophet's (PBUH) commandment that at the point of slaughter, the knife's cutting edge must be well sharpened (Jama'ulFawa'id). In Islam, using bones, claws, teeth, nails, and alike is strictly forbidden. This implies less pain to the animals during slaughter.

> "Use everything to slaughter which allows blood to flow, except for teeth and nails, and all else is permissible" (Sahih Bukhari, p 827; Sunan Abu Dawood).

حَدَثنَا عَمْرُو بْنُ عَلِيٍ ، حَدَثنَا يَحْيَى ، حَدَثنَا سُفْيَانُ ، حَدَثنَا أَبِي ، عَنْ عَبَايَةَ بْنِ رِفَاعَةَ بْنِ رَافِعِ
بْنِ خَدِيجٍ ، عَنْ رَافِعِ بْنِ خَدِيجٍ ، قَالَ قُلْتُ يَا رَسُولَ اللَه إِنَا لاَقُو الْعَدُوِ غَدًا ، وَلَيْسَتْ مَعَنَا
مُدًى فَقَالَ اعْجَلْ أَوْ أَرِنْ مَا أَنْهَرَ الدَم وَذُكِرَ اسْمُ اللَه فَكُلْ ، لَيْسَ السِن وَالظُفرَ ، وَسَأُحَدِثك ،
أَمَا السِن فَعَظْمٌ ، وَأَمَا الظُفرُ فَمُدَى الْحَبَشَةِ وَأَصَبْنَا نَهْبَ إِبِلٍ وَغَنَمٍ فَنَدَ مِنْهَا بَعِيرٌ ، فَرَمَاهُ
رَجُلٌ بِسَهْمٍ فَحَبَسَهُ فَقَالَ رَسُولُ اللَه صلى الله عليه وسلم إِنَ لِهَذِهِ الإِبِلِ أَوَابِدَ كَأَوَابِدِ
الْوَحْشِ ، فَإِذَا غَلَبَكُمْ مِنْهَا شَىْءٌ ، فَافْعَلُوا بِهِ هَكَذَا ،

The incision should be instantaneous, with one uniform directional movement, and achieved without interruptions, uncertainty, or force [28].

Furthermore, the cut must be made from the ventral position of the neck near the lower jaw and just before the spine [28]. This is following hadith that "*the jugular veins and the carotid arteries (wadaja'an), in addition, to the throat (hulqum) and the trachea (mari'), but the vertebral or spinal cord must not be cut.*" The head must not be wholly separated from the remaining body during slaughter.

> "The knife must be razor sharp and without blemishes and damage. For animals with normal necks, the slaughter must begin with an incision on the neck just before the glottis, and for animals with long necks, such as chickens, turkeys, ostriches, camels, etc., the incision must be before the glottis".
>
> " ... must be done once only. The slaughtering implements must not be lifted off the animal during slaughtering. Any lifting is construed as one act of slaughter. Multiple acts of slaughter on one animal are prohibited".

Under Halal, acceptable animals and birds are slaughtered with a razor-sharp knife to have a swift, deep incision cutting the front of the esophagus, trachea, jugular veins, and carotid arteries [29]. In Halal slaughter, proper emphasis is given to the knife's sharpness to facilitate rapid and efficient blood drain and the onset of unconsciousness.

3. The Kosher Perspective of Knife Sharpness

Kosher slaughter/*Shechitah*/*Shechita* is the animal slaughter method followed by the Jewish community derived from a *mitzvà* (commandment) mentioned in the book of Deuteronomy 12:21

> "And ye shall be men of holy calling unto Me, and ye shall not eat any meat that is torn in the field" (Exodus XXII:30)
>
> "...thou shall kill of thy herd and of thy flocks, which the Lord hath given thee, as I have commanded thee..." (Deuteronomy XII:21). "[...] you may slaughter animals from the herds and flocks the Lord has given you, as I have commanded you, and in your towns, you may eat as much of them as you want".
>
> Rapid blood loss and maximum bleed-out is recommended in Kosher slaughter as
>
> "Only be sure that thou eat not the blood: for the blood is life" (Deuteronomy 12:23)

In Kosher meat production, three factors, viz., permitted animals, strict prohibition of blood, and mixing meat with milk, are vital considerations [30]. Wild birds and pigs are considered impure in kosher diets [31]. In Jewish law (*Halacha*), importance is given to compliance with animal welfare and avoiding pain and stress during slaughtering (Ha Levi A. 13th cent, Karo 1563d) [30]. During Shechitah, Jewish law (*Halacha*) emphasizes the suitability of a knife (sharpness and size), immobilization of animals, and neck cut (correct knowledge of anatomy and physiology, and skill [32].

In kosher slaughter, the knife's sharpness, its absence of nicks, a pre-slaughter examination of the knife's sharpness, and its size are all given careful consideration (double the width of the neck of the animal to be slaughtered). It is essential for decreasing stress and suffering during animal slaughter [33]. The sharpness and size of the knife (*Chalaf,* also known as *Chalef/Chalof/Chalif*), along with the rigorous training and inspection by Jewish authorities, ensure a proper supply of kosher meat and its production [34].

Kosher slaughter is completed in five phases, viz., selection of animal (cloven hooves and ability to ruminate as per The New International Version of the Bible 2011, Leviticus XI), health inspection of the animal, slaughter, inspection (*bedika*), and cleaning (*nikkur*) of meat, and koshering (washing and salting meat for removing blood) [27]. *Shechitah* is performed by making a clean incision at the front of the neck and cutting the trachea, esophagus, carotid arteries, and jugular veins by using a *Shechitah* knife (*Chalaf,* derived from the Hebrew word meaning 'to change') with a *Shochet* (authorized slaughter man) as per fundamental commandment, conveyed via the Oral Law and dating back to the time of Moses [18].

The *Chalaf* is designed to have exquisite sharpness and is repeatedly inspected between animals to avoid imperfections [18]. There are five principles of halachic (traditional body of Jewish law) during the *Shechitah* viz. *Shehiyah*/pause (uninterrupted incision), *Derasah*/pressure (no pressing of the blade against the neck), *Halad*/stabbing (adequate length of the blade so it does not get covered with wool, feathers, or hide), *Hagramah*/slanting (severing neck at the appropriate point for neat, clean, and efficient incision), and *Ikkur*/tearing (no tearing of tissues) [18,35]. An animal is deemed unfit if any problem is found with the knife or the neck cut [36].

Chalaf should be perfectly smooth and razor-sharp without any nicks or serrations to facilitate the slaughter process as painlessly as possible. It is twice the length of the neck of the animals going to be slaughtered (poultry: 14–16 cm, sheep and goats: 25 cm, cattle: 40–45 cm) [37]. Before and after *Shechitah, chalaf* must be inspected along with fingernails to ensure compliance with Jewish slaughter regulations during the cut, and any presence of a nick makes the meat unfit for consumption (*terefa/terefah*) or rejection [34]. In chicken slaughter, the *Chalaf* may be inspected following the killing of all animals in a group and checked for nicks, with the provision that all slaughtered birds are considered *terefah*/rejected if nicks are discovered [38].

Further, detecting a nick is tricky as a trained shochet may detect a nick that generally goes unnoticed by the trained sharpener [38]. It is mandatory to check the neck of animals to ensure the absence of any materials that may damage the knife. In case of potential knife-damaging materials such as dirt, dust, soil, etc., the animal should be washed. This process during kosher slaughter slows down the slaughtering process in addition to being a cause of stress. In religious slaughter, washing animals while entering the lairage is widely followed so that animals will have sufficient time to recover from stress [38]. However, proper care should be taken in cold climatic conditions, as it may lead to severe cold stress in animals.

4. Knife Sharpness in Animal Welfare during Slaughter

Sharpness significantly affected the forces generated and energy required during the cutting process, the cutting-edge durability, and the surface finish [39]. In the USA, Section 1902 of the Humane Methods of Livestock Slaughter Act of 1978 stipulates religious slaughter by a method in which "*the animal undergoes unconsciousness by the lack of blood to the brain due to the instantaneous severing of carotid arteries with a sharp instrument and proper handling in accordance with such slaughter*" [40].

The knife design and neck cut process were crucial in preventing the animal from reacting to the cut, ensuring rapid blood flow [28,41]. The knife's sharpness and a clean and uninterrupted cut also influence vasoconstriction, clotting, and ballooning resulting in carotid occlusion/false aneurysm [42] due to constriction of the caudal end of the severed carotid arteries. While the sharpness of the knife influences neck cutting, the method of neck cutting and the number of cuts influence pain perception [19]. A rapid and efficient cut by a razor-sharp knife is crucial for maximum blood loss and early onset of a state of unconsciousness, thus reducing the pain during the whole process. In addition to this, it also improves meat quality by reducing petechial hemorrhages. Stressed and excited animals usually take more time to become unconscious than calm animals, thus not meeting the higher animal welfare standards [9,32].

To preserve animal welfare, religious slaughter without stunning requires improved administration. The animal remains conscious during the neck cut, and it will take some time (varies with species or neck cut) to undergo a state of unconsciousness. Thus, the animal remains sensitive to pain and stress between this neck cut and unconsciousness. It was observed that if religious slaughter without stunning was performed with a proper razor-sharp knife on a calm and rested animal, restrained properly/comfortably, then animals showed very little (flinch) or no reaction to the neck cut [43]. This reaction/flinch was noticed even less than an ear tag punch, a metal clanging noise, air hissing, followed by no further reaction afterwards [43]. Similarly, Barnett et al. [44] also observed mild physical response to neck cutting in some birds (100 birds) whereas no response was observed in the majority of birds (592 birds) during kosher slaughter. The authors [44] also observed the presence of physical response to touching the eye or eyelid in birds up to 5 s after a neck cut, which disappeared after 15 s of the neck cut. The loss of posture and presence of involuntary muscular contraction was noticed after 12–15 s of neck cutting and the loss of 40% of total blood within 30 s of neck severance. However, some animal welfare scientists believe that the pain during religious slaughter could be reduced if religious slaughter is performed correctly [45]. Further, low behavioral responses after a neck cut may not necessarily indicate the pain-free status of the animal [45,46].

As blood is an excellent medium for the growth of various microorganisms, maximum possible drainage is recommended for improving the quality of meat. Further, the presence of blood in carcasses could also make the appearance of the carcass dark, which consumers prefer less. Kosherization has also been reported to improve meat quality by reducing *Salmonella* and coliform counts [47]. The hemoglobin in the blood is a potent lipid prooxidant, and efficient bleeding is also recommended to improve the meat's oxidative stability. Further slaughter and bleeding methods have been reported to affect meat color and sensory properties [48,49]. Koshering (slating and washing) also has an effect on meat quality by significantly reducing shear force and drip loss [50].

4.1. Knife Sharpness on Pain Sensation

Very few studies have evaluated the effect on animal welfare and meat quality associated with the sharpness of slaughter knives [51]. If restrained properly without pressure and gently done, cattle were observed to have little or no reaction to the throat cut in three kosher slaughterhouses [43]. When the blade touched the skin, cattle showed a slight flinch which was less vigorous than the animals' reactions to an ear tag, and cattle were observed to remain calm as the cut proceeded [43]. Further, the wound should be open during the incision to prevent pain. Further, the knife should be of sufficient length, so that its tip remains outside the neck during the cut [52].

The presence of nicks on blades is considered to cause irritation while cutting blood vessels and associated tissues during halal and kosher slaughter. Gibson et al. [16,53–57] noted that the electroencephalogram spectrum varies with neck cutting and could be correlated with pain sensation. However, the size of the knife (10 inches) and its machine sharpening used in the study make it difficult to conclude pain and stress in animals during religious slaughter, particularly kosher slaughter [34,38]. Further, the short knife even sharpened on a grindstone used in these studies could cause the tip to gouge the throat [52].

4.2. Sharpness and Blade Length

Sharp knives need a lower force requirement than blunt knives. This lower brute force lowers the damage to the meat. Knife sharpness is correlated with the forces produced and energy needed during the cutting process, the cutting edge, and the cutting surface [39].

Grandin [58] observed that a short knife used in the halal slaughter of cattle indicates that digging the end of the knife blade into the throat caused intense reaction and pain. Muslim scholars recommended a sufficient length of knife used in Halal slaughter with a minimum of 18 cm for cows and 24 cm for buffaloes. Likewise, Velarde et al. [59] observed that the blade length used for halal slaughter was 29.6 ± 1.79 cm, whereas size and shape were more uniform for kosher slaughter (approx. 40 cm *Chalaf* knife). The blade length of knives used for the halal slaughter of sheep (without stunning) was 22.2 ± 1.82 cm and 25 cm in case of kosher slaughter [59]. Table 1 summarizes the blade length of knives used in halal and kosher slaughtering.

Table 1. Blade length (approx. cm) of knife used for halal and kosher slaughtering.

Species	Halal Slaughter	Kosher Slaughter
Cattle	29.6 ± 1.79 cm (Also varies from 18 cm for cow to 24 cm for buffalo)	40.0
Poultry	–	13.5
Sheep	22.2 ± 1.82	25.4

(Source: [59,60]).

Perfect sharpness and proper height of the knife ensure the severing of all jugular and carotids with a rapid stroke resulting in a sudden drop of arterial pressure to the brain and fast and massive blood loss [18]. In a survey of slaughterhouses in Italy regarding kosher slaughter practices, Bozzo et al. [35] observed the majority of rejection (2.4% of total samples) due to failure to complywith a pause (*Shehiyah*) followed by non-compliance to the pressure (*Derasah*) and stabbing *(Halad)* rule. The authors also did not notice any rejection due to non-compliance to slanting and tearing. This further strengthens the requirement for sharp knives of the appropriate dimension for kosher slaughter.

In addition, Abd El-Rahim [61] advocated that the minimum length of the sharp edge of a knife used to slaughter animals should be at least 12 cm. Further elaborating, Helmut [28] recommended that the knife be constantly sharpened and free from scratches and faults as it may cause pain by dragging and grasping the tissues during a neck cut. According to personal observation by Leffert [45], the blades used for religious slaughter in the USA were of 6–18 inch size with significant variations in the degree of sharpness [45]. This could have affected the variation in the onset of unconsciousness in animals slaughtered by religious methods from 10–60 s [45] in the slaughterhouses visited by the author in the USA.

The shorter knife takes more time to perform neck cuts as its tip may gouge the wound and become stuck in the cut, with sawing motion during the cutting procedure. This would stimulate accessible nerve endings in the skin and at the cut edge of the skin leading to potential pain perception [19]. There have been reports of using blunt knives in halal slaughter in various countries [19,62], consequently requiring more pressure and more attempts during neck incisions, potentially provoking pain. In water buffalo, in lateral recumbency, neck twisting is impossible due to the horns' breadth. In such cases, a blade is inserted into the skin, and after the animal has settled down, the blade is repositioned in the wound, and several vertical cuts (up to 7–19 cuts) are made, as observed by Gregory et al. [20]. There is a need for specific legislation, guidelines, and training programs for handlers to improve the welfare of water buffaloes during slaughter [63].

4.3. Sharpness on Blood Biochemical and Electroencephalogram (EEG)

Imlan et al. [51] evaluated the effect of knife sharpness on stress and pain perfection in Brahman crossbred steers in terms of blood enzymes, plasma catecholamines, and electroencephalogram (EEG) changes. The authors reported a significant difference between the concentration of pre- and post-slaughter glucose, creatine kinase (CK), and lactate dehydrogenase (LDH) concentration in steers slaughtered with a sharp commercial knife (ANAGO sharpness score of 7.8) commonly used in slaughterhouses as compared to those slaughtered with a sharp knife (sharpened with a machine with ANAGO sharpness score of 8.0) [51]. Further, the authors observed a significant increase ($p < 0.0001$) in catecholamines (adrenaline), glucose, CK, and LDH post-slaughter in comparison to pre-slaughter in the commercial sharp knife group as compared to the sharp knife group. Increased LDH and CK in the blood indicate muscle damage, fatigue, and stress in animals [4,6,64,65]. Similarly, the release of catecholamines (epinephrine/adrenaline and nor-epinephrine/nor-adrenaline) in the blood indicates an initial reaction to stress/fear [17,66].

An electroencephalogram presents the electrical activity of neurons rapidly, accurately, and objectively [67,68]. These variables are well interrelated with animals' physiological and biochemical parameters [3]. On analyzing electroencephalogram (EEG) variables, a significant increase was noted in the median frequency (F_{50}) ($p < 0.0001$) and total power (Ptot) ($p < 0.0001$) in the animals slaughtered with the sharp commercial knife as compared to those slaughtered with the sharp knife [51]. These parameters could be correlated with pain and stress observed during neck cuts [17,57,69,70]. The significantly increased alpha, beta, delta, and theta waves in the EEG spectrum in cattle after slaughter increased could be attributed to use of razor-sharp knives resulting in a mild behavioral response in animals being non-painful to the animal [38].

A sharp change in the EEG spectrum upon neck cutting is depicted in Figure 1.

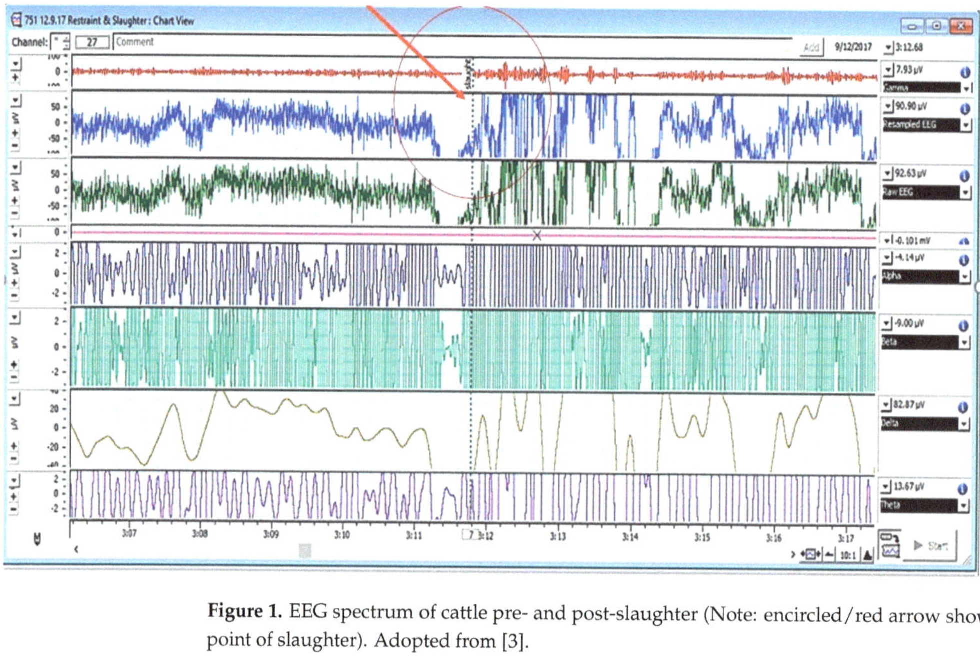

Figure 1. EEG spectrum of cattle pre- and post-slaughter (Note: encircled/red arrow showing the point of slaughter). Adopted from [3].

5. Mechanical Attributes of the Knife

In general, knife sharpness refers to the design, fitness, and quality of cutting-edge design and is measured in terms of the forces needed for cutting [71]. Sharpness can be defined by the level of force exerted by the knife during the cutting of the material [72] or the area of the cutting edge [73]. The main parameters determining the efficiency of all cutting operations depend on the durability, thickness, consistency, and speed accomplished by blades and correlate to the suitable application. In various meat-cutting operations, professional workers have duly acknowledged that the knife blade's sharpness affects the workers' productivity and the product's quality [72]. Further, the grip forces and cutting actions were greater with blunt but workable knife blades vs. recently sharpened knives [72] with a 20–30% decline in grip force, cutting actions, and time with sharper contrast with blunter blades.

It is highly preferable to use a sharp knife with a durable edge. The sharpness of the blade is determined by several factors, such as the properties of the steel, the relative movements of the blade and target material, the curvature and edge angle of the blade, grinding, and finish [74]. A good sharpness condition needs a cutting force of 25 N and a blunt condition with a 75 N cutting force. Moreover, McGorry et al. [74] concluded that while a sharp blade finish lowers the force needed for elongated cuts through different tissues, remarkably, the finish has a comparable effect for shorter cuts through muscle only as a tendency.

6. Quantifying the Sharpness

The objective quantification of the sharpness of the knife is quite challenging. During religious slaughter, it is carried out by the slaughterman prior to religious slaughter. It is also recommended to sharpen the knife and inspect its sharpness before halal or kosher slaughter by Muslims and Jewish authorities. As sharpness is directly linked with animal welfare compliance and meat quality, there is, thus, an urgent need to objectively quantify the sharpness of knives used in religious slaughter. The objective quantification of the sharpness of the knife is quite challenging. During religious slaughter, it is carried out by the slaughterman prior to the religious slaughter. Thus, it is quite challenging to standardize the sharpness measurement under various slaughtering conditions. The current measurement of knife sharpness is basically done by visual inspection/assessment, which is very subjective and varies with individual experience, training, and expertise. Commercial slaughterhouses have very high slaughter rates, so the step of knife sharpness assessment and maintaining the desired level of knife sharpness could slow down the slaughter rate and warrants extra labor and capital investment.

The traditional method to test the knife sharpness in slaughterhouses is based on a paper test under which a piece of paper is held by one corner, and if the knife can cut the hanging-dangling part of the paper, it is considered to be sharp, and if it fails to do so, then blunt. A sharp knife (in a dry state) should slice a standard paper (A4 printer paper, 80 g weight) hanging by one corner. It is a rapid and cost-saving method but can only detect the sharpness in the center of the knife, leaving untested the other commonly used area of the knife during slaughter operations [75]. Under kosher slaughter after sharpening and in between the slaughter of animals, the shochet carefully checks the knife for roughness and nicks by running his fingernails up and down on the edge of the *Chalaf* (Figure 2). Keeping in mind the sharpness of the *Chalaf*, this should be carried out with the utmost precautions and after proper training.

Figure 2. *Chalaf* sharpness test by shochet during kosher slaughter. Source: [60]. Copyright permission was obtained from the figure owner on 18 May 2023.

In present-day slaughterhouses with high throughput, it is quite challenging to quantify the sharpness of knives every time. ANAGO®, New Zealand has developed an instrument called an ANAGO® sharpness tester for the objective of accurate and reliable assessment of the sharpness of knives as well as to assist in effectively controlling and improving the sharpness to the desired level. This helps in measuring knife sharpness and achieving the desired sharpness [76]. The equipment can be applied to ensure sharpness in slaughterhouses on a large scale. This instrument can also detect the presence of nicks on the blade.

The ANAGO® sharpness score presents a sharpness profile of the knife from the tip to the heel based on the relative force required to cut. The sharpness score ranges from 2.0 (42 times as much force required) to 10.0 (no force required while cutting). For measuring the knife sharpness score, sections of 20 mm blocks in length on the blade are divided from the tip to the end [76]. A sharpness score of 8.0 could be judged as in the sharp category, while most of the knives used in commercial slaughterhouses were reported to have an average of 7.80 [51]. Additionally, to maintain the sharpness level to a score of 8.0 or above, it is necessary to sharpen the knife by machine after every slaughter. In contrast, manual sharpening of the knife after every slaughter can achieve a maximum score of 7.80 [51] as presented in Figure 3.

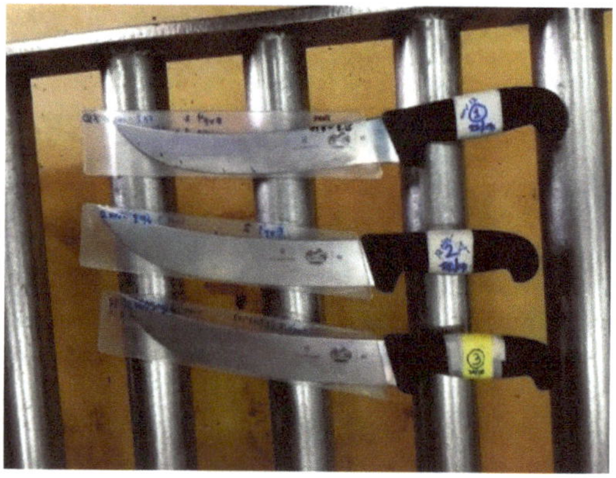

Figure 3. Knives used for the halal slaughter of cattle. Source: [77]. Copyright permission was obtained from the Universiti Putra Malaysia.

Figure 4 presents Chalaf used for the kosher slaughter of animals.

Figure 4. Chalaf used for the kosher slaughter of animals. Source: [78]. Copyright permission was obtained from the figure owner on 19 May 2023.

7. Determinants of Knife Sharpness

The overall impact and outcome of the knife sharpness are affected by several factors, viz., training and expertise of the slaughterers, neck cut position, restraints, and slaughter position. These factors improve the efficiency of knife sharpness, neat-clean neck cut, and improve bleeding efficiency, thereby maximizing animal welfare compliance while alleviating pain and distress.

7.1. Neck Cut Position

The neck incision under kosher slaughter is usually done at the position of cervical vertebra C2–C4. However, if done at C1, there are fewer incidences of false aneurysm formation, premature blockage of blood loss, and accumulation of blood in the upper and lower respiratory tract, later associated with unpleasant sensory signals [19]. Neck cut position is a key determinant in the onset of a false aneurysm [79] usually appearing between 7–21 s of the neck cut [80]. However, the case of a higher cut (C1) may potentially result in a cut to the larynx, including the associated bones, which may damage the knife's sharpness and result in the rejection of the meat due to non-compliance with religious practices, particularly in Kosher slaughter [38]. Furthermore, cattle slaughtered with C1 neck cuts produced a different sound than those with the neck cut at C2–C4 [19]. Similarly, Gibson et al. [81] also noted a lower time to final collapse in anima, reducing suffering, upon performing a high neck cut. It could be due to the increased branching of carotid arteries at a higher neck position (above tracheal ring 2), thereby minimizing or preventing the retraction of carotid arteries within the connective tissue sheath [41,81].

The neck severance at the C1 would further reduce the potential risk of irritation associated with blood infusion into the respiratory tract due to the cutting of laryngeal nerves (transferring signals from the upper respiratory tract) and the vagus nerve (transmitting signals from the lungs and inferior trachea) [19,82]. In a controlled trial with captive bolt stunning of cattle, Gregory et al. [80] observed that making a neck cut at the C3 position had four times higher chances of early arrested blood flow and a 2.5 times higher frequency of false aneurysm formation compared to neck cutting at the C1 position. Gregory et al. [80] attributed the benefits of a sharp and clean cut at C1 using a very sharp knife with rapid blood loss and lower incidences of the formation of the false aneurysm to the following factors, viz.,

(1) Higher branching of the carotid artery at the C1 position lowers the chances of sealing all carotid artery branches.
(2) The C1 neck cut needs stretching of the chin, thereby stretching the artery with less chance of their retraction within the connective tissue.
(3) The presence of less connective tissue at the C1 position.

7.2. Skill and Training of Slaughterers

Under kosher slaughter, it is mandatory for a shochet to undergo rigorous training and education to obtain a license to slaughter animals. However, such training and skill requirements are not mandatory in halal slaughter. A trained slaughterer cuts both carotid arteries and jugular veins more effectively by making a fast swift cut close to the jaw, thereby ensuring efficient bleeding and early onset of unconsciousness. Grandin [83] observed a very short collapse time for cattle (5 s) if kosher slaughtered by a good shochet whereas a collapse time of up to 1 min in the case of a throat cut by a poor shochet. Further, a proper cut also ensures lower incidences of false aneurysms.

During neck cutting, the average numbers of cutting movements vary with the skill of the slaughterer and the restraint system used for immobilizing the animal. Cenci-Goga et al. [84] observed variations in the number of cutting movements between halal and kosher-shackled and slaughtered sheep due to the difference in the skill of the slaughterers during neck cutting. The authors [84] in a survey on religious slaughter in Italy observed 25.2 cutting movements for halal-slaughtered cattle in the upright position, 2.9 for halal-slaughtered sheep shackled and the neck cut, and 1.25 for kosher-slaughtered sheep shackled and the neck cut. In mechanically turned restraints, fewer cutting movements were reported by the authors in the same study. The higher number of cuts directly influences the pain felt by animals during slaughter. Gregory [85] reported a sudden nociceptor discharge lasting for 4 s upon throat cut.

Thus, proper training/technical knowledge for gentle handling of animals to handlers, less stressful restraint devices during immobilization, elimination of distractions, and appropriate neck cutting with extremely sharp knives would all maintain high animal welfare standards while producing high-quality meat with fewer instances of petechial hemorrhages [9,32].

7.3. Restraints and Slaughter Position

Gentle stretching of the neck during neck cutting with proper restraint improves the quality of the neck cut. The wounds should be left open to alleviate pain and facilitate bleeding. After the throat cut, the animal should be released from restraint. An inverting-conscious animal could aggravate the fear and distress in animals as well as the aspiration of ruminal fluids coupled with compression of internal thoracic organs exerting pressure and inhibiting respiration [86]. During neck cutting, the neck should be properly within reach of slaughterers for neck cutting in a swift movement. Velarde et al. [59] also observed fewer cuts (three cuts) performed during the slaughter of cattle at 90° position on their sides as compared to higher cuts (five cuts) in cattle slaughtered at 180° position on their back as well as nine cuts performed in slaughter in the upright position.

During kosher and halal slaughter in Italy, cattle are restrained in the upright position [84]. Stressful restraint causes struggling, thereby masking the behavioral response to neck cuts. The restraints should not provide excess pressure and avoid jerky movements, which could make animals exited and agitated, thereby affecting the throat cut and slaughter process. A calm animal loses consciousness earlier than an agitated animal [83]. Animals should not be suspended by limbs except for poultry and there must not be any injury inflicted to restrain animals such as the cutting of tendons [1]. Small ruminants suspended on a shackle were reported as having significantly higher struggling as compared to manual restraints on their side [87]. Act of abusive and rough handling such as beating, poking with pointed sticks, dragging, leg clamping, shackling, and hoisting is also strictly prohibited under animal welfare and slaughter legislation [9,22].

The restraint equipment used for halal and kosher slaughter should hold the animal in a comfortable upright position before and during the slaughtering process. The same standards should be applied if a rotating box is used for restraining the animals [88]. Velarde et al. [59] observed that cattle in halal slaughter without stunning were restrained by four different methods, namely, by turning the animal at 45°, turning on their side (at 90°), turning on their back (180°), and upright positions with modified ASPCA pens. The authors [59] observed vocalization in fewer cattle in upright restraint boxes as compared to inverted restraints/turned on their backs. The cattle restrained by turning on their side lost posture early followed by cattle that were turned 45° and slaughtered in the upright position, with the longest time to loss of posture recorded in cattle turned on their back [59].

8. Prospects and Challenges

Maintaining the proper level of knife sharpness in modern slaughterhouses with higher slaughter rates is quite challenging. In the case of religious slaughter, it is required to slaughter as per the specific requirement of animal handling, knife specification, and good neck cut as the recommended practice of halal or kosher to make it safe and fit for eating to that particular community. For maintaining the desirable slaughter line speed, more infrastructure and human resources are required in religious slaughter. This will result in increasing the cost of production of meat.

The knife size and sharpness sometimes do not match the prescribed guidelines in cases of religious slaughter. There are more incidences reported in various slaughterhouses regarding the use of shorter and poorly sharpened knives during the slaughter of animals than using the desired ones [62]. One survey conducted over ten years duration reported that over 80% of abattoir workers used an inappropriately sharpened knife [89]. The slaughterers performing repetitive tasks continuously for hours could also result in the cutting of the corners of the blades [45]. Using blunt/less sharpened knives has been associated with lower production and increased injuries due to greater force during meat cutting [90].

A razor-sharp knife requires lower grip force, cutting time, and cutting moments, whereas a blunt knife was projected to require 25% more cutting moments and musculoskeletal disorders (MSDs) [91]. Further, Grandin [92] described how using an inadequately sharp knife results in a higher force applied by the operator and an increase in cutting time. Further, a sharp knife is more productive by increasing the cutting speed by 1.5 times [92]. Thus, a small increment in knife sharpness could result in a significant gain in overall productivity by improving the speed of production and the safety and quality of the meat.

There is an urgent need to make people aware of the importance of knife sharpness and the role played by a sharp knife in improving and maintaining high standards, particularly in religious slaughter without stunning. There should be proper emphasis on regular training and refresher courses for the personnel involved in this meat production. A study by Claudon et al. [91] with 196 respondents in France revealed that 42% of the 196 butchers complained that the blade being used was not sharp enough and only 16% stated having been trained in sharpening and maintenance of knives. There should be national-level training programs for livestock handlers, slaughterhouse workers, and staff. The personnel employed in meat production should have a certificate of competence.

Islam and Judaism have given utmost importance to knife sharpness, but the specific detailing of the knife is not mentioned for halal slaughter, as that mentioned in kosher slaughter. Thus, it is recommended to develop proper specifications for the knife used for halal slaughter [27]. There is a need to make slaughter workers and butchers aware of the importance of knife sharpness and its crucial role in ensuring the slaughter of animals with minimal pain. This follows the intrinsic principles of animal welfare prescribed in Islam and Jewish.

Further, slaughterhouse management should be more concerned with proper facilities and the gentle handling of animals during slaughter. The butchers/shochet should undergo regular training or refresher course to update them about the importance of gentle handling, knife sharpness, and a neat-clean neck cut. This will protect from scotoma (factory blindness) and compassionate fatigue in slaughterers due to a monotonous work profile. There is a need for regular auditing and inspection of slaughterhouses for proper compliance with knife sharpness and other factors that affect animal welfare during slaughter.

There is a requirement for proper monitoring of the whole process of kosher and halal slaughter. Various outcome-based measures or variables should be continuously monitored to improve the process. The high levels of vocalization (moo or bellow) during handling and restraint are associated with physiological stress due to aversive conditions such as excessive pressure from a restraint device, the sharp edge of a restraint sticking into an animal, or the use of electric goads [93–95]. Cattle undergoing ritual slaughtering in a Weinberg pen (in which the animal is inverted during slaughtering) were recorded to spend eight times longer time and have significantly higher cortisol and hematocrit values as compared to cattle slaughtered conventionally or in an ASPCA (American Society for Prevention of Cruelty to Animals) pen (in which the animal remains standing during slaughtering) [96]. Time to collapse, loss of posture, or eye rollback is an important indicator of animal welfare and should be continuously monitored [88]. The loss of consciousness should occur within 30–40 s after the neck cut [88]. In addition, there is a need for constant supervision and monitoring of knife sharpness by the slaughterhouse management as people tend to become sloppy when monitoring and oversight by management are reduced.

The authors believe there should be improved knife sharpening devices, and slaughter persons should be trained in knife sharpening. The slaughter persons should have basic knowledge of animal sentience, pain, and distress associated with slaughter without stunning and various ways to alleviate the pain and distress during slaughter. Sufficient human resources with expertise in slaughter techniques and proper knowledge of religious values should be employed to reduce the work pressure in slaughterhouses. Alternatively, the slaughterhouse management may provide a set of sharp knives for slaughtering a group of animals in a day or a shift depending on usage and need. At a point of slaughter, more than one slaughterer should be appointed so that when one person is slaughtering, another will inspect and sharpen the knife and, later, vice versa. The management should also practice regular interviews/interactions with workers to assess their compassion and empathy towards animals and the intrinsic value of various guidelines prescribed in religious slaughter. Whenever needed, they should intervene accordingly.

9. Conclusions

Knife sharpness plays a crucial role in rapid and clean neck severance, alleviating pain and stress in animals for producing good quality meat following religious practices. With the ever-increasing demand for halal and kosher meat due to its authenticity, nutritive value, and animal welfare compliance, it is of utmost importance to emphasize knife sharpness during slaughter as per the prescribed religious practices. Manufacturing design of slaughter knives for ergonomics, maintenance, sharpness, and assessment must be conducted for food safety and animal welfare, as well as for the sensory appeal of meat such as the palate knives and perceived healthiness.

Author Contributions: Conceptualization, Y.-M.G., Z.I. and A.Q.S.; data curation U.K., P.K., A.A.A., M.A.A. and J.C.I. writing—original draft, P.K., A.A.A., M.A.A. and J.C.I. writing—review and editing Y.-M.G., U.K., Z.I. and A.Q.S. All authors have read and agreed to the published version of the manuscript.

Funding: This research received no external funding.

Institutional Review Board Statement: Not applicable.

Informed Consent Statement: Not applicable.

Data Availability Statement: Data will be made available on request.

Acknowledgments: The first author, Pavan Kumar, is thankful to the Indian Council of Agricultural Research, New Delhi, India, for providing the Netaji-Subhas ICAR International Fellowship for pursuing his doctoral study at Universiti Putra Malaysia.

Conflicts of Interest: The authors declare no conflict of interest.

References

1. World Organization for Animal Health Founded as OIE. Introduction to the Recommendations for Animal Welfare. In *Terrestrial Animal Health Code*; 12 rue de Prony: Paris, France, 2018.
2. Wickham, S.L.; Collins, T.; Barnes, A.L.; Miller, D.W.; Beatty, D.T.; Stockman, C.; Blache, D.; Wemelsfelder, F.; Fleming, P.A. Qualitative Behavioral Assessment of Transport-Naïve and Transport-Habituated Sheep. *J. Anim. Sci.* **2012**, *90*, 4523–4535. [CrossRef] [PubMed]
3. Kumar, P.; Abubakar, A.A.; Sazili, A.Q.; Kaka, U.; Goh, Y.-M. Application of Electroencephalography in Preslaughter Management: A Review. *Animals* **2022**, *12*, 2857. [CrossRef] [PubMed]
4. Kumar, P.; Ahmed, M.A.; Abubakar, A.A.; Hayat, M.N.; Kaka, U.; Ajat, M.; Meng, G.Y.; Sazili, A.Q. Improving Animal Welfare Status and Meat Quality through Assessment of Stress Biomarkers: A Critical Review. *Meat Sci.* **2022**, *197*, 109048. [CrossRef] [PubMed]
5. Kumar, P.; Abubakar, A.A.; Verma, A.K.; Umaraw, P.; Nizam, M.H.; Mehta, N.; Ahmed, M.A.; Kaka, U.; Sazili, A.Q. New Insights in Improving Sustainability in Meat Production: Opportunities and Challenges. *Crit. Rev. Food Sci. Nutr.* **2022**, *12*, 1–29. [CrossRef] [PubMed]
6. Kumar, P.; Abubakar, A.A.; Ahmed, M.A.; Hayat, M.N.; Kaka, U.; Pateiro, M.; Sazili, A.Q.; Hoffman, L.C.; Lorenzo, J.M. Pre-Slaughter Stress Mitigation in Goats: Prospects and Challenges. *Meat Sci.* **2022**, *195*, 109010. [CrossRef]
7. Clonan, A.; Wilson, P.; Swift, J.A.; Leibovici, D.G.; Holdsworth, M. Red and Processed Meat Consumption and Purchasing Behaviours and Attitudes: Impacts for Human Health, Animal Welfare and Environmental Sustainability. *Public Health Nutr.* **2015**, *18*, 2446–2456. [CrossRef]

8. Bozzo, G.; Barrasso, R.; Grimaldi, C.A.; Tantillo, G.; Roma, R. Consumer Attitudes towards Animal Welfare and Their Willingness to Pay. *Vet. Ital.* **2019**, *55*, 289–297.
9. Grandin, T. Making Slaughterhouses More Humane for Cattle, Pigs, and Sheep. *Annu. Rev. Anim. Biosci.* **2013**, *1*, 491–512. [CrossRef]
10. Verma, A.K.; Umaraw, P.; Kumar, P.; Mehta, N.; Sazili, A.Q. Processing of Red Meat Carcasses. *Postharvest Postmortem Process. Raw Food Mater.* **2021**, *2*, 243–280.
11. Kumar, P.; Verma, A.K.; Umaraw, P.; Mehta, N.; Sazili, A.Q. Processing and Preparation of Slaughtered Poultry. *Postharvest Postmortem Process. Raw Food Mater.* **2021**, *2*, 281–314.
12. International Association for the Study of Pain IASP IASP Announces Revised Definition of Pain. Available online: https://www.iasp-pain.org/publications/iasp-news/iasp-announces-revised-definition-of-pain/?ItemNumber=10475#: ~\{\}:text=Jul%2016,%202020&text=The%20defini-tion%20is:%20%E2%80%9CAn%20unpleasant,pain%20for%20further%20 valuable%20context (accessed on 8 July 2022).
13. Tilbrook, A.J.; Clarke, I.J. Neuroendocrine Mechanisms of Innate States of Attenuated Responsiveness of the Hypothalamo-Pituitary Adrenal Axis to Stress. *Front. Neuroendocrinol.* **2006**, *27*, 285–307. [CrossRef] [PubMed]
14. Nielsen, S.S.; Alvarez, J.; Bicout, D.J.; Calistri, P.; Depner, K.; Drewe, J.A.; Garin-Bastuji, B.; Gonzales Rojas, J.L.; Schmidt, C.G.; Michel, V.; et al. Welfare of Cattle at Slaughter. *EFSA J.* **2020**, *18*, e06275. [CrossRef] [PubMed]
15. Grandin, T. Evaluating Methods of Restraint for Holding Animals during Kosher and Halal Slaughter. In *Preparation and Processing of Religious and Cultural Foods*; Woodhead Publishing: Shaxton, UK, 2018; pp. 349–358. [CrossRef]
16. Gibson, T.; Johnson, C.; Murrell, J.; Hulls, C.; Mitchinson, S.; Stafford, K.; Johnstone, A.; Mellor, D. Electroencephalographic Responses of Halothane-Anaesthetised Calves to Slaughter by Ventral-Neck Incision without Prior Stunning. *N. Z. Vet. J.* **2009**, *57*, 77–83. [CrossRef] [PubMed]
17. Sabow, A.B.; Goh, Y.M.; Zulkifli, I.; Sazili, A.Q.; Kaka, U.; Kadi, M.Z.A.A.; Ebrahimi, M.; Nakyinsige, K.; Adeyemi, K.D. Blood Parameters and Electroencephalographic Responses of Goats to Slaughter without Stunning. *Meat Sci.* **2016**, *121*, 148–155. [CrossRef] [PubMed]
18. Rosen, S.D. Physiological Insights into Shechita. *Vet. Rec.* **2004**, *154*, 759–765. [CrossRef] [PubMed]
19. Gregory, N.; von Wenzlawowicz, M.; von Holleben, K.; Fielding, H.; Gibson, T.; Mirabito, L.; Kolesar, R. Complications during Shechita and Halal Slaughter without Stunning in Cattle. *Anim. Welf.* **2012**, *21*, 81–86. [CrossRef]
20. Gregory, N.G.; von Wenzlawowicz, M.; Alam, R.M.; Anil, H.M.; Yeşildere, T.; Silva-Fletcher, A. False Aneurysms in Carotid Arteries of Cattle and Water Buffalo during Shechita and Halal Slaughter. *Meat Sci.* **2008**, *79*, 285–288. [CrossRef]
21. Velarde, A.; Dalmau, A. Slaughter without Stunning. In *Advances in Agricultural Animal Welfare: Science and Practice*; Woodhead Publishing: Shaxton, UK, 2018; pp. 221–240. [CrossRef]
22. American Veterinary Medical Association. *AVMA Guidelines for the Humane Slaughter of Animals*; American Veterinary Medical Association: Schaumburg, IL, USA, 2016.
23. Baldwin, B.A.; Bell, F.R. The Anatomy of the Cerebral Circulation of the Sheep and Ox. The Dynamic Distribution of the Blood Supplied by the Carotid and Vertebral Arteries to Cranial Regions. *J. Anat.* **1963**, *97*, 203.
24. Nielsen, S.S.; Alvarez, J.; Bicout, D.J.; Calistri, P.; Depner, K.; Drewe, J.A.; Garin-Bastuji, B.; Gonzales Rojas, J.L.; Gortázar Schmidt, C.; Miranda Chueca, M.Á.; et al. Slaughter of Animals: Poultry. *EFSA J.* **2019**, *17*, e05849. [CrossRef]
25. Italian National Bioethics Committee Macellazioni Rituali e Sofferenza Animale. *Macellazioni Ritual. E Sofferenza Anim.* **2003**, 91–96.
26. Nakyinsige, K.; Man, Y.B.C.; Sazili, A.Q. Halal Authenticity Issues in Meat and Meat Products. *Meat Sci.* **2012**, *91*, 207–214. [CrossRef]
27. Farouk, M.M.; Al-Mazeedi, H.M.; Sabow, A.B.; Bekhit, A.E.D.; Adeyemi, K.D.; Sazili, A.Q.; Ghani, A. Halal and Kosher Slaughter Methods and Meat Quality: A Review. *Meat Sci.* **2014**, *98*, 505–519. [CrossRef] [PubMed]
28. Helmut, P. *Review of Stunning and Halal Slaughter*; Meat & Livestock Australia: North Sydney, Australia, 2010; pp. 1–26.
29. Aghwan, Z.A.; Regenstein, J. mac Slaughter Practices of Different Faiths in Different Countries. *J. Anim. Sci. Technol.* **2019**, *61*, 111–121. [CrossRef] [PubMed]
30. della Corte, V.; del Gaudio, G.; Sepe, F.; Nevola, G. Kosher Meat and Production Issues: A Bibliometric Analysis. *Trends Food Sci. Technol.* **2021**, *116*, 749–754. [CrossRef]
31. Rosenblum, J.D. *The Jewish Dietary Laws in the Ancient World*; Cambridge University Press: Cambridge, UK, 2016; ISBN 9781316106655.
32. Pozzi, P.S.; Waner, T. Shechita (Kosher Slaughtering) and European Legislation. *Vet. Ital.* **2017**, *53*, 5–19. [CrossRef]
33. Gregory, N.G.; Grandin, T. *Animal Welfare and Meat Production*; Gregory, N.G., Grandin, T., Eds.; CABI: Wallingford, UK, 2007; ISBN 9781845932152.
34. Zivotofsky, A.Z. Government Regulations of Shechita (Jewish Religious Slaughter) in the Twenty-First Century: Are They Ethical? *J. Agric. Environ. Ethics.* **2012**, *25*, 747–763. [CrossRef]
35. Bozzo, G.; di Pinto, A.; Bonerba, E.; Ceci, E.; Mottola, A.; Roma, R.; Capozza, P.; Samoilis, G.; Tantillo, G.; Celano, G.V. Kosher Slaughter Paradigms: Evaluation of Slaughterhouse Inspection Procedures. *Meat Sci.* **2017**, *128*, 30–33. [CrossRef]
36. Regenstein, J.M.; Chaudry, M.M.; Regenstein, C.E. The Kosher and Halal Food Laws. *Compr. Rev. Food Sci. Food Saf.* **2003**, *2*, 111–127. [CrossRef]

37. Goga, B.C.; Anil, A.; Fermani, A.G. *La Macellazione Religiosa: Protezione Degli Animali e Produzione Igienica Delle Carni*; Point Veterinaire Italie: Milano, Italy, 2010; ISBN 8895995791.
38. Regenstein, J.M. The Religious Slaughter of Animals. In *The Halal Food Handbook*; Wiley: New York, NY, USA, 2020; pp. 93–119.
39. McCarthy, C.T.; Hussey, M.; Gilchrist, M.D. On the Sharpness of Straight Edge Blades in Cutting Soft Solids: Part I—Indentation Experiments. *Eng. Fract. Mech.* **2007**, *74*, 2205–2224. [CrossRef]
40. Unites States Department of Agriculture FSIS Directive 6900.2 Revision 3. Humane Handling and Slaughter of Livestock. 2020. Available online: https://www.fsis.usda.gov/policy/fsis-directives/6900.2 (accessed on 17 May 2023).
41. Sazili, A.Q.; Adeyemi, K.D.; Sabow, A.B.; Nakyinsige, K. Meat Quality and Animal Welfare: Religious and Scientific Perspectives. In *Preparation and Processing of Religious and Cultural Foods*; Elsevier: Amsterdam, The Netherlands, 2018; pp. 359–375.
42. Anil, M.H. Religious Slaughter: A Current Controversial Animal Welfare Issue. *Anim. Front.* **2012**, *2*, 64–67. [CrossRef]
43. Grandin, T.; Regenstein, J.M. Religious Slaughter and Animal Welfare: A Discussion for Meat Scientists. *Meat Focus Int.* **1994**, *3*, 115–123.
44. Barnett, J.L.; Cronin, G.M.; Scott, P.C. Behavioural Responses of Poultry during Kosher Slaughter and Their Implications for the Birds' Welfare. *Vet. Rec.* **2007**, *160*, 45–49. [CrossRef] [PubMed]
45. Leffert, S. Is Ongoing Ritual Slaughter of Livestock Justifiable in Modern America? *J. Appl. Anim. Res.* **2021**, *49*, 492–522. [CrossRef]
46. von Holleben, K.; von Wenzlawowicz, M.; Gregory, N.; Anil, H.; Velarde, A.; Rodriguez, P.; Cenci Goga, B.; Catanese, B.; Lambooij, B. Report on Good and Adverse Practices-Animal Welfare Concerns in Relation to Slaughter Practices from the Viewpoint of Veterinary Sciences. *Dialrel Deliv.* **2010**, *1.3*. Available online: https://www.dialrel.net/dialrel/welcome.html (accessed on 8 July 2022).
47. Hajmeer, M.N.; Marsden, J.L.; Crozier-Dodson, B.A.; Basheer, I.A.; Higgins, J.J. Reduction of Microbial Counts at a Commercial Beef Koshering Facility. *J. Food Sci.* **1999**, *64*, 719–723. [CrossRef]
48. Alvarado, C.Z.; Richards, M.P.; O'Keefe, S.F.; Wang, H. The Effect of Blood Removal on Oxidation and Shelf Life of Broiler Breast Meat. *Poult. Sci.* **2007**, *86*, 156–161. [CrossRef]
49. Sabow, A.B.; Zulkifli, I.; Goh, Y.M.; Zainal, M.; Kadir, A.A.; Kaka, U.; Imlan, J.C.; Abubakar, A.A.; Adeyemi, K.D.; Sazili, A.Q. Bleeding Efficiency, Microbiological Quality and Oxidative Stability of Meat from Goats Subjected to Slaughter without Stunning in Comparison with Different Methods of Pre-Slaughter Electrical Stunning. *PLoS ONE* **2016**, *12*, e0178890. [CrossRef]
50. Żurek, J.; Rudy, M.; Duma-Kocan, P.; Stanisławczyk, R.; Gil, M. Impact of Kosher Slaughter Methods of Heifers and Young Bulls on Physical and Chemical Properties of Their Meat. *Foods* **2022**, *11*, 622. [CrossRef]
51. Imlan, J.C.; Kaka, U.; Goh, Y.-M.; Idrus, Z.; Awad, E.A.; Abubakar, A.A.; Ahmad, T.; Nizamuddin, H.N.Q.; Sazili, A.Q. Effects of Slaughter Knife Sharpness on Blood Biochemical and Electroencephalogram Changes in Cattle. *Animals* **2020**, *10*, 579. [CrossRef]
52. Grandin, T. *Welfare during Slaughter without Stunning (Kosher or Halal) Differences between Sheep and Cattle*; Department of Animal Science, Colorado State University: Fort Collins, CO, USA, 2011.
53. Mellor, D.; Gibson, T.; Johnson, C. A Re-Evaluation of the Need to Stun Calves Prior to Slaughter by Ventral-Neck Incision: An Introductory Review. *N. Z. Vet. J.* **2009**, *57*, 74–76. [CrossRef]
54. Johnson, C.; Gibson, T.; Stafford, K.; Mellor, D. Pain Perception at Slaughter. *Anim. Welf.* **2012**, *21*, 113–122. [CrossRef]
55. Gibson, T.; Johnson, C.; Murrell, J.; Mitchinson, S.; Stafford, K.; Mellor, D. Electroencephalographic Responses to Concussive Non-Penetrative Captive-Bolt Stunning in Halothane-Anaesthetised Calves. *N. Z. Vet. J.* **2009**, *57*, 90–95. [CrossRef] [PubMed]
56. Gibson, T.; Johnson, C.; Murrell, J.; Chambers, J.; Stafford, K.; Mellor, D. Components of Electroencephalographic Responses to Slaughter in Halothane-Anaesthetised Calves: Effects of Cutting Neck Tissues Compared with Major Blood Vessels. *N. Z. Vet. J.* **2009**, *57*, 84–89. [CrossRef] [PubMed]
57. Gibson, T.; Johnson, C.; Murrell, J.; Chambers, J.; Stafford, K.; Mellor, D. Amelioration of Electroencephalographic Responses to Slaughter by Non-Penetrative Captive-Bolt Stunning after Ventral-Neck Incision in Halothane-Anaesthetised Calves. *N. Z. Vet. J.* **2009**, *57*, 96–101. [CrossRef] [PubMed]
58. Grandin, T. Progress and Challenges in Animal Handling and Slaughter in the U.S. *Appl. Anim. Behav. Sci.* **2006**, *100*, 129–139. [CrossRef]
59. Velarde, A.; Rodriguez, P.; Dalmau, A.; Fuentes, C.; Llonch, P.; von Holleben, K.V.; Anil, M.H.; Lambooij, J.B.; Pleiter, H.; Yesildere, T.; et al. Religious Slaughter: Evaluation of Current Practices in Selected Countries. *Meat Sci.* **2014**, *96*, 278–287. [CrossRef] [PubMed]
60. Professional Shechita Knives. Available online: https://thekoshercut.com/shop/shechita-knives/professional-shechita-knives/ (accessed on 17 May 2023).
61. Abd El-Rahim, I.H.A. Recent Slaughter Methods and Their Impact on Authenticity and Hygiene Standards. In *The Halal Food Handbook*; Wiley: New York, NY, USA, 2020; pp. 81–91.
62. Grandin, T. Animal Welfare and Society Concerns Finding the Missing Link. *Meat Sci.* **2014**, *98*, 461–469. [CrossRef]
63. de la Cruz, L.; Gibson, T.J.; Guerrero-Legarreta, I.; Napolitano, F.; Mora-Medina, P.; Mota-Rojas, D. The Welfare of Water Buffaloes during the Slaughter Process: A Review. *Livest. Sci.* **2018**, *212*, 22–33. [CrossRef]
64. Zhang, C.; Wang, L.; Zhao, X.H.; Chen, X.Y.; Yang, L.; Geng, Z.Y. Dietary Resveratrol Supplementation Prevents Transport-Stress-Impaired Meat Quality of Broilers through Maintaining Muscle Energy Metabolism and Antioxidant Status. *Poult. Sci.* **2017**, *96*, 2219–2225. [CrossRef]

65. Siqueira, T.S.; Borges, T.D.; Rocha, R.M.M.; Figueira, P.T.; Luciano, F.B.; Macedo, R.E.F. Effect of Electrical Stunning Frequency and Current Waveform in Poultry Welfare and Meat Quality. *Poult. Sci.* **2017**, *96*, 2956–2964. [CrossRef]
66. Salehpoor, F.; Bazzazi, A.M.; Estakhri, R.; Zaheri, M.; Asghari, B. Correlation between Catecholamine Levels and Outcome in Patients with Severe Head Trauma. *Pak. J. Biol. Sci.* **2010**, *13*, 738–742. [CrossRef] [PubMed]
67. Kaka, U.; Cheng, C.H.; Meng, G.Y.; Fakurazi, S.; Kaka, A.; Behan, A.A.; Ebrahimi, M. Electroencephalographic Changes Associated with Antinociceptive Actions of Lidocaine, Ketamine, Meloxicam, and Morphine Administration in Minimally Anaesthetized Dogs. *Biomed Res. Int.* **2015**, *2015*, 305367. [CrossRef] [PubMed]
68. Kaka, U.; Goh, Y.M.; Chean, L.W.; Chen, H.C. Electroencephalographic Changes Associated with Non-Invasive Nociceptive Stimulus in Minimally Anaesthetised Dogs. *Pol. J. Vet. Sci.* **2016**, *19*, 675–683. [CrossRef]
69. Gibson, T.; Johnson, C.; Stafford, K.; Mitchinson, S.; Mellor, D. Validation of the Acute Electroencephalographic Responses of Calves to Noxious Stimulus with Scoop Dehorning. *N. Z. Vet. J.* **2007**, *55*, 152–157. [CrossRef] [PubMed]
70. Raghazli, R.; Othman, A.H.; Kaka, U.; Abubakar, A.A.; Imlan, J.C.; Hamzah, H.; Sazili, A.Q.; Goh, Y.M. Physiological and Electroencephalogram Responses in Goats Subjected to Pre-and during Slaughter Stress. *Saudi. J. Biol. Sci.* **2021**, *28*, 6396–6407. [CrossRef] [PubMed]
71. Reilly, G.A.; McCormack, B.A.O.; Taylor, D. Cutting Sharpness Measurement: A Critical Review. *J. Mater. Process. Technol.* **2004**, *153–154*, 261–267. [CrossRef]
72. McGorry, R.W.; Dowd, P.C.; Dempsey, P.G. Cutting Moments and Grip Forces in Meat Cutting Operations and the Effect of Knife Sharpness. *Appl. Ergon.* **2003**, *34*, 375–382. [CrossRef]
73. Evans, A.R. Connecting Morphology, Function and Tooth Wear in Microchiropterans. *Biol. J. Linn. Soc.* **2005**, *85*, 81–96. [CrossRef]
74. McGorry, R.W.; Dowd, P.C.; Dempsey, P.G. The Effect of Blade Finish and Blade Edge Angle on Forces Used in Meat Cutting Operations. *Appl. Ergon.* **2005**, *36*, 71–77. [CrossRef]
75. Farouk, M.M.; Pufpaff, K.M.; Amir, M. Industrial Halal Meat Production and Animal Welfare: A Review. *Meat Sci.* **2016**, *120*, 60–70. [CrossRef]
76. ANAGO Understanding Sharpness. Available online: https://anago.co.nz/how-it-works/understanding-sharpness/ (accessed on 17 May 2023).
77. Imlan, J.C. Meat Quality Attributes and Physiological, Metabolic and Electroencephalographic Responses to Slaughter Position and Knife Sharpness in Cattle. Ph.D. Thesis, Universiti Putra Malaysia, Serdand, Malaysia, 2019.
78. Jewish Knives Collection. Available online: https://jewishknives.com/collections/chalefim%20 (accessed on 18 May 2023).
79. Bozzo, G.; Bonerba, E.; Barrasso, R.; Roma, R.; Luposella, F.; Zizzo, N.; Tantillo, G. Evaluation of the Occurrence of False Aneurysms During Halal Slaughtering and Consequences on the Animal's State of Consciousness. *Animals* **2020**, *10*, 1183. [CrossRef] [PubMed]
80. Gregory, N.G.; Schuster, P.; Mirabito, L.; Kolesar, R.; McManus, T. Arrested Blood Flow during False Aneurysm Formation in the Carotid Arteries of Cattle Slaughtered with and without Stunning. *Meat Sci.* **2012**, *90*, 368–372. [CrossRef] [PubMed]
81. Gibson, T.J.; Dadios, N.; Gregory, N.G. Effect of Neck Cut Position on Time to Collapse in Halal Slaughtered Cattle without Stunning. *Meat Sci.* **2015**, *110*, 310–314. [CrossRef] [PubMed]
82. Gregory, N.G.; von Wenzlawowicz, M.; Holleben, K. von Blood in the Respiratory Tract during Slaughter with and without Stunning in Cattle. *Meat Sci.* **2009**, *82*, 13–16. [CrossRef] [PubMed]
83. Grandin, T. Improving Religious Slaughter Practices in the U.S. *Anthropol. Food* **2006**, *5*, 1–8. [CrossRef]
84. Cenci-Goga, B.T.; Mattiacci, C.; De Angelis, G.; Marini, P.; Cuccurese, A.; Rossi, R.; Catanese, B. Religious Slaughter in Italy. *Vet. Res. Commun.* **2010**, *34*, 139–143. [CrossRef] [PubMed]
85. Gregory, G.N. *Physiology and Behavior of Animal Suffering*; Blackwell Publishing Company: Hoboken, NJ, USA, 2004; pp. 25–50.
86. Imlan, J.C.; Kaka, U.; Goh, Y.-M.; Idrus, Z.; Awad, E.A.; Abubakar, A.A.; Ahmad, T.; Quaza Nizamuddin, H.N.; Sazili, A.Q. Effects of Slaughter Positions on Catecholamine, Blood Biochemical and Electroencephalogram Changes in Cattle Restrained Using a Modified Mark IV Box. *Animals* **2021**, *11*, 1979. [CrossRef] [PubMed]
87. Novelli, S.; Sechi, P.; Mattei, S.; Iulietto, M.F.; Cenci, G.B.T. Report on Religious Slaughter Practices in Italy. *Vet. Ital.* **2016**, *52*, 5–11.
88. Grandin, T. Maintaining Acceptable Animal Welfare during Kosher or Halal Slaughter. Available online: https://www.grandin.com/ritual/maintain.welfare.during.slaughter.html (accessed on 17 May 2023).
89. Claudon, L.; Marsot, J. Effect of Knife Sharpness on Upper Limb Biomechanical Stresses—A Laboratory Study. *Int. J. Ind. Ergon.* **2006**, *36*, 239–246. [CrossRef]
90. Karltun, J.; Vogel, K.; Bergstrand, M.; Eklund, J. Maintaining Knife Sharpness in Industrial Meat Cutting: A Matter of Knife or Meat Cutter Ability. *Appl. Ergon.* **2016**, *56*, 92–100. [CrossRef]
91. Claudon, L. Influence on Grip of Knife Handle Surface Characteristics and Wearing Protective Gloves. *Appl. Ergon.* **2006**, *37*, 729–735. [CrossRef] [PubMed]
92. Marsot, J.; Claudon, L.; Jacqmin, M. Assessment of Knife Sharpness by Means of a Cutting Force Measuring System. *Appl. Ergon.* **2007**, *38*, 83–89. [CrossRef] [PubMed]
93. Grandin, T. Objective Scoring of Animal Handling and Stunning Practices at Slaughter Plants. *J. Am. Vet. Med. Assoc.* **1998**, *212*, 36–39. [PubMed]
94. Grandin, T. Cattle Vocalizations Are Associated with Handling and Equipment Problems at Beef Slaughter Plants. *Appl. Anim. Behav. Sci.* **2001**, *71*, 191–201. [CrossRef] [PubMed]

95. Bourguet, C.; Deiss, V.; Tannugi, C.C.; Terlouw, E.M.C. Behavioural and Physiological Reactions of Cattle in a Commercial Abattoir: Relationships with Organisational Aspects of the Abattoir and Animal Characteristics. *Meat Sci.* **2011**, *88*, 158–168. [CrossRef]
96. Dunn, C.S. Stress Reactions of Cattle Undergoing Ritual Slaughter Using Two Methods of Restraint. *Vet. Rec.* **1990**, *126*, 522–525. [PubMed]

animals

MDPI

Article

Pre-Slaughter Stunning of Farmed Atlantic Halibut in CO_2-Saturated Seawater: Assessment of Unconsciousness by Electroencephalography (EEG)

Daniel Santiago Rucinque [1,*], Hans van de Vis [2], Henny Reimert [2], Bjørn Roth [3], Atle Foss [4], Cesar Augusto Taconeli [5] and Marien Gerritzen [2]

1 Faculdade de Zootecnia e Engenharia de Alimentos, Laboratório de Aquicultura, Universidade de São Paulo, Av. Duque de Caxias Norte 222, São Paulo 13635-900, Brazil
2 Wageningen Livestock Research, P.O. Box 338, 6700 AH Wageningen, The Netherlands
3 Department of Processing Technology, Nofima, 4068 Stavanger, Norway
4 Akvaplan-niva, Fram Centre, 9296 Tromsø, Norway
5 Department of Statistics, Federal University of Paraná, Curitiba 81531-980, Brazil; cetaconeli@gmail.com
* Correspondence: dsrucinqueg@alumni.usp.br

Simple Summary: The World Organization for Animal Health recommends the use of stunning methods before the slaughter of farmed fish destined for human consumption. The use of carbon dioxide (CO_2) is not recommended by the World Organisation for Animal Health for the stunning of farmed fish at pre-slaughter. However, its use continues to be common in the world. CO_2 stimulates chemoreceptors, causing aversion and escape attempts when fish are immersed in CO_2-saturated water. The use of CO_2 in seawater with oxygenated water is allowed for pre-slaughter of halibut in Norway. Hence, there are no studies assessing loss of consciousness in halibut exposed to this stunning method. The best approach to determine unconsciousness is by measuring brain electrical activity through EEG.

Abstract: As fish welfare becomes a growing concern, it is important to ensure humane treatment during slaughter. This study aimed to assess the onset of unconsciousness in Atlantic halibut immersed in CO_2-saturated seawater through electroencephalography (EEG). Of the 29 fish studied, 10 exhibited escape attempts, indicating aversion to CO_2-saturated water despite its oxygenation. EEG signals showed four distinct phases: transitional, excitation (high amplitude–high frequency), suppressed, and iso-electric phases. The onset of the suppressed phase, indicative of unconsciousness, occurred on average 258.8 ± 46.2 s after immersion. The spectral analysis of the EEG signals showed a progressive decrease in median frequency, spectral edge frequency, and high frequency contribution, which corresponded to the gradual loss of consciousness. The study concludes that CO_2-saturated water is not recommended for pre-slaughter handling of halibut due to the extended time required for the onset of unconsciousness and the observed aversive behaviour. Ensuring humane treatment during slaughter is important for addressing public concern and safeguarding fish welfare in all stages of production.

Keywords: fish welfare; humane slaughter; unconsciousness; EEG

Citation: Rucinque, D.S.; van de Vis, H.; Reimert, H.; Roth, B.; Foss, A.; Taconeli, C.A.; Gerritzen, M. Pre-Slaughter Stunning of Farmed Atlantic Halibut in CO_2-Saturated Seawater: Assessment of Unconsciousness by Electroencephalography (EEG). *Animals* **2023**, *13*, 1993. https://doi.org/10.3390/ani13121993

Academic Editors: Gareth Pearce and Elbert Lambooij

Received: 19 April 2023
Revised: 25 May 2023
Accepted: 12 June 2023
Published: 15 June 2023

1. Introduction

Global production from aquaculture of fish, crustaceans and molluscs continues to grow and reached 88 million tonnes in 2020. The capture production was 90.3 million tonnes, a decrease of 2.06 percent compared with 2019. The global finfish aquaculture production was 57.5 million tonnes in 2020 [1]. According to the Federation of European Aquaculture Producers the European production of finfish farming production was 2.5 million tonnes in 2020. In 2020, 1500 tonnes of farmed Atlantic halibut (*Hippoglossus hippoglossus*) were produced in Norway [2]. Atlantic halibut is a species of high value in the market that is likely to play a more important role in European aquaculture production [3].

Concern about fish welfare is increasing in Western countries. A study of 2.147 Norwegian citizens showed that they are concerned about fish welfare [4]. Even though levels of public knowledge and concern about fish welfare issues are lower when compared to terrestrial species, society calls for improvement of animal farming practices. For example, Norway has specific regulations for the humane slaughter of farmed fish destined for human consumption [5,6]. However, the use of carbon dioxide (CO_2) in seawater is allowed for the stunning of halibut, but not for Atlantic salmon. Additionally, the World Organisation for Animal Health—OIE recommends practices to safeguard farmed fish welfare at slaughter [7]. To protect animals at slaughter, they should be rendered unconscious and insensible by stunning to avoid pain, fear or distress prior to slaughter, which is a general provision in the EU legislation to protect animals at slaughter [8]. Hence, to protect fish at slaughter, two steps should be applied. The first is stunning, which leads to loss of consciousness (LOC) and sensibility. The second step is killing by bleeding out or the application of another method to induce death while the fish is still unconscious. It should be noted that both steps may be achieved by the application of only one method. Nevertheless, Annex 1 of the European relation regarding the protection of animals at the time of killing does not mention any specifical requirement to stun/kill fish, which differs from the specific requirements to use head-only electrical stunning in cattle, ovine, caprine, chickens and turkeys, as well as for electrical waterbath stunning of poultry.

Methods to avoid or minimize pain, fear and distress at slaughter of fish comprise electrical stunning, percussion or the use of permitted anaesthetics in water [9]. The European Food Safety Authority [10] recommends assessing stunning and killing methods through electroencephalography (EEG) [11]. EEG registrations in fish are supplemented by behavioural observations [12]. The electrical activity recorded on EEG can be classified into Delta (<4 Hz), Theta (4 to 8 Hz), Alpha (8 to 13 Hz), Beta (13 to 32 Hz) and Gamma (32 to 45 Hz) frequency bands. When the main power contribution of the EEG is in the Alpha, Beta and Gamma frequency bands (8–45 Hz), the animal is considered to be conscious, whereas when the power contribution shifts to lower frequencies, Theta and Delta, the animal is considered to be unconscious [13]. It should be noted that no firm conclusions on LOC can be drawn based on observation of behaviour only [9]. Fourier transformation boils down to spectral analysis, as EEG signals are dissected in its component spectra. In principle, such a method allows for the detection of underlying sinus waves that in their sum result in the complex wave detected [14]. Further, spectral analysis on EEG traces is useful to identify different states of consciousness. Spectral variables can be calculated from EEG as total power (Ptot), median frequency (F50) and spectral edge frequency (SEF). Ptot is defined as the total area under the power spectrum curve; F50 represents the median frequency of the power spectrum curve and SEF represents the frequency where 95% of the power spectrum curve is located [14]. F50 is more sensitive to changes in lower frequencies, whereas SEF is more sensitive to shifts towards higher frequencies. It is well recognized that increases in EEG Ptot and associated decreases in both F50 and SEF are correlated with clinical signs of LOC and anaesthesia [15–17].

One of the first reports of the use of carbon dioxide (CO_2) in water for the purposes of anaesthesia and transport of fish was in carp (*Cyprinus carpio*) [18]. The authors observed drops in the water pH from 7.5 to 5.0 when the tank was flushed with CO_2 and active movements were observed at the initial phase of anaesthesia [18]. In other experiments, CO_2 in anoxic water was used for the stunning of Atlantic salmon (*Salmo salar*) [11,19,20] and rainbow trout (*Oncorhynchus mykiss*) [21]. Atlantic salmon and rainbow trout showed strong aversion for at least 30 s after immersion in carbon dioxide, and even durations of more than three minutes have been recorded [22]. For this reason, CO_2 stunning without aeration or oxygenation has been banned in Norway. In Norway, halibut is stunned in well-oxygenated water flushed with CO_2 until a pH of approximately 5 is obtained. In practice, the halibuts are netted or pumped into a tank with CO_2-saturated water and are left in the tank until all movement stops. They are then removed from the water and exsanguinated [19].

At present, assessments of effectiveness of CO_2 stunning of halibut in well-oxygenated seawater via registration of EEGs have not been performed. Hence, it is not known whether the method applied results in loss of consciousness with avoidable pain, fear or distress in halibut. Therefore, the aim of this study was to assess the onset of unconsciousness through EEG assessment of Atlantic halibut (*Hippoglossus hippoglossus*) immersed in CO_2 saturated seawater.

2. Materials and Methods

2.1. Animals and Immersion in CO_2

This study was approved on 16 May 2015 by the Norwegian authority (Mattilsynet), licence number 2015/108619.

Thirty farmed halibut with an average weight of 2.33 $\pm$ 0.56 kg were provided by a commercial halibut producer in Norway, in autumn. Prior to the experiment, fish were kept in holding tanks containing aerated seawater with a pH of 7.8 and at a temperature of 8.7 °C.

Prior to immersion, the tank containing seawater with a temperature of 8.6 °C was flushed with carbon dioxide until a pH of 5.2 was obtained. Typically, commercial stunning tanks operate at pH levels of about 5.5–6.0, corresponding to CO_2 levels of 200–450 mg L^{-1} [19,20]. Subsequently, water was flushed with oxygen to avoid anoxic water. When an oxygen concentration of 8.5 mg/L was obtained, halibut were immersed in the water for stunning. After 446 $\pm$ 32 s of immersion in the CO_2, twenty-eight fish were bled by applying a double gill-cut after carbon dioxide stunning. One fish was not bled to observe the consistency of unconsciousness.

One fish was euthanized for head dissection to determine the exact position of the brain. The needle-shaped electrodes (2 cm length, 1.5 cm diameter, 55% silver, 21% copper and 24% zinc) were placed percutaneously across the skull and remained on the surface of the brain lateral to *opticum tetum*. For handling, prior to electrode insertion, fish were restrained on a rectangular wooden plate (25 $\times$ 40 cm) using adjustable straps around the body. The first electrode was inserted at 4 cm caudal from an imaginary line between the eyes. The next electrode was inserted at 0.5 cm ventral to the first electrode. The third intramuscular earth electrode was inserted in the tail. Cyanoacrylate under the skin was used to fix the electrodes. This methodology is similar to that reported by [23].

2.2. EEG Analyses

EEG was recorded using a Bioamplifier connected to a PowerLab 8/35 from AD instruments, with 1000 Hz sampling frequency. Baseline EEG was recorded during 60 s before stunning in a tank with fresh seawater (50 $\times$ 35 $\times$ 23 cm) (n = 29 fish). After the baseline recording, fish were placed in the stunning tank with CO_2-saturated seawater, which was aerated prior to stunning of halibut. In addition to the EEG recording, behaviour was recorded using a camcorder. Behaviour of the restrained fish was used to identify escape attempts and movement artefacts to support interpretation of the EEG traces. Escape attempts were characterized by strong muscle contractions in an attempt to escape after immersion in the CO_2 tank. The proportion of fish with escape attempts, latency (s) for onset movements from the immersion in the CO_2 tank and duration of movements were recorded.

EEG recordings were monitored, saved and analysed using LabChart 7 Pro (version 7.3.7, AD Instruments, Cologne, Germany). Periods with clear artefacts lasting for more than a few seconds were identified and discarded in the subsequent analysis. Thereafter, on LabChart, the EEG recordings were filtered (band pass: 0.1 to 45 Hz) for visual analysis such as filtering delate background noise. The screen was split in two in order to compare each EEG portion from the baseline of each fish (Figures 1–5). Visual inspection of the traces was used to assign portions of the EEG to one of the 4 phases with the following characteristics:

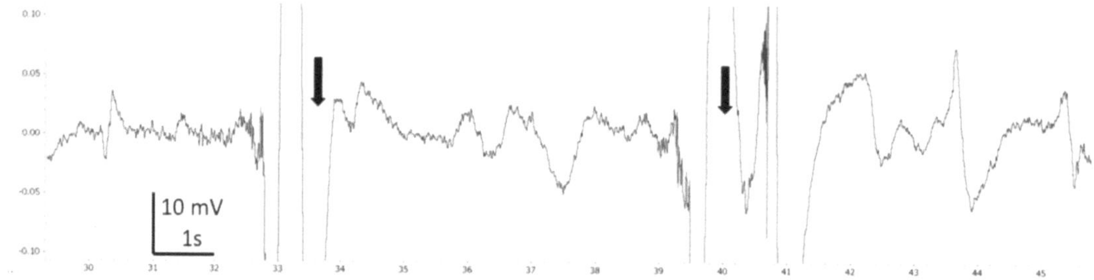

Figure 1. Example of baseline (normal EEG) observed in halibut before immersion in CO_2-saturated seawater. Band pass filter (0.1–45 Hz). Arrows represent breathing interference.

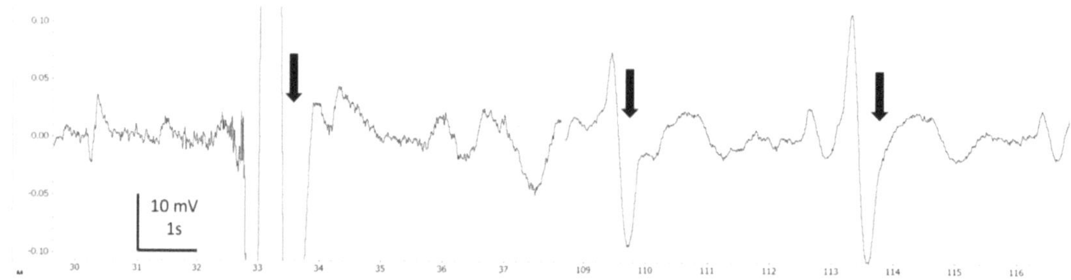

Figure 2. Example of comparison of baseline (left side) vs. transitional phase (right side) observed in halibut from immersion in CO_2-saturated seawater. Band pass filter (0.1–45 Hz). Arrows represent breathing interference.

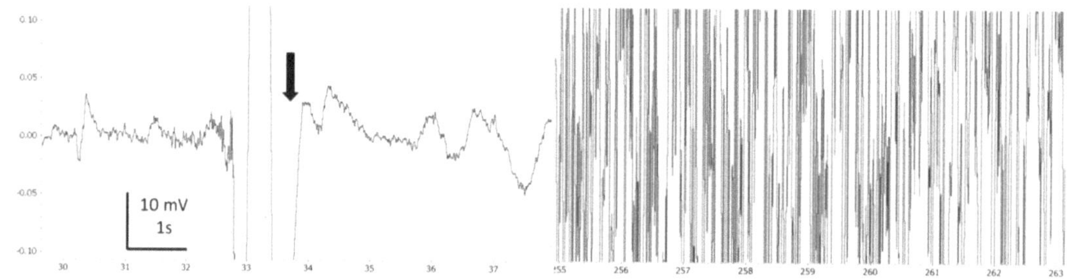

Figure 3. Example of comparison of baseline (left side) vs. excitation phase (right side) observed in halibut from immersion in CO_2-saturated seawater. Band pass filter (0.1–45 Hz). Arrows means breathing interference.

'Normal': activity identical to the baseline but after immersion in the CO_2 tank;

'Transitional': increase in the amplitude and similar frequency to baseline phase;

'Suppressed': a greatly suppressed EEG but containing low amplitude and low frequency wave activity; such a phase can be identified by the reduction of more than 40% on the amplitude in comparison with the baseline;

'Iso-electric': residual low-level noise indicating lack of EEG activity.

Normal, transitional, suppressed and iso-electric patters have been reported previously in studies with chickens stunned with CO_2 [16,24].

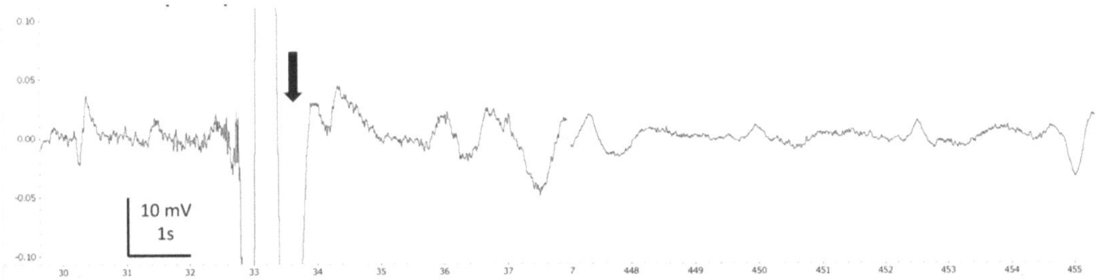

Figure 4. Example of comparison of baseline (left side) vs. suppressed phase (right side) observed in halibut from immersion in CO_2-saturated seawater. Band pass filter (0.1–45 Hz). Arrows represent breathing interference.

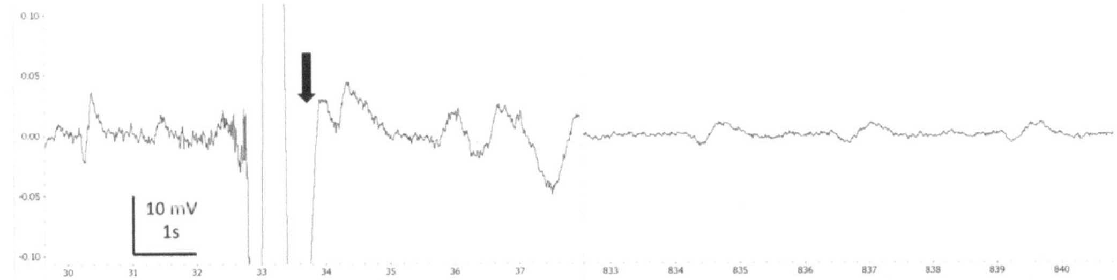

Figure 5. Example of comparison of baseline (left side) vs. iso-electric phase (right side) observed in halibut from immersion in CO_2 saturated seawater. Top of the figure with band pass filter (0.1–45 Hz).

Representative 2 s epochs free of noise and/or artefacts were selected from EEG trace for spectral analysis. Six fish (6/29) were discarded due to artefacts on the signal, preventing the selection of several 2 s epochs on the baseline. Hence, spectral analysis was performed on 23 fish registers. Spectral analyses were performed on filtered recordings (band pass: 4 to 45 Hz) was set: FFT 1K; data window Hamming and window overlap 93.75%. Such filter diluted the interference from breathing. The percentage of total power was determined from 2 s epochs for the following wave bands of contribution to total power: Delta (0–4 Hz), Theta (4–8 Hz), Alpha (8–12 Hz) and Beta (12–32 Hz). In addition, Total power (v^2) (Ptot), median frequency (Hz) (F50) and spectral edge frequency 95% (Hz) (SEF) were calculated from each 2 s epoch and classified according to the EEG phases identified on visual analysis. Epochs of less than 2 s were not included.

Five criteria were considered for fish to define unconscious: decrease of Ptot, decrease of F50, decrease of SEF, increase of low frequencies (Delta and Thera) (0–8 Hz) and decrease of high frequencies (Alpha and Beta) (8–32 Hz). Such criteria have been used in birds [25], lambs [26], cattle [27,28] and fish [13,29]. However, these studies did not evaluate all the above-cited criteria used in the present study. Changes of the EEG signal to a transitional, suppressed or iso-electric phase were based on visual comparison of the signal with the baseline.

2.3. Statistical Analyses

The variables of escape attempts (proportion of fish with escape behaviour), latency (s) for onset movements from immersion in CO_2 tank and duration (s) of movements are shown with descriptive statistics. The effect of time process on Ptot, F50, SEF, low frequencies (LF) and high frequencies (HF) were analysed through linear mixed models. In this case, visual phase was included as a fixed factor with five levels (baseline, transitional, excitation, suppressed and iso-electric), and a normal distributed random effect of fish was included to adjust the variance between different fish and correlation between measurements from a single fish. Results for each fish in each time were first summarized by the sample median, and it was considered as the response variable for the linear mixed models. The adequacy of fitted models was assessed through residual analysis. Ptot values were log-transformed due to high skewness, and back-transformed when the model results were presented. The results provided by the fitted model were summarized through estimated means and corresponding 95% confidence intervals. Furthermore, the variance of random effects and corresponding intra-class correlation coefficient (ICC) are presented as well. Statistical significance ($p < 0.05$) was verified for all response variables, and the analysis was succeeded by a Tukey multiple comparison post hoc test, where the conclusions are based on a 5% global significance level. All analyses were performed using the R environment for statistical computation, version 4.1.2 [30]. The lme4 library [31] was used in fitting the regression models.

3. Results

During immersion in CO_2-saturated seawater, a series of consistent changes in the appearance of the EEG were observed: baseline (Figure 1), a transitional phase (Figure 2), a phase with high-amplitude–high-frequency signal (excitation phase) (Figure 3), a suppressed phase (Figure 4) and finally an iso-electric phase (Figure 5). Escape attempts were noted in 10 out of 29 (34.5%) fish with a median latency of 12.5 ± 17.5 s from immersion in CO_2 tank and duration of 14 ± 10.5 s (median ± Interquartile Range IQR) (Figure 6).

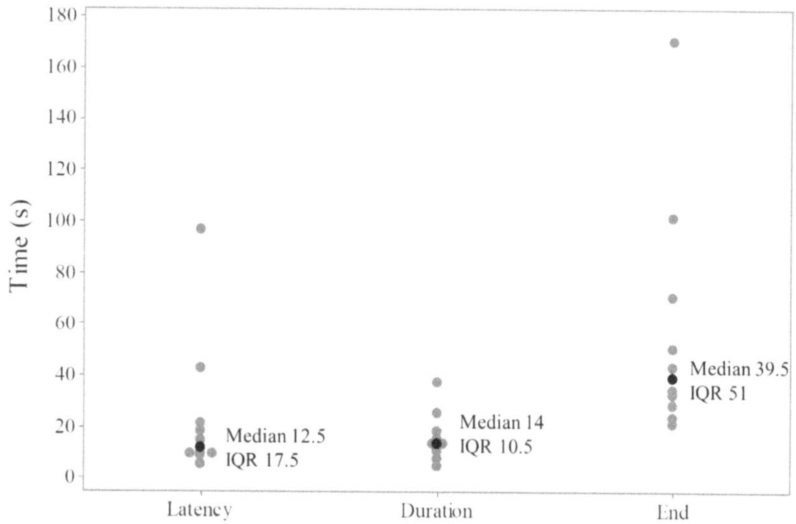

Figure 6. Individual time (s) for escape attempts from immersion in CO_2-saturated seawater. Latency for first movement, duration of movements and end of last movement. The green point indicates individual values; the black points represent the median value. Interquartile range IQR.

Timings for onset and duration of the different phases observed in EEG recordings are presented in Table 1. Both suppressed and iso-electric phases were not observed in all fish after bleeding. Furthermore, one fish did not show suppressed phase before bleeding; for this reason, the number of fish was not the same in all phases. The characteristics of EEG from each fish can be observed in Figure 7. The transitional phase was the first change observed on EEG and was followed by an excitation phase. It is possible that consciousness was not lost at the start of the excitation phase. The mean time for the onset of the suppressed phase was 258.8 ± 46.2 s. The visual characteristics of the suppressed phase suggest that halibut is unconscious at this point.

Table 1. Onset and duration (s) (mean ± SD) of different EEG phases observed in 29 halibuts from immersion in CO_2-saturated seawater.

	Transitional	Excitation Phase	Suppressed	Suppressed *	Iso-Electric *
Onset	13.7 ± 10.3	126.2 ± 55.0	258.8 ± 46.2	482.6 ± 32.4	567.4 ± 89.1
Duration	107.9 ± 54.3	134.8 ± 36.5	180.2 ± 65.6	251.4 ± 83.7	237.5 ± 94.1
n	29	29	28	19	17

* After bleeding (time after CO_2 immersion).

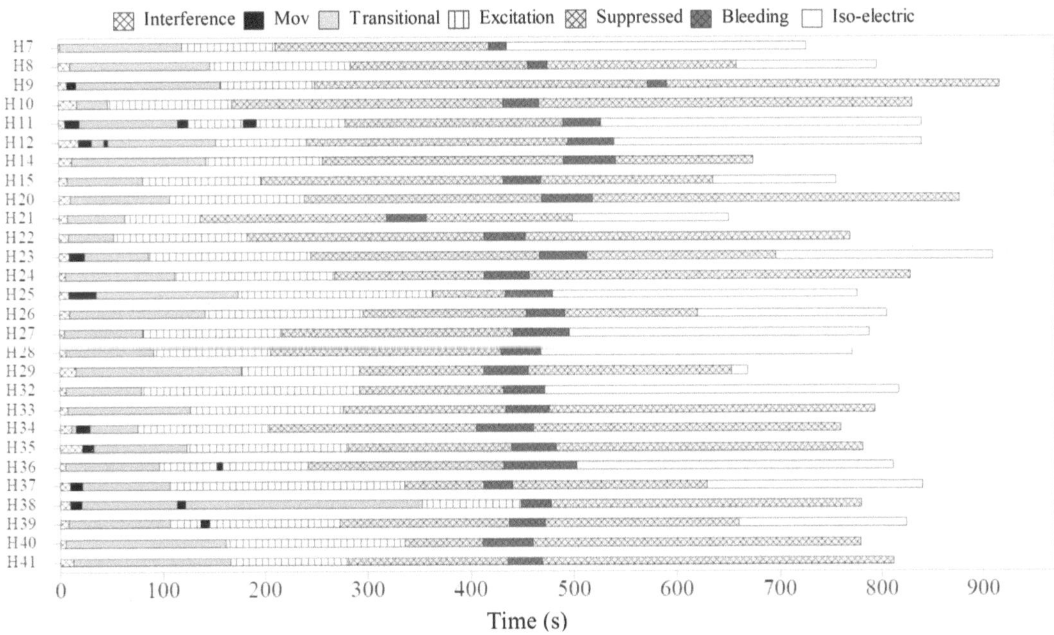

Figure 7. Characteristics of the EEG trace in halibut (n = 28) immersed in CO_2-saturated seawater.

The results from the model used to understand the spectral analysis can be observed in Table 2. The values on the excitation phase were higher in Ptot, F50 and HF due to both high amplitude and high frequency on the trace (See Figure 3). Compared to the baseline values, Ptot value was lower at the iso-electrical phase but not at the suppressed phase. However, the values of F50, SEF and HF in the suppressed phase were significantly lower in contrast to baseline, in concordance with an unconscious animal. The LF values were significantly higher at both suppressed and iso-electric phases, compared to baseline. F50, SEF, LF and HF values were similar between the suppressed and iso-electric phase.

Table 2. Summary of the fitted linear mixed models for the spectral analysis (Total Power: Ptot; Median Frequency: F50; Spectral Edge Frequency: SEF, contribution from Low Frequencies (0–8 Hz): LW; contribution from High Frequencies (8–32 Hz): HF, presented as estimated means (95% confidence interval) in halibut from immersion in CO_2-saturated seawater. Different letters at the same column indicate different means according to the Tukey multiple comparison post hoc test.

Phase	Ptot (μV^2)	F50 (Hz)	SEF (Hz)	LF (%)	HF (%)
Baseline	21.4 (11.7; 39.0) ^{B}	10.5 (9.2; 11.7) ^{B}	31.6 (29.5; 33.7) ^{A}	38.9 (31.9; 45.8) ^{B}	55.4 (49.3; 61.6) ^{B}
Transitional	32.98 (18.08; 60.2) ^{B}	10.0 (8.7; 11.2) ^{BC}	32.3 (30.2; 34.4) ^{A}	41.9 (34.9; 48.8) ^{B}	51.9 (45.7; 58.1) ^{BC}

Table 2. *Cont.*

Phase	Ptot (μV²)	F50 (Hz)	SEF (Hz)	LF (%)	HF (%)
Excitation	5729.9 (3141.22; 10,452.1) A	12.1 (10.9; 13.4) A	32.7 (30.6; 34.8) A	25.5 (18.5; 32.5) C	68.4 (62.2; 74.5) A
Suppressed	28.42 (15.58; 51.8) B	8.1 (6.8; 9.3) D	24.2 (22.0; 26.3) B	52.7 (45.7; 59.7) A	45.0 (38.8; 51.2) C
Iso-electric	5.11 (2.39; 10.9) C	8.3 (6.9; 9.8) CD	26.3 (23.7; 28.9) B	53.2 (44.9; 61.5) A	42.7 (35.4; 50.0) C
Random effect	21.4 (11.7; 39.0) B	10.5 (9.2; 11.7) B	31.6 (29.5; 33.7) A	38.9 (31.9; 45.8) B	55.4 (49.3; 61.6) B

4. Discussion

The results from spectral analysis, such as the reductions in F50, SEF and HF and the increase in LF in the suppressed phase, are compatible with the unconsciousness state, with both LOC and loss of pain sensibility. The variables of spectral and visual analysis can be used together to determine the onset of unconsciousness in fish. The main challenge in recording baselines in conscious fish was a predominance of high frequencies, which compromises the spectral analysis. Refining the handling protocol, using local anaesthesia before implanting electrodes, and reducing stress during handling can improve the quality of baseline records.

After the excitation phase, fish showed a suppression phase (Figure 4). The suppressed phase can be explained by marked depression caused by high concentrations of CO_2 in the cerebrospinal fluid [32]. We interpret the suppressed phase as typical for loss of consciousness, as the spectral analysis showed a significant decrease in F50, SEF and HF, in comparison to baseline values compatible with an unconscious animal as reported in broilers. Broilers submitted to general anaesthesia presented a reduction in F50 and SEF in comparison to baseline values [17]. In hens killed using CO_2, the suppressed EEG which followed the transitional phase had a very distinctive appearance and spectral properties were strongly associated with an unconscious state [24]. Fish stunned by an electrical current or by percussion showed a period of strong depression of the electrical activity on the EEG, after the epileptiform insult labelled in this study as "exhaustion phase" [33]. Hence, the subsequent phases observed in the EEG analysis of halibut in our study exposed to saturated CO_2 showed a pattern consistent with a progressive induction of unconsciousness, and the LOC can be estimated on the onset of suppressed phase due to changes on spectral variables.

Bowman et al. [34] found that for rainbow trout (*Oncorhynchus mykiss*) immersed in CO_2-saturated water at different acclimation temperatures of 14, 8 and 2 °C, it took significantly longer to lose equilibrium after submersion in CO_2-saturated water at 2 °C compared to 14 °C ($p < 0.05$). Furthermore, the eye-roll reflex was lost shortly after equilibrium, taking (mean, min-max value) 1.30 (1–2) min, 3.15 (2–4.5) min and 4.15 (3–5.5) min at 14, 8 and 2 °C, respectively, and the time taken for the eye-roll reflex to be lost was significantly different between all temperature groups [34]. Breathing ceased at 3.61 (2.90–4.55) min at 14 °C, 5.74 (4.01–7.33) min at 8 °C and 9.00 (7.33–11.33) min at 2 °C, and the time taken for breathing to cease was significantly different between all temperature groups [34].

In pigs exposed to 80% or 95% CO_2, time to loss of consciousness was 46 s and 33 s, respectively [35]. In poultry, McKeegan et al. [36] reported a suppressed pattern in broiler chickens exposed to 40% CO_2 + 60%. Nitrogen was observed to occur at 23 ± 4 s. Suppressed EEG occurred on average at 60 ± 23 s in broiler chickens exposed to CO_2 in a concentration below 40% in a multistage system [37]. Such times indicate a faster induction process in birds and pigs than fish, when CO_2 or CO_2 mixtures are used. Concentrations of 80% and 90% CO_2 in air for stunning are equivalent to 800,000 parts per million (ppm) and 900,000 ppm, respectively [38]. In the atmosphere, CO_2 is normally found at concentration between 300 and 700 ppm. In seawater, CO_2 concentrations of 200 to 400 mg L^{-1} for stunning are equivalent to 200 to 400 ppm, respectively. Additionally, pigs and birds are exposed to high CO_2 concentrations and hypoxic environments (<12% O_2 in air), different from our experiment in which the O_2 concentrations in water were at normal farm system levels. Therefore, fish stunning with CO_2 using both lower CO_2 concentrations in water and a normoxic environment is contrary to CO_2 stunning in pigs or birds. This might explain the reason why the onset of unconsciousness is faster in pigs and birds than in fish.

Another possible explanation for the longer times to loss of consciousness compared to broilers is related to the Bohr effect. The Bohr effect generally describes the effect of pH on the blood–O_2-binding affinity. In most teleost fish, the Bohr effect is twice as strong as in mammals because up to four additional protons are bound, thereby stabilizing the low-O_2-affinity T-state [39]. T-state haemoglobin (Tense haemoglobin) form is the most stable haemoglobin form in the absence of oxygen [40]. The presence of a Bohr effect allows more-efficient blood–O_2 transport because under the acidic conditions in the tissue capillaries, where CO_2 or lactic acid are released into the blood, the Hb–O_2 affinity is decreased. The right-shift of the O_2-binding curves allows increased offloading of O_2 without compromising venous PO_2; thereby the O_2-diffusion gradient from the blood to the tissues can be maintained [39]. The affinity of haemoglobin for O_2 is quantified with P_{50}, which is the PO_2 at 50% saturation. In most birds, and in contrast to mammals, there is no independent effect of CO_2 on P_{50}. CO_2 forms carbamino compounds with haemoglobin in mammals, and these cause small increases in P_{50}. In some birds, such as sparrows and burrowing owls, the Bohr effect is greater when pH is changed with CO_2 compared with fixed acid. Therefore, carbamino formation does occur and can decrease O_2 affinity in stripped avian haemoglobin [41]. Therefore, as this mechanism means that fish are more resistant to environments with high concentrations of CO_2, perhaps the maintenance of the pH of the cerebrospinal fluid may be more efficient in fish, explaining the slower induction of unconsciousness in comparison with birds or mammals.

The iso-electric phase was observed after bleeding (Figure 5). This phase corresponds to low-level noise reflecting a lack of EEG activity as was noted in broilers under low-atmospheric-pressure stunning [25] and during exposure to gas mixtures [42]. The iso-electric phase, in conjunction with other indicators such as the absence of reflexes and cardiac arrest, are indicators of brain death in birds under anaesthesia [17]. High concentrations of CO_2 in the brain, in addition to hypovolemia and ischemia caused by cutting the gills, suggest that the iso-electric phase is an indicator of brain death.

Given that the visual analysis from the EEG is a qualitative and subjective analysis, we performed the spectral analysis on the EEG data in Halibut. In other species, the spectral analysis from EEG traces was performed with animals under anaesthesia [17,43,44]. This approach allows the administration of various drugs that enhance analgesia, anaesthesia, muscle relaxation, hypnosis and amnesia responses, thus obtaining EEG recordings free from any interference or artefacts.

Regarding Ptot, we did not observe significant differences among the baseline and suppressed phases; the difference was observed at the iso-electrical phase. In rainbow trout (*Oncorhynchus mykiss*) submerged in CO_2 at pH < 5.0 and 10 °C, EEG signal amplitude (Ptot) increased immediately after transfer to the treatment tank compared to pre-treatment amplitude measurements. Time until amplitude declined to <50% pre-treatment amplitude was 7.44 (4–15) min after transfer [35].

Sandercock et al. [17] reported an increase in Ptot (uV^2) during sedation and deep anaesthesia, using dexmedetomidine 80 μg/kg IM and sevoflurane 8% vaporizer, respectively. Anaesthesia using propofol in turkeys induced a reduction in Ptot from EEG [44]. The progressive increase in Ptot and subsequent fall was also reported in broilers during exposure to low-atmospheric-pressure stunning [25] and hens under anaesthesia. In rainbow trout (*Oncorhynchus mykiss*) subjected to anaesthesia with MS-222 [29], a decrease in Ptot was observed from 4 min of immersion in the anaesthetic tank.

Values to F50 and SEF were significantly lower at the suppressed phase in comparison with values from baseline, a similar result reported by Hernandez et al. [45] and Sandercock et al. [17] for poultry under anaesthesia. Bowman et al. [29] did not find significant differences on F50 over the time of anaesthesia of rainbow trout when compared to baseline.

Bowman et al. [34] did not observe a significant difference between pre-treatment mean and treatment values from EEG regarding F50 in rainbow trout (*Oncorhynchus mykiss*) submerged in CO_2 water at pH < 5.0 and 10 °C over 40 min. Hence, F50 and SEF can be indicators in future studies with spectral analysis from EEG in fish. The reduction in F50 and SEF is comparable to the shift from the conscious to unconscious state, when the high frequencies (8–32 Hz) are reduced and the low frequencies (0–8 Hz) are predominant. F50 and SEF seem to be more robust for identifying unconsciousness in fish. Hence, the characteristics of the visual inspection (subjective analysis) are consistent with the observed differences in the variables F50 and SEF from spectral analysis (objective analysis) in halibut exposed to CO_2-saturated seawater.

Low frequencies (<8 Hz) waves were specifically examined, as they are associated with sleep, anaesthesia and LOC. Such increases of low frequency activity was also reported in broilers under low-atmospheric-pressure stunning [25]. In rainbow trout (*Oncorhynchus mykiss*) submerged in CO_2 water at pH < 5.0 and 10 °C over 40 min, Bowman et al. [34] showed no significant difference between the pre-treatment means and the relative power of high frequencies (Beta/Alpha) and low frequencies (Theta/Delta) over treatment. Bowman et al. [29]) did not find significant differences between low or high frequencies over the time of anaesthesia of rainbow trout when compared to baseline. In eels, the use of a captive needle pistol immediately induced low-frequency waves tending to no brain activity recorded on EEG [46,47]. Hence, low-frequency activity on suppressed and iso-electric phases suggests a deep state of unconsciousness in halibut from immersion in CO_2-saturated seawater. Results of frequency bands as determined by spectral analysis suggest that halibut immersed in CO_2-saturated seawater lose consciousness at 258.8 ± 46.2 s from the immersion. In our results, halibut showed an increase in low frequencies during the suppressed phase in comparison with the baseline, indicating a state of unconsciousness at the suppressed phase.

Further, high-frequency (8–32 Hz) activities were reduced in the suppressed phase in halibut in contrast with the baseline. Results from spectral analysis of F50, SEF, low-frequency and high-frequency values on the suppressed phase are in concordance with the shift of frequencies from an unconscious animal. Hence, in halibut, the onset for unconsciousness was at 258.8 ± 46.2 s from immersion in CO_2-saturated seawater.

Regarding the escape attempts, fish were restrained during the stunning procedure; such conditions are not suitable to assess behaviour, but strong muscle contractions can still be observed. The number of escape attempts would be probably higher in unrestrained fish. In terrestrial animals, inhalation of concentrations above 30% CO_2 by volume in atmospheric air causes aversion, irritation of membranes, pain and respiratory distress during the induction phase in pigs and poultry [48]. Trout, salmon, carp and eel exposed to CO_2 displayed vigorous aversive reactions, rapid swimming and attempts to escape [22]. Rainbow trout (*Oncorhynchus mykiss*) submerged for 12 min in CO_2-saturated water at pH < 5.0, and at different acclimation temperatures of 14, 8 and 2 °C, all showed strong aversive behaviours in the form of repeated attempts to escape the tank and vigorous swimming. The observations made in halibut in the current study had a similar pattern of reactions to high CO_2 concentrations.

Transitional, suppressed and iso-electric phases have been reported in hens, broilers, ducks and turkeys exposed to N_2- or CO_2-filled foam [16] and hens killed by CO_2 [24]. Such different phases were also described in experiments with pigs exposed to high CO_2 levels [35]. In halibut, shortly after immersion in the stunning tank, movements and muscular contractions related to the escape attempts and breathing movements from gills cause artefacts on the recorded signals and therefore influence the assessment of the EEG signal especially on the transitional phase. Similar to our results, the time for the onset of the transitional phase in hens exposed to N_2 foam was 10 s [16].

However, the duration of the transitional phase in hens was only 20 s, which is considerably shorter than in our results (108.8 ± 54.9 s). The transitional phase (Figure 2) is probably a transition trace between the baseline and the excitation phase. The EEG visual analysis alone is not sufficient to determine the onset of unconsciousness. Therefore, spectral analysis helped in confirming the onset of unconsciousness at the beginning of the suppressed phase.

After the transitional phase, a period of consistent high amplitude and high frequency (Figure 3) was observed. This is called the excitation phase, since the characteristics observed for this phase were very similar to those of a general epileptiform or grand mal insult in studies on fish stunned by electric shock or percussion [33,49]. It is possible that consciousness was not lost at the start of the excitation phase. Hence, results from the spectral analysis suggest that halibut lost consciousness at the beginning of the suppressed phase, 258.8 ± 46.2 s, due to reduction in F50, SEF and % contribution from HF. Epileptic-like EEG patterns or general epileptiform insults or excitation phase were not reported in hens, broilers, ducks and turkeys exposed to N_2- or CO_2-filled foam [16], hens killed by CO_2 [24] or broilers killed by gas mixtures [48]. The gas stunning process induces unconsciousness faster in pigs and birds, <50 s; thus, the excitation phase or general epileptiform insult were probably not observed in these species. The excitation phase could be masked by movement artefact, for example.

Similar to our results, after the transitional phase, Mckeegan et al. [16] reported a suppressed phase at 30 s after the hens were immersed in the N_2 foam, which is a reliable indicator of LOC. Broilers exposed to 2% residual oxygen in a mixture of 20% carbon dioxide and 80% argon in air showed rapid loss of spontaneous electrical activity in the first 10 s of exposure, leading to a profoundly suppressed EEG (LOC) at the onset of loss of posture at 12 s. The magnitude of EEG suppression occurring prior to the onset of convulsions at 17 s is indicative of a pathological brain state that is incompatible with consciousness and sensibility [50].

Previous studies on EEG in fish immersed in CO_2-saturated water are scarce [11,34], and the available studies used the time to loss of visual evoked responses as a measure of (un)consciousness rather than EEG recordings. Bowman et al. [34] studied rainbow trout (*Oncorhynchus mykiss*) submerged in CO_2 water at pH < 5.0 and 10 °C over 40 min, showing that there is a poor relationship between the loss of visual evoked responses and loss of visual indicators of consciousness, which suggests that when visual indicators alone are used, fish risk being misjudged as insensible before sensibility is lost. To our knowledge, there is no manuscript measuring EEG activity in halibut immersed in CO_2-saturated water and comparing the visual and the spectral variables. Our results can help to produce specifical requirements to stun/kill farmed fish according to the Annex 1 of Regulation 1099/2099 in Europe.

5. Conclusions

Stunning of halibut in CO_2-saturated seawater, which is well-oxygenated, induces vigorous movements and escape attempts. The progressive induction of unconsciousness showed four different phases on EEG. Spectral analysis from EEG traces allowed confirming the phases observed by visual inspection, namely the transitional, excitation, suppressed and iso-electric phases. The onset of unconscious was determined at the suppressed phase. Visual EEG analyses in conjunction with spectral analysis allowed a more accurate approach to assess unconsciousness.

In order to decrease suffering at the time of slaughter, the induction of unconsciousness must be immediate (<1 s) or progressive, but not aversive. In our experiment, halibut took 258.8 ± 46.2 s to lose consciousness. As movements and clear escape attempts, indicative as being aversive, were also observed, the use of CO_2 in seawater is not recommended for stunning of Atlantic halibut.

Author Contributions: Conceptualization, H.v.d.V., B.R., A.F. and M.G.; Methodology, H.v.d.V., H.R., B.R. and A.F.; Software, M.G.; Formal analysis, D.S.R., H.R. and C.A.T.; Investigation, H.R.; Data curation, C.A.T.; Writing—original draft, D.S.R. and H.R.; Writing—review and editing, D.S.R., H.v.d.V., H.R., B.R., A.F., C.A.T. and M.G.; Supervision, H.v.d.V. and M.G.; Funding acquisition, H.v.d.V. and M.G. All authors have read and agreed to the published version of the manuscript.

Funding: Financial support was given by the Norwegian Research Council (project no. 226060). São Paulo Research Foundation (FAPESP) provided a grant with scholarship to Daniel Santiago Rucinque (process number 2018/01584-0) during his stay in The Netherlands.

Institutional Review Board Statement: This study was approved on 16 May 2015 by the Norwegian authority (Mattilsynet), licence number 2015/108619.

Informed Consent Statement: Not applicable.

Data Availability Statement: Data are contained within the article.

Acknowledgments: The authors thank Marc Bracke for reviewing the manuscript. The project group thanks Sterling White Halibut for their collaboration.

Conflicts of Interest: The authors declare no conflict of interest.

References

1. FAO (Food & Agriculture Organisation). *The State of World Fisheries and Aquaculture 2022. Towards Blue Transformation*; FAO: Rome, Italy, 2022.
2. FEAP (Federation of European Aquaculture Producers). *European Aquaculture Production Report—2021*; FEAP: Biñan, Philippines, 2021.
3. Imsland, A. The Flatfishes (Order: Pleuronectiformes). In *Finfish Aquaculture Diversification*; le François, N.R., Jobling, M., Carter, C., Blier, P.U., Eds.; CAB International: Wallingford, UK, 2010; pp. 450–496. ISBN 1845934946.
4. Ellingsen, K.; Grimsrud, K.; Nielse, H.M.; Mejdell, C.; Olesen, I.; Honkanen, P.; Navrud, S.; Gamborg, C.; Sandøe, P. Who Cares about Fish Welfare?: A Norwegian Study. *Br. Food J.* **2015**, *117*, 257–273. [CrossRef]
5. Norway Animal Welfare Act. Available online: https://www.regjeringen.no/en/dokumenter/animal-welfare-act/id571188/ (accessed on 30 September 2015).
6. Röcklinsberg, H. Fish Consumption: Choices in the Intersection of Public Concern, Fish Welfare, Food Security, Human Health and Climate Change. *J. Agric. Environ. Ethics* **2015**, *28*, 533–551. [CrossRef]
7. OIE Aquatic Animal Health Code. Available online: https://www.oie.int/en/what-we-do/standards/codes-and-manuals/aquatic-code-online-access/ (accessed on 18 August 2015).
8. EU (European Union). COUNCIL REGULATION (EC) No 1099/2009 of 24 September 2009 on the Protection of Animals at the Time of Killing. Available online: http://eur-lex.europa.eu/LexUriServ/LexUriServ.do?uri=OJ:L:2009:303:0001:0030:EN:PDF (accessed on 17 April 2016).
9. van de Vis, H.; Abbink, W.; Lambooij, B.; Bracke, M. Stunning and Killing of Farmed Fish: How to Put It into Practice? In *Encyclopedia of Meat Sciences*; Dikeman, M., Devine, C., Eds.; Elsevier: Amsterdam, The Netherlands, 2014; p. 1712. ISBN 9780123847348.
10. EFSA (European Food Safety Authority). Guidance on the Assessment Criteria for Applications for New or Modified Stunning Methods Regarding Animal Protection at the Time of Killing. *EFSA J.* **2018**, *16*, 5343. [CrossRef]
11. Robb, D.H.F.; Wotton, S.B.; McKinstry, J.L.; Sørensen, N.K.; Kestin, S.C. Commercial Slaughter Methods Used on Atlantic Salmon: Determination of the Onset of Brain Failure by Electroencephalography. *Vet. Rec.* **2000**, *147*, 298–303. [CrossRef] [PubMed]
12. Kestin, S.; Van de Vis, J.; Robb, D. Protocol for Assessing Brain Function in Fish and the Effectiveness of Methods Used to Stun and Kill Them. *Vet. Rec.* **2002**, *150*, 302–307. [CrossRef] [PubMed]
13. Lambooij, B.; Bracke, M.; Reimert, H.; Foss, A.; Imsland, A.; van de Vis, H. Electrophysiological and Behavioural Responses of Turbot (Scophthalmus Maximus) Cooled in Ice Water. *Physiol. Behav.* **2015**, *149*, 23–28. [CrossRef]
14. Tonner, P.H.; Bein, B. Classic Electroencephalographic Parameters: Median Frequency, Spectral Edge Frequency Etc. *Best Pract. Res. Clin. Anaesthesiol.* **2006**, *20*, 147–159. [CrossRef]
15. Martín-Cancho, M.F.; Lima, J.R.; Luis, L.; Crisóstomo, V.; López, M.A.; Ezquerra, L.J.; Carrasco-Jiménez, M.S.; Usón-Gargallo, J. Bispectral Index, Spectral Edge Frequency 95% and Median Frequency Recorded at Varying Desflurane Concentrations in Pigs. *Res. Vet. Sci.* **2006**, *81*, 373–381. [CrossRef]
16. McKeegan, D.E.F.; Reimert, H.G.M.; Hindle, V.A.; Boulcott, P.; Sparrey, J.M.; Wathes, C.M.; Demmers, T.G.M.; Gerritzen, M.A. Physiological and Behavioral Responses of Poultry Exposed to Gas-Filled High Expansion Foam. *Poult. Sci.* **2013**, *92*, 1145–1154. [CrossRef]
17. Sandercock, D.A.; Auckburally, A.; Flaherty, D.; Sandilands, V.; Mckeegan, D.E.F. Avian Reflex and Electroencephalogram Responses in Different States of Consciousness. *Physiol. Behav.* **2014**, *133*, 252–259. [CrossRef]
18. Yoshikawa, H.; Ishida, Y.; Ueno, S.; Mitsuda, H. The Use of Sedation Action of CO_2 for Long-Term Anesthesia in Carp. *Bull. Jpn. Soc. Sci. Fish.* **1988**, *54*, 545–551. [CrossRef]
19. Erikson, U. Live Chilling and Carbon Dioxide Sedation at Slaughter of Farmed Atlantic Salmon: A Description of a Number of Commercial Case Studies. *J. Appl. Aquac.* **2008**, *20*, 38–61. [CrossRef]
20. Erikson, U. Assessment of Different Stunning Methods and Recovery of Farmed Atlantic Salmon (Salmo Salar): Isoeugenol, Nitrogen and Three Levels of Carbon Dioxide. *Anim. Welf.* **2011**, *20*, 365–375. [CrossRef]

21. Bernier, N.; Randall, D. Carbon Dioxide Anaesthesia in Rainbow Trout: Effects of Hypercapnic Level and Stress on Induction and Recovery from Anaesthetic Treatment. *J. Fish Biol.* **1998**, *52*, 621–637. [CrossRef]
22. Robb, D.H.F.; Kestin, S.C. Methods Used to Kill Fish: Field Observations and Literature Reviewed. *Anim. Welf.* **2002**, *11*, 269–282. [CrossRef]
23. Lambooij, E.; Grimsbø, E.; van de Vis, J.W.; Reimert, H.G.M.; Nortvedt, R.; Roth, B. Percussion and Electrical Stunning of Atlantic Salmon (Salmo Salar) after Dewatering and Subsequent Effect on Brain and Heart Activities. *Aquaculture* **2010**, *300*, 107–112. [CrossRef]
24. McKeegan, D.E.F.; Sparks, N.H.C.; Sandilands, V.; Demmers, T.G.M.; Boulcott, P.; Wathes, C.M. Physiological Responses of Laying Hens during Whole-House Killing with Carbon Dioxide. *Br. Poult. Sci.* **2011**, *52*, 645–657. [CrossRef] [PubMed]
25. McKeegan, D.E.F.; Sandercock, D.A.; Gerritzen, M.A. Physiological Responses to Low Atmospheric Pressure Stunning and Their Implications for Welfare. *Poult. Sci.* **2013**, *92*, 858–868. [CrossRef]
26. Velarde, A.; Ruiz-de-la-Torre, J.L.; Roselló, C.; Fàbrega, E.; Diestre, A.; Manteca, X. Assessment of Return to Consciousness after Electrical Stunning in Lambs. *Anim. Welf.* **2002**, *11*, 333–341. [CrossRef]
27. Gibson, T.J.; Johnson, C.B.; Murrell, J.C.; Mitchinson, S.L.; Stafford, K.J.; Mellor, D.J. Electroencephalographic Responses to Concussive Non-Penetrative Captive-Bolt Stunning in Halothane-Anaesthetised Calves. *N. Z. Vet. J.* **2009**, *57*, 90–95. [CrossRef]
28. Gibson, T.J.; Johnson, C.B.; Murrell, J.C.; Mitchinson, S.L.; Stafford, K.J.; Mellor, D.J. Amelioration of Electroencephalographic Responses to Slaughter by Non-Penetrative Captive-Bolt Stunning after Ventral-Neck Incision in Halothane-Anaesthetised Calves. *N. Z. Vet. J.* **2009**, *57*, 96–101. [CrossRef]
29. Bowman, J.; Hjelmstedt, P.; Gräns, A. Non-invasive Recording of Brain Function in Rainbow Trout: Evaluations of the Effects of MS-222 Anaesthesia Induction. *Aquac. Res.* **2019**, *50*, 3420–3428. [CrossRef]
30. R Core Team. *R: A Language and Environment for Statistical Computing*; R Core Team: Vienna, Austria, 2021; Available online: https://www.R-project.org/ (accessed on 25 April 2022).
31. Bates, D.; Maechler, M.; Bolker, B.M.; Walker, S. Fitting Linear Mixed-Effects Models Using Lme4. *J. Stat. Softw.* **2015**, *67*, 1–48. [CrossRef]
32. Woodbury, D.M.; Karler, R. The Role of Carbon Dioxide in the Nervous System. *Anesthesiology* **1960**, *21*, 686–703.
33. Lambooij, E.; Pilarczyk, M.; Bialowas, H.; van den Boogaart, J.G.M.; van de Vis, J.W. Electrical and Percussive Stunning of the Common Carp (*Cyprinus carpio* L.): Neurological and Behavioural Assessment. *Aquac. Eng.* **2007**, *37*, 171–179. [CrossRef]
34. Bowman, J.; van Nuland, N.; Hjelmstedt, P.; Berg, C.; Gräns, A. Evaluation of the Reliability of Indicators of Consciousness during CO_2 Stunning of Rainbow Trout and the Effects of Temperature. *Aquac. Res.* **2020**, *51*, 5194–5202. [CrossRef]
35. Verhoeven, M.; Gerritzen, M.; Velarde, A.; Hellebrekers, L.; Kemp, B. Time to Loss of Consciousness and Its Relation to Behavior in Slaughter Pigs during Stunning with 80 or 95% Carbon Dioxide. *Front. Vet. Sci.* **2016**, *3*, 38. [CrossRef] [PubMed]
36. McKeegan, D.E.F.; McIntryre, J.A.; Demmers, T.G.M.; Lowe, J.C.; Wathes, C.M.; van den Broek, P.L.C.; Coenen, A.M.L.; Gentle, M.J. Physiological and Behavioural Responses of Broilers to Controlled Atmosphere Stunning: Implications for Welfare. *Anim. Welf.* **2007**, *16*, 409–426. [CrossRef]
37. Gerritzen, M.A.; Reimert, H.G.M.; Hindle, V.A.; Verhoeven, M.T.W.; Veerkamp, W.B. Multistage Carbon Dioxide Gas Stunning of Broilers. *Poult. Sci.* **2013**, *92*, 41–50. [CrossRef]
38. OSHA (Occupational Safety and Health Administration). Carbon Dioxide In Workplace Atmospheres. Available online: https://www.osha.gov/dts/sltc/methods/inorganic/id172/id172.html (accessed on 1 October 2018).
39. Berenbrink, M. Evolution of the Bohr Effect. In *Encyclopedia of Fish Physiology from Genome to Enviroment*; Farrell, A.P., Ed.; Elsevier Inc.: San Diego, CA, USA, 2011; pp. 921–928. ISBN 9780080923239.
40. Nikinmaa, M. Hemoglobin. In *Encyclopedia of Fish Physiology from Genome to Enviroment*; Farrell, A.P., Ed.; Elsevier Inc.: San Diego, CA, USA, 2011; pp. 887–892. Available online: https://www.elsevier.com/books/encyclopedia-of-fish-physiology/farrell/978-0-12-374545-3 (accessed on 1 October 2018).
41. Scanes, C.G.; Dridi, S. *Sturkie's Avian Physiology*; Academic Press: Cambridge, MA, USA, 2022.
42. Raj, A.B.M.; Wotton, S.B.; McKinstry, J.L.; Hillebrand, S.J.W.; Pieterse, G. Changes in the Somatosensory Evoked Potentials and Spontaneous Electroencephalogram of Broiler Chickens during Exposure to Gas Mixtures. *Br. Poult. Sci.* **1998**, *39*, 686–695. [CrossRef]
43. Gibson, T.J.; Johnson, C.B.; Stafford, K.J.; Mitchinson, S.L.; Mellor, D.J. Validation of the Acute Electroencephalographic Responses of Calves to Noxious Stimulus with Scoop Dehorning. *N. Z. Vet. J.* **2007**, *55*, 152–157. [CrossRef]
44. Hernandez, E.; James, F.; Torrey, S.; Widowski, T.; Schwean-Lardner, K.; Monteith, G.; Turner, P.V. Electroencephalographic, Physiologic and Behavioural Responses during Cervical Dislocation Euthanasia in Turkeys. *BMC Vet. Res.* **2019**, *15*, 132. [CrossRef] [PubMed]
45. Hernandez, E.; James, F.; Torrey, S.; Widowski, T.; Schwean-Lardner, K.; Monteith, G.; Turner, P.V. Evaluation of Brain Death in Laying Hens During On-Farm Killing by Cervical Dislocation Methods or Pentobarbital Sodium Injection. *Front. Vet. Sci.* **2019**, *6*, 297. [CrossRef] [PubMed]
46. Lambooij, E.; Van de Vis, J.W.; Kloosterboer, R.J.; Pieterse, C. Evaluation of Captive Needle Stunning of Farmed Eel (*Anguilla anguilla* L.): Suitability for Humane Slaughter. *Aquaculture* **2002**, *212*, 141–148. [CrossRef]
47. Velarde, A.; Raj, M. *Animal Welfare at Slaughter*, 1st ed.; Velarde, A., Raj, M., Eds.; 5m Publishing: Sheffield, UK, 2016.

48. Gerritzen, M.A.; Lambooij, B.; Reimert, H.; Stegeman, A.; Spruijt, B. On-Farm Euthanasia of Broiler Chickens: Effects of Different Gas Mixtures on Behavior and Brain Activity. *Poult. Sci.* **2004**, *83*, 1294–1301. [CrossRef]
49. Lambooij, E.; Gerritzen, M.A.; Reimert, H.; Burggraaf, D.; van de Vis, J.W. A Humane Protocol for Electro-Stunning and Killing of Nile Tilapia in Fresh Water. *Aquaculture* **2008**, *275*, 88–95. [CrossRef]
50. Raj, M.; O'Callaghan, M.; Thompson, K.; Beckett, D.; Morrish, I.; Love, A.; Hickman, G.; Howson, S. Large Scale Killing of Poultry Species on Farm during Outbreaks of Diseases: Evaluation and Development of a Humane Containerised Gas Killing System. *Worlds Poult. Sci. J.* **2008**, *64*, 227–243. [CrossRef]

animals

MDPI

Article

How Well Do Dogs Cope with Air Travel? An Owner-Reported Survey Study

Katrin Jahn [1,*], Jacqui Ley [2], Theresa DePorter [3] and Kersti Seksel [4]

1 German Veterinary Clinic, Villa 112, 39th Street, Khalifa City A, Abu Dhabi P.O. Box 34867, United Arab Emirates
2 Melbourne Veterinary Specialists Centre, 70 Blackburn Rd, Glen Waverley, VIC 3150, Australia; behaviour@melbvet.com.au
3 Oakland Veterinary Referral Services, 1400 S Telegraph Road, Bloomfield Hills, MI 48302, USA
4 Kersti Seksel and Associates Ltd., Seaforth, NSW 2092, Australia
* Correspondence: drkatrin@germanvet.ae; Tel.: +971-2-556-2024

Simple Summary: Pet air travel has increased in the last decade, and 6% of pets in the US board a plane every year. Dogs have been reported to make up 58% of pets travelling worldwide. Despite this, there is little data available about air travel in dogs. A total of 635 questionnaires were collected from dog owners whose dogs had travelled by air in the last 12 months to obtain initial data on international dog air travel and how well dogs cope with air travel physically, mentally, and emotionally. Data on how dogs were being prepared for air travel, specifically regarding stress management, were also collected. Results showed that most dogs cope with and recover well from air travel but that there is a group of individuals who suffer physical, mental, and emotional ill health consequences during or after air travel, including death. Most dog owners planned air travel themselves, and over half did not seek professional advice. Stress management products such as anxiolytic medication and supplements or pheromones were only used in one-quarter of canine travellers. This study presents opportunities for all stakeholders of pet air travel, including owners, veterinarians, airlines, airports, and pet shippers, to improve pet welfare during air travel.

Abstract: It is estimated that 2 million domestic animals travel on commercial flights every year in the US alone and that dogs make up 58% of pets travelling worldwide. There has been little research on the welfare effects of air travel on dogs. The purpose of this owner-reported study was to understand how well dogs cope with and recover from air travel from a physical, mental, and emotional health perspective. An online survey questionnaire was distributed globally to pet owners whose dogs had travelled by air in the last 12 months, and the results were collected and analysed. Information was received about dog and owner demographics, logistics, and preparation for travel, as well as the dog's experience of air travel. Results showed that most dogs cope with and recover well from air travel but that there is a group of individuals who suffer physical, mental, and emotional ill health consequences during or after air travel, including death. Stress management products such as anxiolytic medication, supplements, and pheromones were underutilised and, in some instances, actively discouraged. More education of all stakeholders of pet air travel is needed to improve the physical, mental, and emotional health and welfare of canine air travellers.

Keywords: air travel; travel; stress management; welfare; human–animal bond; anxiolytics; pheromones

Citation: Jahn, K.; Ley, J.; DePorter, T.; Seksel, K. How Well Do Dogs Cope with Air Travel? An Owner-Reported Survey Study. *Animals* **2023**, *13*, 3093. https://doi.org/10.3390/ani13193093

Academic Editor: Elbert Lambooij

Received: 25 September 2023
Revised: 29 September 2023
Accepted: 1 October 2023
Published: 4 October 2023

1. Introduction

Pet travel has increased by 19% in the last decade [1], and it is estimated that 2 million domestic animals travel on commercial flights every year in the US alone [2]. Moreover, 6% of pets in the US board a plane every year, and dogs have been reported to make up 58% of pets travelling worldwide [2]. Considering these large numbers, there is generally little data available about air travel in dogs and specifically how dogs cope with air travel, and there have been only a few publications looking at air travel and stress in dogs. One study [3] investigating the physiological signs and behaviour of 24 Beagles during air transport concluded that air transportation is stressful for dogs and that sedation with acepromazine did not affect physiological and behavioural stress responses. A further study [4] looked at the relationship between kennel size and stress in 12 greyhounds travelling between Ireland and England by air and concluded that some greyhounds found the experience extremely stressful regardless of kennel size. Studies in other species show that horses experience a sharp increase in heart rate and changes in behavioural activities during air transport, especially during the transitional stages such as the aircraft ascending and descending [5] and in two Giant Pandas transported by air from China to the US, their urinary cortisol measures were highest during the time of the flight compared to the remainder of the 30-day period post-transport [6].

Regarding animal deaths, injuries, or loss on flights, data are easily accessible only from the Department of Transportation (DOT) in the US. The US DOT Air Travel Consumer Report for 2022 shows 7 animal deaths, injuries to 1 other animal, and 1 lost animal, for a total of 9 incidents, down from the 19 incident reports filed for the pre-pandemic calendar year 2019. For the calendar year 2022, 188,223 animals were transported by US airlines, for a rate of 0.48 incidents per 10,000 animals transported [7]. In pre-pandemic 2019, 404,556 animals were transported, for a rate of 0.47 incidents per 10,000 animals transported [8]. While this seems like a low incidence, it is important to consider that this report applies only to US carriers, is not a reflection of the global situation and that incidents occurring immediately after a flight, for example, in quarantine facilities, are not accounted for.

The purpose of this survey study was to collect some initial data on international dog air travel and to obtain information on how well dogs cope with air travel from a physical, mental, and emotional health perspective. Furthermore, we wanted to collect data pertaining to how dogs were being prepared for air travel and what measures were commonly being taken to manage their stress during air travel. It is our hypothesis that most dogs cope well with and recover quickly from air travel but that there are a certain number that suffer physical, mental, and emotional ill health consequences during and after air travel. We also hypothesise that dog owners worry about their dogs traveling by air and that, currently, stress management practices for dogs are underutilised or even actively discouraged.

2. Materials and Methods

2.1. Subjects

The inclusion criteria for participants of this web survey were dog owners from any country around the world whose dog had travelled by air in the last 12 months. We chose the timeframe of 12 months as the questionnaire asked several descriptive questions relating to the dog's air travel experience, and we wanted to ensure good recollection of this on the part of the participant whilst giving a large enough timeframe to capture a representative number of participants. If the participant had more than one dog who had travelled by air in the last 12 months, we asked them to choose the dog whose name started closest to the letter "A".

2.2. Survey Questionnaire and Data Collection

A web-based, electronic survey was designed in English on Survey Monkey© and initially circulated to a pilot cohort of 18 people to test functionality and ease of completion. Feedback was collected from the pilot cohort, and some minor changes were made to the survey based on this feedback.

The survey comprised 51 questions divided into three sections:

(1) **Demographic data—dog**: This section included questions about the dog's age, gender and neuter status, body weight and size, breed, hair coat, whether they were brachy-

cephalic, age at which and how the dog was acquired, purpose of having the dog, physical and mental/emotional health, training history and previous flight experience.

(2) **Demographic data—owner**: This section included questions about the owner's age, gender, education, nationality, country of residence, and how stressed they felt at the thought of their dog travelling by air.

(3) **Data about the dog's air travel process and experience**: This was the most comprehensive section and was further divided into questions about

   (a) Logistics and preparation for air travel;
   (b) The dog's experience of air travel.

This section included questions about the reason for air travel, the dog's age at the time of air travel, the route and length of air travel, the number of transit stops during air travel, where in the plane the dog travelled (for example, in the cabin or the hold of the aircraft), preparation for air travel, signs of stress/distress at different times during the air travel process and physical, mental, and emotional health after air travel.

The questions took a variety of forms, including multiple choice questions (e.g., the dog's age), ticking checklists (e.g., how owners prepared their dog for air travel), open-ended questions (e.g., which country the dog's flight started from), and Likert scale questions (e.g., how stressed the owner felt prior to their dog travelling by air). An "other" option was included wherever appropriate to allow participants to provide further information. The amount of time needed for completion was estimated at 15–20 min. The full survey questionnaire is posted in the Supplementary Materials.

In the introduction of the survey, participants were informed that their participation was completely voluntary and that they were able to stop the survey and their participation at any time if they wished. They were further informed that the survey was anonymous and that they could not be identified during any stage of the process. Consent was received from participants for data being collected and stored as a result of their participation in the survey and for this data to be published. The primary author's email address was provided as a way of contact and for any questions or concerns from participants, and participants were given the option for their data to be withdrawn at any point by contacting the same email address.

The survey was posted on various Facebook groups (including dog owner, dog breeder, pet travel, pet air travel, and global expatriate Facebook groups) as well as the primary author's Instagram and LinkedIn pages entitled "How well did your dog cope with air travel?" and sent via email to pet shippers belonging to the IPATA (International Pet and Animal Transportation Association) organisation for circulating amongst their client lists from November 2022 to February 2023. To further generate participation, we encouraged people to share links to the survey on their own social media accounts and to tag people whose dogs had travelled by air in the last 12 months. Additionally, dog owners who had completed the questionnaire were eligible to receive a free PDF designed by the first author entitled "Top tips for air travel with pets". To maintain anonymity, participants were asked to email the first author directly or share their email address on the Facebook Messenger application to receive the PDF.

2.3. Statistical Analysis

A total of 771 surveys were received; of these, 136 were discarded due to not being fully completed or being completed for a different species (cat), and a total of 635 questionnaires were statistically analysed. Data collected from the online questionnaires were initially downloaded into a Microsoft Excel spreadsheet, where it was coded. Descriptive statistics were performed using the statistics program IBM SPSS Statistics Version 29. All the answers for each criterion were plotted, and the percentage was calculated for each category.

3. Results

3.1. Demographic Data—Dog

At the time of air travel and survey completion, most dogs were adults between 2 and 8 years of age. At the time of air travel, only 10.1% were under 6 months of age, and 16.4% were over 8 years of age (see Table 1).

Table 1. Dog's age at time of air travel.

Age	N	%
Under 6 months	64	10.1
6–24 months	128	20.2
2–4 years	146	23
4–6 years	98	15.4
6–8 years	95	15
8–12 years	63	9.9
Over 12 years	41	6.5

Most dogs (59.1%) were acquired as puppies at the age of 8–16 weeks, and 20.9% of dogs were acquired at the age of 4–24 months. Just under half the dogs in the survey were acquired from private breeders, and around one-quarter of dogs in the survey were acquired from a shelter or rescue (See Figure 1).

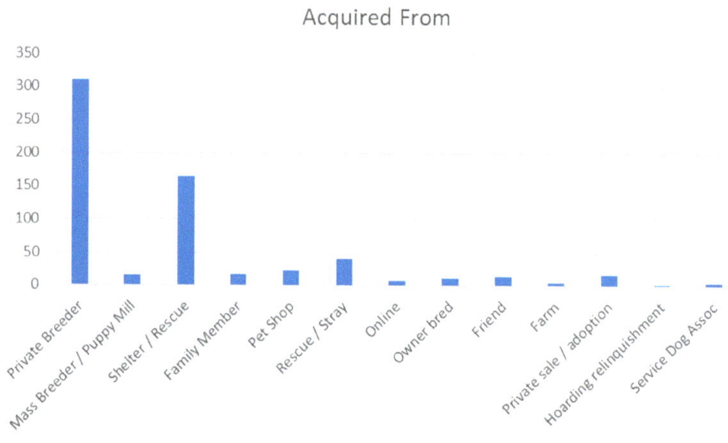

Figure 1. Where dogs were acquired.

A total of 80% of dogs in the survey were acquired as companions and around 25% as family dogs. Approximately 25% each were acquired for breeding purposes or as sports dogs.

At the time of survey completion, most dogs were neutered (67.4%), with an almost equal distribution of male and female dogs.

As shown in Figure 2, most dogs were medium-sized dogs (10–25 kg bodyweight) or small dogs (5–10 kg). Moreover, 19.7% were large dogs (25–40 kg), 13.9% were miniature dogs (under 5 kg), and only 2% were giant breed dogs (over 40 kg).

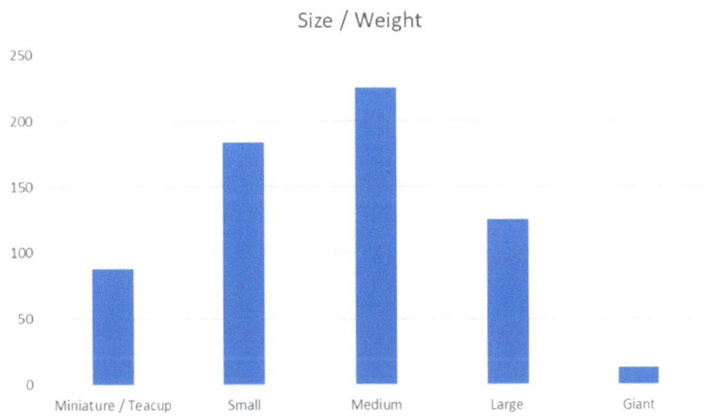

Figure 2. Size/Weight of dogs.

A total of 58.6% were purebred dogs, with the most common breed groups (according to the seven UK Kennel Club breed groups) being pastoral—also known as working- or herding breeds—(17.8%) and toy (17.6%) breeds (see Figure 3). Hybrid breeds are defined in this survey as modern designer breed dogs such as Labradoodles and Cockapoos and made up 5.8% of dogs in this study. Moreover, 8.3% (53 dogs) were brachycephalic breed dogs.

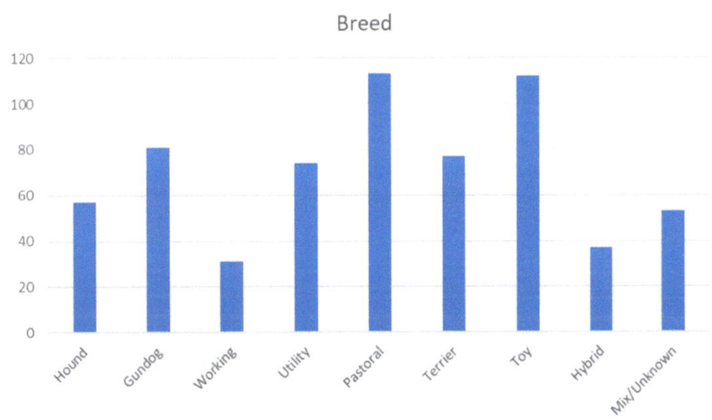

Figure 3. Distribution of breeds (according to UK Kennel Club breed groups).

Most dogs had either short or medium hair coats, 11.5% had long hair coats, and 9.4% had dense or thick hair coats.

Moreover, 67.4% of dogs did not have any owner-reported physical illnesses at the time of the flight, and where owners did report physical illnesses, 13.2% of dogs suffered from food- or environmental allergies, and 7.6% of dogs suffered from osteoarthritis or musculoskeletal disease (see Table 2).

Moreover, 44.3% of owners reported that their dog suffered from a mental/emotional illness at the time of the flight. In addition, 14.6% reported that their dog suffered from separation anxiety, and 13.5% reported that their dog suffered from more than one of the behaviour disorders mentioned for selection in the questionnaire (see Table 3).

Table 2. Owner-reported physical illness at time of air travel.

Illness	N	%
None	428	67.4
Allergies (environmental and food)	84	13.2
Osteoarthritis and musculoskeletal	48	7.6
Gastro-intestinal	17	2.7
Heart	13	2
Sensory impairment	10	1.6
Endocrine/liver/infectious	7	1.1
Respiratory tract	5	0.8
Neurological	4	0.6
Kidney	1	0.2
Neoplasia	1	0.2
3 or more concurrent conditions	17	2.7

Table 3. Owner–reported mental/emotional illness at time of air travel.

Illness	N	%
None	354	55.7
Separation anxiety	93	14.6
Noise phobia	24	3.8
Fear/phobia other	37	5.8
Aggressive behaviours	19	3
Anxiety disorder	18	2.8
Compulsive disorder	4	0.6
More than one behaviour disorder	86	13.5

Most dogs (81.1%) were not taking any regular medications at the time of air travel, and of the dogs taking medication, 5.5% were taking psychotropic medication, 3.3% were taking non-prescription supplements, 3.1% were taking regular parasite control medications, 2.2% were taking allergy control medications, and 1.9% were taking non-steroidal anti-inflammatory medications.

The distribution of frequency of air travel of the dogs in this survey is shown in Figure 4.

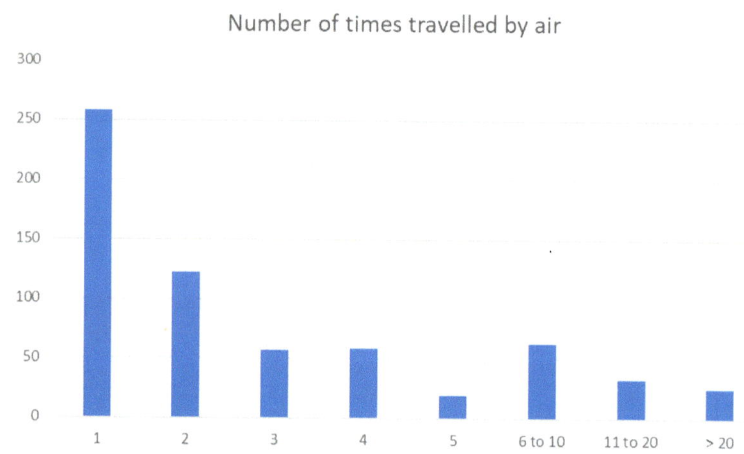

Figure 4. Number of times travelled by air.

3.2. Demographic Data—Owner

A total of 90.6% of participants were female, and 78% of participants were between the ages of 25 and 54.

Moreover, 63.5% of participants held either an undergraduate or postgraduate degree.

A total of 49% of participants resided in North America, 26.8% in Europe, and 11.2% in the Middle East/Africa region. Moreover, 5.7% resided in Asia, 4.4% in Oceania (Australia, New Zealand, and Pacific Islands), and 2.8% in Latin America. The nationality of participants was as follows: 48% were North American, 28.7% European, between 4 and 5% of participants were Asian or from Oceania, respectively, or had dual nationality, and between 3 and 4% of participants, respectively, were of the Middle East/African or Latin American nationalities, and 2.7% preferred not to answer this question.

When asked how stressed the participants felt at the thought of their dog travelling by air, 41.4% were either extremely stressed or very stressed, 34% were somewhat stressed, 17.5% were not very stressed, and only 7.1% were not at all stressed (see Figure 5).

Figure 5. Owner's stress level at the thought of their dog travelling by air.

3.3. Data about the Dog's Air Travel Process and Experience

3.3.1. Logistics and Preparation for Air Travel

The primary reason for dogs travelling by air in this survey was relocation (43.5%). This was followed by taking the dog on vacation (24.3%). Moreover, 8.5% of participants received their dog via air transportation as puppies, 7.9% of dogs travelled as service dogs, and 6.9% of dogs travelled to dog shows or competitions (see Figure 6).

Most flights both originated (52.8%) and terminated (44.4%) in North America, 22% originated and 31% terminated in Europe, 9.1% originated and 10.1% terminated in the Middle East/Africa and 8.5% originated and 5.4% terminated in Asia.

Most flights (60.9%) were direct flights, 33.1% of flights had 1 stop/layover, and 6% of flights had multiple stops/layovers.

35.4% of trips were termed long trips (10–24 h), 26.1% were medium-length trips (6–10 h), 24.3% were short trips (1–6 h), 9.4% were multi-leg with multiple stopovers, and only 4.7% were long/multi-leg trips with boarding and/or quarantine stays.

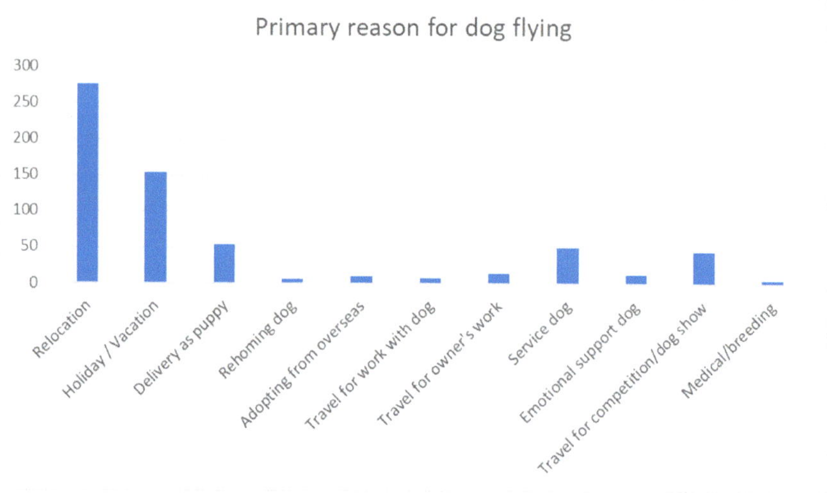

Figure 6. Primary reason for the dog flying.

Participants were asked where their dogs were cared for during different stages of the flight and were asked specifically about dedicated airport animal lounges, where dogs are kept in separated, temperature-controlled environments, are possibly taken out of their crates and offered food and water, and possibly examined by a veterinarian or non-veterinary care assistant. Dedicated airport animal lounges, if present, are usually part of the cargo shipment area and would not be used by dogs travelling in the cabin. A total of 8.4% said that their dogs did not visit a dedicated airport animal lounge, 24.9% of participants said that yes, their dog was cared for at a dedicated airport animal lounge at least at one point of air travel and 16.7% of participants did not know what a dedicated airport animal lounge was or did not know where their dog was cared for during air travel.

A total of 51.2% of dogs travelled in the hold of the aircraft, and 48.5% of dogs travelled in the cabin, either on a leash or in a carrier and either with the participant or another accompanying person.

A total of 13 dogs (2%) were said to be emotional support dogs, and of these, 11 travelled in the cabin with the participant.

Moreover, 23.1% of participants used a pet shipping company to plan the entire process of pet air travel, 66.8% planned and prepared the trip completely by themselves, and 10.1% used a pet shipping company for some aspects of the trip preparation and planned other aspects themselves.

Most participants (21.6%) started preparing their dogs for air travel 1–4 weeks before the travel date, with the second largest group (18.1%) beginning preparation 1–3 months before travel. A smaller number of owners started preparing more than 6 months before travel (14.3%), and surprisingly, 13.9% only started preparing their dogs 24 h to 1 week before travel. Around 10% each started 3–6 months before travel, 1 day before travel, or did not prepare their dogs for travel themselves (see Figure 7).

Over half of the participants (55.3%) stated that they did not seek any advice from professionals, and 6.5% used their own previous experience to prepare their dogs for air travel.

Of the participants that sought advice from professionals, 39.5% sought advice from airlines, 37.3% from IATA (International Air Transport Association), 31.8% from books, 29.8% from veterinarians, 24.9% from the Internet/Google and 15.9% from pet shippers. Advice was sought to a lesser degree from dog trainers, other pet professionals, family members, pet shipping organisations such as IPATA (International Pet and Animal

Transportation Association) and ATA (Animal Transportation Association), social media, breeders, and government agencies (see Table 4).

Figure 7. Preparation time before air travel.

Table 4. Advice sought from professionals (multiple answers possible).

Type of Professional	N	%
None	351	55.3
Airline	251	39.5
IATA	237	37.3
Books	202	31.8
Veterinarian	189	29.8
Internet/google	158	24.9
Pet Shipper	101	15.9
IPATA/ATA	61	9.6
Trainer	47	7.4
Family/Friends	41	6.5
Own experience	41	6.5
Social media	30	4.7
Breeder	13	2
Government agencies	4	0.6
Other pet professionals	116	18.3

When asked about the preparation of their dog for air travel, 67.4% of participants crate-trained their dogs prior to air travel, 74.8% visited a veterinarian to perform a physical examination on their dog, and 79.2% visited a veterinarian to perform required vaccinations, blood tests, microchip placement or parasite treatments prior to air travel.

When asked about the use of stress management or calming products, 13.4% of participants used anxiolytic medications, 12.8% used calming supplements, 9.8% used pheromone products, and 2% used a calming diet in preparation for or during the flight. A total of 5.8% of participants used CBD/hemp products, and 3.1% used aromatherapy products. A total of 76.7% of participants did not use any kind of stress management products (medications, pheromones, supplements, and diets), and seven participants (1.1%) were pro-actively advised not to use any kinds of calming or anxiolytic products before or during air travel. When anxiolytic medication or supplements were used, 49 participants (7.7%) used trazodone, 34 (5.4%) used gabapentin, 11 (1.7%) used alpha-casozepine (Zylkene®, Vetoquinol—Lure, France), 10 (1.6%) used a benzodiazepine, 6 (0.9%) each used YuCalm® (Lintbells—Hitchin, UK) and the anti-nausea medication maropitant, 3 (0.5%) each used acepromazine and Benadryl® and even fewer used clonidine, L- Theanine (Anxitane®, Virbac—Carros, France), Composure® (VetriScience—VT, USA) and Soliquin® (Nutramax—SC, USA). No participants used oromucosal dexmedetomidine gel (Sileo®, Zoetis—NJ, USA) or pregabalin.

A total of 39 participants (6.1%) used behaviour modification techniques such as desensitisation and familiarisation with certain aspects of the air travel process, four participants put familiar items in the dog's crate, and three participants used a Thundershirt® during air travel.

Other methods of preparation included the dog losing weight (1 participant), putting a tile tracker (a Bluetooth device that can be attached to an object to track its location) on the dog's collar (1 participant), and chartering a private jet for air travel (2 participants).

3.3.2. The Dog's Experience of Air Travel

Participants were asked how distressed they felt their dogs were at the following points during air travel: 1. When handed over to the pet shipper or airport staff before the flight (if the dog travelled in the hold of the aircraft), 2. At the airport or during the flight (if the dog flew in a cabin with the owner or an accompanying person), 3. At the time of arrival/collection at the airport or when the dog was delivered to the participant after the flight, the results can be seen in Table 5.

Table 5. Dog distress levels at various points of air travel.

Distress Level	Before Flight	At Airport or during Flight	Arrival/Collection/ Delivery after Flight
Not at all distressed	26%	20.5%	30.4%
Not so distressed	22.7%	18.9%	23.5%
Moderately distressed	23.8%	10.7%	19.2%
Very or extremely distressed	10.2%	6.1%	13.1%
N/A	17.3%	43%	15.1%

Participants were also asked whether they saw their dog displaying certain body language signs associated with stress at the three points during air travel: (1) when handed over to the pet shipper or airport staff before the flight (if the dog travelled in the hold of the aircraft), (2) at the airport or during the flight (if the dog flew in a cabin with the owner or an accompanying person), and (3) at the time of arrival/collection at the airport or when the dog was delivered to the participant after the flight (see Table 6).

Table 6. Dog body language signs of stress at various points of air travel.

Body Language Signs of Stress	Before Flight	At Airport or during Flight	Arrival/Collection/ Delivery after Flight
Panting	14.2%	13.1%	17.6%
Trembling	13.1%	9.4%	11.5%
Whining	11.5%	8.8%	14.5%
Tense	10.4%	7.2%	9%
Pawing	6%	7.7%	9.8%
Barking	6.9%	3.8%	8.8%
Cowering	4.6%	2.2%	5%
Hiding	5.8%	3.3%	5.5%
Pacing	3.6%	2.5%	6%
Urination	0.6%	1.7%	5%

A total of 70.1% of dogs showed at least one sign of the above-mentioned stress signs at handover before the flight, 72.4% of dogs showed at least one of the above-mentioned stress signs during the flight, and 32% of dogs showed at least one of the above-mentioned signs of stress at arrival/handover after the flight.

Moreover, 3% of respondents stated that their dog had shown "other" signs of stress before, 4.1% during and 4.9% after the flight (open-ended question box). Some of these responses included being wide-eyed, the tail being tucked, being restless, unable to settle and hyper-alert to the surroundings, biting on the water bowl and mesh covering of the travel crate, chewing the leash tied onto the crate to pieces, having a raw nose from rubbing it on the crate door, being desperate to urinate and defecate, being thirsty, being tired, not wanting to eat food or treats, not moving, having trouble standing and walking, trying to escape from crate, pulling on the leash whilst walking through the airport, lunging from inside the crate when approached, and, sadly, one dog died in quarantine after the flight.

Most dogs (83.6%) did not receive medical attention from a veterinarian immediately after air travel. Of the dogs that did receive medical attention after the flight, 83 dogs (13.1%) received a routine veterinary examination, 18 dogs were seen unrelated to air travel (for example, to refill regular medication), 8 dogs were seen to treat a major health issue, 1 dog was seen to treat a minor health issue, 1 dog was seen to treat a mental/emotional/behavioural condition, and 2 dogs were seen to confirm death.

Participants were asked to comment on the dog's stress level at various times following air travel (48 h, 3–7 days, 8–30 days, and over 30 days after air travel), and the responses can be seen in Table 7.

Table 7. Dog stress levels at various times after the flight.

Stress Level	48 h after Air Travel	3–7 Days after Air Travel	8–30 Days after Air Travel	Over 30 Days after Air Travel
Not at all or not much more stressed than usual	71.4%	84.4%	91.8%	96.4%
Moderately more stressed than usual	19.2%	10.4%	6%	3%
Much more or extremely more stressed than usual	9.4%	5.2%	2.2%	0.6%

Participants were asked about the dog's behaviours and signs of stress at various times following air travel (48 h, 3–7 days, 8–30 days, over 30 days). Between 34% (2 days after air travel) and 39.4% (over 30 days after air travel) of participants stated that their dogs did not show any behaviours or signs associated with stress after air travel. Of the dogs that did show behaviours or signs associated with stress, the most commonly seen are listed in Table 8.

A total of 30 participants (4.7%) responded that their dog had shown "other" behaviours or signs of stress after air travel. These included howling and crying when being left alone, especially for longer periods, being more alert and jumpier, being unwilling to go for a walk, being "overheated", vomiting, and being unsettled with another dog in the household.

A total of 13.8% of participants stated that their dog developed a behaviour problem within 3 months after the flight. Moreover, 4.3% of dogs developed separation anxiety, 3% of dogs were generally more anxious, and 3.6% of dogs developed more than one of the behaviour disorders mentioned for selection in the questionnaire (see Table 9).

Table 8. Behaviours and stress signs seen at various times after the flight.

Stress Behaviours	48 h after Air Travel	3–7 Days after Air Travel	8–30 Days after Air Travel	Over 30 Days after Air Travel
More clingy	31.8%	15.4%	6.1%	3.5%
Anxious	18.1%	9.1%	4.4%	2.7%
Sleeping more	17.3%	-	-	-
Unsettled	14%	5.8%	-	-
Low appetite	13.9%	-	-	-
Less interactive	13.2%	-	-	-
Thirsty	11.7%	-	-	-
House soiling	7.4%	-	-	-
Sleeping less/poor quality sleep	6.8%	-	-	-
Fearful of triggers in the environment	6.6%	5.5%	3%	2%

Table 9. Development of an owner–reported behaviour problem within 3 months after the flight.

Behaviour Problem	N	%
None	545	86.2
Separation anxiety	27	4.3
Noise phobia	5	0.8
Increased anxiety	19	3
Fear of specific triggers	6	0.9
Aggressive behaviours	6	0.9
Compulsive disorder	1	0.2
More than one behaviour problem	23	3.6

A total of 8.2% of participants thought that the behaviour change was a result of the flight, 12.1% thought it was due to a change in the dog's physical environment, and 9% thought it was due to a change in the dog's social environment (more than one answer was possible).

A total of 10.4% of participants stated that their dog experienced worsening of an already existing behaviour problem within 3 months after the flight. Moreover, 4.1% of dogs experienced a worsening in separation anxiety, 1.1% experienced a worsening in aggressive behaviours, and 3.6% of dogs experienced a worsening of more than one of the behaviour disorders mentioned for selection in the questionnaire (see Table 10).

Table 10. Worsening of an owner–reported behaviour problem within 3 months after the flight.

Behaviour Problem	N	%
None	568	89.6
Separation anxiety	26	4.1
Noise phobia	1	0.2
Increased anxiety	6	0.9
Fear of specific triggers	2	0.3
Aggressive behaviours	7	1.1
Compulsive disorder	1	0.2
More than one behaviour problem	23	3.6

A total of 5.8% of participants thought that the worsening of the behaviour problem/s was a result of the flight, 7.4% thought it was due to a change in the dog's physical environment, and 6.3% thought it was due to a change in the dog's social environment (more than one answer was possible).

A total of 8.3% of participants stated that their dog experienced worsening of an already existing physical illness within 3 months after the flight. Moreover, 2.7% of dogs experienced a worsening in food or environmental allergies, 1.9% experienced a worsening in osteoarthritis or musculoskeletal disease, and 1.3% of dogs experienced a worsening in gastro-intestinal disease (see Table 11).

Table 11. Worsening of an already existing owner–reported physical illness within 3 months after the flight.

Illness	N	%
None	582	91.7
Allergies (environmental and food)	17	2.7
Osteoarthritis and musculoskeletal	12	1.9
Gastro-intestinal	8	1.3
Heart	1	0.2
Sensory impairment	1	0.2
Endocrine/liver/infectious	4	0.6
Respiratory tract	4	0.6
Neurological	1	0.2
Kidney	1	0.2
Neoplasia	2	0.3
3 or more concurrent conditions	1	0.2

A total of 2.8% of participants thought that the worsening of the physical illness was a result of the flight, 5.7% thought it was due to a change in the dog's physical environment, and 2.2% thought it was due to a change in the dog's social environment (more than one answer was possible).

3.4. *The Two Dogs That Died*

Questionnaire 28 was a 4–6-year-old female neutered, ex-racing greyhound weighing 25–40 kg. She was obtained from a shelter between 4–24 months of age. There were no known pre-existing physical medical illnesses; however, her owner–reported pre-existing mental/emotional illnesses, namely separation anxiety and fears and phobias of specific triggers. She was not taking any medication at the time of the flight and had never travelled by air before. Her owner was very stressed about the thought of her dog travelling and used a pet shipping company to arrange the entire travel process.

Travel was to be from the UK to Australia (a long, multi-leg trip including quarantine), with one transit in Singapore; however, the dog did not make it to Australia and died in the quarantine facility in Singapore. The dog travelled in the hold of the aircraft.

In preparation for travel, the dog underwent a physical examination by a veterinarian and received vaccinations and blood tests for travel. No further preparation or stress management was performed, and the owner was told that "dogs are not allowed any calming medications" for air travel, which was advice received from a veterinarian and the pet shipping agent.

The dog was very distressed at handover, and behavioural signs of stress included trembling, panting, cowering, hiding, being tense, scratching, and "shake off".

Questionnaire 330 was a 4–6-year-old male entire, French Bulldog of medium size (10–25 kg). He was obtained at 8–16 weeks of age from an animal farm/shop in Dubai as a family companion and for his breed and look/aesthetics. He had a known owner-reported pre-existing physical disease in the form of skin allergies and food hypersensitivities but suffered from no known mental/emotional disease and was not taking any medications. He had travelled by air five times before.

His owner was extremely stressed at the thought of her dog travelling and used a pet shipping company to arrange the entire travel process. Travel was for the purpose of relocating to live in another country, and the route was from Spain to Singapore (long trip—10–24 h) with multiple transits/stops. The dog flew in a "special animal freight plane", and the owner does not know if this was a cargo or passenger plane or if the dog flew in the hold or cabin of the aircraft.

In preparation for travel, the dog underwent a physical examination by a veterinarian and received vaccinations and blood tests for travel. Preparation started 3–6 months prior to travel and included travel crate familiarisation. No further preparation or stress management was performed.

The dog was reported to be "not so distressed" at handover before the flight, and sadly, he was found dead on arrival in his crate in Singapore.

4. Discussion

4.1. Subjects and Data Collection

The validity of behavioural survey reporting by pet owners as a form of observational study has been confirmed in several papers, and questionnaires are a commonly used method of collecting behavioural data from companion animals [9]. However, despite their frequent use, they are somewhat subjective, as they do not collect data straight from the animals but from their owners [10]. In addition, observational studies bring some challenges, including confounding and recall bias, which may influence the quality of the data. The significant female response bias (90.6%) is well-recognised in both veterinary and human surveys. This may be due to the increased willingness of women to express opinions via a survey, female caregiver status, greater engagement in social media, or wider exposure to survey recruitment strategies [11].

A limitation of the study design is the lack of a control group of non-air travelling dogs to distinguish whether findings are specific to air travel or can occur during other stressful events or, in general, daily canine life. A further limitation of the study design is that the questionnaire was only available in English and, if the study were to be repeated for either dogs or other species, translation of the questionnaire into other languages, such as Spanish, should be considered to reach a wider and possibly more representative audience.

4.2. Demographic Data—Dog

The dogs in this survey are mostly "owned" companions or "pets", likely have one or two dedicated caretakers, and are considered part of the family. This is validated by the fact that most dogs in this survey were adults or seniors (69.8%) at the time of air travel, and the two primary purposes for air travel were relocation (43.5%) and taking the dog on vacation (24.3%). Moreover, 80% of participants stated that their dog was obtained for companionship and 23.5% for different types of dog sports, indicating a strong bond. A strong human–animal bond may benefit animal welfare [12], and animals that are part of a human-animal bond are more likely to be treated as individuals and with special concern for their welfare [13], and it is, therefore, possible that the dogs in this survey were prepared for air travel with a greater level of concern for their welfare. However, a strong human-animal bond may also be the source of compromised welfare [12], and it is possible that, in some cases, air travel was chosen for dogs that were perhaps not suitable, either from a physical or a mental/emotional health perspective to undergo air travel. There is a second air travelling dog group, namely puppies or pregnant bitches, which are shipped either legally or illegally for the purposes of "puppy-trafficking" or resale and are faced with different or additional challenges to their health and welfare [14].

A total of 64.4% of dogs in this survey were either medium or small-size dogs (between 5 kg and 25 kg). It is possible that there are more small- and medium-sized dogs in the general dog population or that these sizes are more suitable for air transport, both from a logistical and financial point of view, and that owners who saw themselves relocating or travelling by air in the future chose small- or medium-sized dogs as companions to facilitate this.

Only 8.3% of dogs in this survey were brachycephalic breeds, which, given current brachycephalic breed popularity [15], seems underrepresented and contrasts with the number of brachycephalic breed dogs shipped through London Heathrow Airport between

2012 and 2017, which was 26% of all dogs [16]. It is possible that more brachycephalic breeds were transported by air between 2012 and 2017 and that this has now reduced due to perceived higher risks of air travel as well as airline reluctance to transport these breeds. It is also possible that there are regional differences and that more brachycephalic breed dogs were transported into and out of the UK compared to globally. Finally, it is possible that the high number of brachycephalic breed dogs being shipped through the UK from 2012–2017 were indeed large numbers of brachycephalic puppies being shipped for resale due to the popularity of these breeds. One omission in the current survey questionnaire was a question regarding the Brachycephalic Obstructive Airway Syndrome (BOAS) assessment of brachycephalic breed dogs, whether this was performed, and whether this had any consequence on decision-making and their ability to travel by air.

The most commonly reported physical ill health presentations at the time of air travel were environmental and food allergies (13.2%); however, only 2.2% of dogs were taking anti-allergic medication at the time of air travel. A total of 7.6% of dogs were reported to have osteoarthritis or another musculoskeletal disease; however, only 1.9% were being treated with non-steroidal anti-inflammatory medications.

A total of 44.3% of dogs were reported to be suffering from one or more mental health diseases, but only 5.5% of dogs were being treated with psychotropic medications. We need to consider whether physical diseases were diagnosed and being managed by a veterinarian and whether mental/emotional ill health disorders were diagnosed and being managed by a veterinary behaviourist or veterinarian with a special interest in behavioural medicine, who would most likely have recommended medical treatment, or whether these conditions were self-diagnosed by the participant. In one study on canine food allergies [17], 60% of dog owners suspected that their dog had a food allergy before their veterinarian, and it may be possible that pet owners reported diseases in their dogs in this survey without veterinary diagnosis or treatment. The internet is the most widely used source for health information, and given the fact that most pet owners think of their pets as family members and are just as vested in their pets' health and wellbeing as other family members or themselves, it is reasonable to assume that many pet owners are using the internet to help educate and guide them in making decisions about the health of their pet, which may lead to self-diagnosis [18]. Obtaining an accurate diagnosis and initiating veterinary-prescribed treatment of both physical and mental/emotional health disorders before air travel may be an important factor in managing a pet's physical, mental, and emotional well-being and welfare during and after air travel.

Almost 60% of dogs in this survey had travelled by air once or twice. Canine air travel can be complex to organise, financially expensive, and, according to this survey, is either extremely or very stressful for over 40% of pet owners and these may be some of the reasons for the infrequency of air travel for most dogs in this study. Under 10% of dogs were frequent flyers with over 10 air travel events. Further exploration of the data are necessary to find correlations and will be part of a future paper.

Two further omissions in Section 1 (dog demographic factors) of the questionnaire were noted at the time of data analysis: (1) information regarding whether dogs had been groomed (shaved or clipped) before air travel to help with temperature regulation and (2) the question regarding the dog's neuter status was addressed to the time of survey completion and not at the time of the air travel. If the survey were to be repeated, these questions may want to be changed or included.

4.3. Demographic Data—Owner

When looking at nationality and country of residence of the participants, around half were North American, around one quarter were European and the remaining one quarter of participants were divided between Middle East/Africa, Asia, Oceania (Australia, New Zealand, and Pacific countries), and Latin America. This correlates approximately with the regions of both flight departures and arrivals for the dogs travelling by air (45–52% North America, 22–31% Europe, 10% Middle East/Africa, and 5–8% Asia) in this survey.

It is likely that canine, domestic air travel within North America is both easier and more affordable due to reduced requirements for documentation (import- and export certificates), quarantine, and complex veterinary preparation such as blood tests and parasite control. This may explain why 50% of participants in this survey were North American and around 50% of flights both originated and terminated in North America. It may also be that air travel is just a more commonly accepted way of domestic travel for people in North America and that dog owners find this a "normal", easy, and acceptable way of taking their dogs on trips with them. The same may apply to a lesser degree within the European Union. Canine air travel to Australia, New Zealand, and some countries in Asia involves much more financial commitment and preparation, sometimes months in advance, in terms of blood testing and documentation, and requires a mandatory quarantine stay for all dogs arriving, and this may account for the reduced numbers of participants from these countries and travelling to these countries.

As mentioned, over 40% of participants were either very or extremely stressed, and a further 34% were somewhat stressed by the thought of their dog travelling by air. This is much higher than the participant-reported stress levels of the dogs themselves at various points of air travel and in the first month after air travel. There could be several reasons for this. Owners may experience a lack of control, which can be stressful for many people [19]. Another factor adding to this may be the lack of transparency when it comes to pet air travel, especially when dogs travel in the hold of the aircraft. Owners are forced to relinquish control once the dog is handed over to the pet shipper or airport staff, and they often do not know where their pets, which may be regarded as "family members", will be kept before, during, and after the flight. When asked about the care of dogs travelling in the hold during air travel, 16.7% of owners stated that they did not know where their dogs were cared for during air travel, and most owners knew that this would not be in a dedicated airport animal lounge. Dedicated airport animal lounges are only available in a small number of international airports, including Frankfurt, Amsterdam, London Heathrow, and JFK, and this may be an area of consideration when thinking about welfare improvements in international pet air travel in the future.

Despite many owners reporting feeling stressed, the participants did not rate their dogs as showing levels of stress in the same numbers as themselves. A reason for the lower incidence of reported stress levels in dogs could be a lack of education when it comes to the recognition of canine fear, anxiety, and stress signs. For example, some dog owners may not recognise a dog "freezing" and being very still due to fear and may interpret this as the dog being calm. In contrast, they would recognise and be able to report more accurately on feelings of stress experienced by themselves. It is also possible that dog owners do not know where and how to access professional advice and support when it comes to preparing their dogs for air travel, which may contribute to feelings of stress and being overwhelmed. A total of 66.8% of participants planned and prepared their dog's air travel themselves, without the support of a pet shipper, and over half of the participants did not seek any advice from professionals to prepare their dogs for air travel. Making professional support, advice, and education regarding canine air travel more widely and easily available may contribute to reducing dog owner stress.

Finally, there are a number of alarming reports in the media of dogs dying both in the cabin and the hold of aircraft during flights [20,21], which can understandably be seen as a large source of stress for dog owners whose dogs are about to travel by air. While these events are horrific, they are not the norm for pet air travel, and most pets arrive alive and well.

4.4. Data about the Dog's Air Travel Process and Experience

4.4.1. Logistics and Preparation for Air Travel

Over 60% of flights in this survey were direct flights. As mentioned in the introduction, studies looking at air travel in horses [5] noted sharp increases in heart rate and changes in behavioural activities, especially during the transitional stages, such as the aircraft ascending and descending. Although these data do not exist in dogs, it may still be reasonable to assume that multiple transit phases with multiple ascents and descents may be more stressful for dogs, and therefore, the most direct route of air travel should be chosen wherever possible.

Interestingly, there was an almost equal distribution of dogs travelling in the cabin and the hold of the aircraft in this study. In the authors' experience, travel in a cabin is restricted to small dogs (under 8 kg weight) for most international flights, and therefore, it is possible that the large number of North American domestic flights, where medium- and large-sized dogs are also permitted to travel in the cabin may have influenced the more equal distribution of these data.

As discussed previously, two-thirds of participants planned their dog's air travel journey by themselves, and over half did not seek any professional advice when preparing their dogs for air travel. Reasons for this may include cost, not knowing where or how to easily access advice and support, and the fact that the logistics of domestic flight planning within North America is straightforward and may not require support from professionals. Further data analysis is necessary to see correlations between route/region and preparation for air travel and will be included in a future paper.

Where advice was sought from professionals, most was sought from the airlines and IATA, with veterinarians only accounting for around 30% and pet shippers accounting for only around 16% of advice given. This means that there is a great opportunity for veterinarians and pet shippers to be more involved with further education, especially when it comes to safe stress management methods for pet air travel. A recent paper [22] describes a number of safe stress management options for air travel in cats, many of which can be transferred to dogs. Owners and pet care professionals, including pet shippers, may not be aware of these options, and further widespread education on this topic is necessary and should be encouraged and embraced by all professionals in the pet shipping industry.

The most widely used air travel preparation methods were crate familiarisation (67.4%), physical examination by a veterinarian (around 75%), and medical procedures such as vaccinations, blood tests, microchip placement, and parasite treatments (around 80%). One of the most reproducible methods for reducing a pet's stress levels during air transport is for the pet to be acclimatised to the carrier and trained to use it at home [23]. In the home environment, the carrier can become a 'safe' zone and/or a place that the pet associates with pleasant experiences, such as feeding, so that on the day of travel, the pet is less likely to be concerned about being inside the carrier [22]. Enough time must, therefore, be allocated for familiarisation with the carrier. The veterinary examination and other veterinary procedures are usually legal requirements for air travel, which is likely why these numbers were high. In contrast to this, almost 80% of participants stated that they did not use any stress management products such as anxiolytics, supplements, or pheromones, and seven participants were actively advised not to use any stress management products.

There is a lot of controversy about medicating pets for air travel, and much of this stems from the traditional use of sedatives and tranquilisers such as acepromazine. Almost half of airline transport deaths in animals from 1990 to 1995 resulted from sedation [24]. According to the IATA (International Air Transport Association) Live Animal Regulations (LAR) [25], sedating animals either before or during a flight is deemed a considerable risk to the animal and is, therefore, not recommended.

Historically, acepromazine has been prescribed to alleviate stress during air transport. However, tranquilisation with acepromazine (0.5 mg/kg) does not affect the physiological or behavioural stress responses in dogs during air transport [3].

Acepromazine is a phenothiazine derivative that has minimal impact on the animal's emotional state of fear and/or anxiety; in fact, the tranquillising effect is instead dependent on motor inhibition mechanisms [26]. Phenothiazines have poor anxiolytic properties, cause sedation, and may increase startle reactions to noise [27], making them a poor choice for the management of fear, anxiety, and stress during air travel. In the authors' opinion, it is likely that the historical use of acepromazine and its unfavourable effects have influenced opinion and the willingness to use other kinds of medication to manage stress in pets during air travel. With the advent of modern anxiolytics and other modalities for anxiolysis and stress reduction in the last years, it is important to consider anxiolytic medications to mitigate stress and, therefore, positively impact welfare in dogs travelling by air. What we should be considering now are medications that can be classed as "True Anxiolytics", and that can alleviate fear whilst preserving the animal's ability to function relatively normally, both emotionally and physically [28].

Short-acting anxiolytic medications that are commonly used for pet-related anxiety include gabapentin and pregabalin, trazodone, benzodiazepines, and maropitant for motion sickness [22]. There are some potential side effects to using anxiolytic medications, including (but not limited to) ataxia, sedation, and paradoxical hyperactivity [22], all of which are not desired during air travel and, therefore, anxiolytic medications should always be trialled before air travel and should always be used under the guidance of a qualified veterinarian familiar with these medications [22]. Effective anxiolytic supplements include those with the active ingredients alpha-casozepine (Zylkene®, Vetoquinol—Lure, France) [29] and L-Theanine (Anxitane®, Virbac – Carros, France, Composure®, VetriScience—VT, USA, Solliquin®, Nutramax—SC, USA) [30]. Canine appeasing pheromone products such as Adaptil® (Ceva – Marseille, France), sprayed in all eight corners of the travel crate 10–15 min before putting the dog into the crate prior to travel [22], can also be very useful for managing stress during air travel [30–33].

4.4.2. The Dog's Experience of Air Travel

When looking at owner-reported canine distress levels before, during, and after the flight, an average of 10% of owners thought their dogs were very or extremely distressed at these times. Owners reported that they saw the highest number of very or extremely distressed dogs (13.1%) after the flight and the lowest number of very or extremely distressed dogs (6.1%) during the flight. The lowest number of distressed dogs overall during all 3 phases of air travel were reported to be those that travelled in the cabin with their caretakers, suggesting that being able to travel with your caretaker in the cabin is less stressful than travelling alone in the hold of the aircraft.

Body language signs of stress before, during, and after the flight were reported consistently, with panting, trembling, and whining being the three most commonly seen stress signs. It is possible that "panting" was a true stress sign; however, panting could also be a temperature regulation behaviour by the dog. The questionnaire did not specify whether "pawing" related to pawing at the crate door or possibly pawing at people, so it is difficult to know what this behaviour looked like exactly and whether it was an attempt to escape or get out of the crate or perhaps appeasement or attention-seeking behaviour. A further observation was that only a very small number of dogs urinated either during or after the flight, which may be due to a reluctance of well-house-trained dogs to urinate in an inappropriate location such as their crate or could be a sign of dehydration due to reduced water intake and panting. From a welfare point of view and considering that most of the flights described in this survey were long- (10–24 h) or medium-length (6–10 h) trips, providing better elimination opportunities during the process of air travel is something that should be looked at. Anecdotally, and from the authors' experience at airports, there are not many suitable elimination areas that are hygienic or attractive to dogs, and this could be a suggested area of improvement for airports.

Around 10% of dogs were much more or extremely stressed than normal 48 h after the flight; however, this reduced over time to 0.6% 30 days after air travel. The number of dogs that were moderately stressed after air travel also reduced from 19.2% 48 h after air travel to 2% 30 days after air travel. The most common stress behaviours seen after air travel were "being more clingy", being anxious, sleeping more, and being unsettled, all of which were reduced by 30 days after air travel.

One omission on the questionnaire in this section was a question about the state of repair of the dog's travel crate upon arrival. This may have given further information about distress behaviours during the flight, such as pawing at or pushing into the crate door, as well as any traumatic experiences the dog may have had during the flight due to the crate being damaged externally, for example, from it being dropped or crushed.

The stress and distress behaviours evaluated in this survey were all owner or caretaker-reported behaviours, which poses some problems. Dog owners are often not educated in canine body language and may have overlooked or misinterpreted some behaviours. Again, dogs may have been "frozen" or still as an expression of their fear or stress, which may have been interpreted by the owner as the dog being "calm", and this group of stressed dogs may therefore have been overlooked.

On the other hand, most dog owners know their dogs well, especially if they have been bonded for many years, and being in a stressful situation may make them more vigilant and observant of their dog's behaviour.

Most dogs (85–90%) did not develop or did not experience a worsening of owner–reported behaviour problems after air travel. Of the dogs that did, the largest number showed signs of separation anxiety or showed more than one behavioural presentation. As mentioned previously, the owner-reported separation anxiety might not be a diagnosis in the true sense of the classification used in veterinary behavioural medicine but may rather be an expression of hyper-attachment shown by the dog after a stressful event. Hyper-attachment is characterised by the dog constantly looking for contact with their owners and being more likely to follow their owners around [34], a behaviour that may be consistent with the description of "more clingy" in the questionnaire. Hyper-attachment can result from several mechanisms including neoteny, which is the retention of infantile characteristics into adulthood and is a consequence of the domestication process and may have increased the tendency of some dogs to develop a strong attachment to their owners [34]. The development or worsening of more than one behavioural sign may be explained by the fact that stress affects overall mental health and increases the risk of generalised anxiety, separation anxiety, phobias, compulsive behaviours, and posttraumatic stress disorders [35].

Over 90% of dogs did not experience a worsening of owner-reported physical health signs after air travel, and of those that did, allergies and osteoarthritis/musculoskeletal problems were the most common presentations.

In all instances, dog owners felt that the development or worsening of behavioural or physical conditions was as likely to be attributed to changes in the physical or social environment as to the flight itself.

Sadly, two dogs in this survey passed away: one dog during and one directly after air travel in the quarantine facility. Although this is too small a number of casualties to detect any statistical significance, there are some characteristics that are noteworthy when considering these dogs. One dog was a brachycephalic breed dog, and the other dog suffered from owner-reported pre-existing mental/emotional illnesses, namely separation anxiety and fears and phobias of specific triggers, which were not being treated. Neither dog received any stress management products for air travel, and one dog was not familiarised with the travel crate. Further research is essential to understand which dogs are at greatest risk of losing their lives during air travel, and stakeholders must work together to prevent this.

The results in this final section of the questionnaire fit with our hypothesis that most dogs cope well with and recover quickly after air travel but that there are a certain number that suffer physical, mental, and emotional ill health consequences during and after air travel.

Finally, it is important to consider the limitations of this study and that the findings are based on owner-reported data, which may be biased or inaccurate. In addition, the number of participants in the study was relatively small compared to the number of dogs that travel by air every year, and data may, therefore, not be representative of all dogs that travel by air.

5. Conclusions

The results of this survey suggest that most owned pet dogs that undergo air travel cope with and recover well from air travel, but that there is a group of individuals who suffer physical, mental, and emotional ill health consequences during or after air travel, including death. Travelling in the cabin with the owner or a caretaker seems to be less stressful than travelling in the hold of the aircraft, as rated by the owners.

Most participants in this survey planned their dog's air travel themselves, and over half did not seek professional advice for their dog's preparation. Most dogs did not receive stress management products such as anxiolytic medication, supplements, or pheromones, and some owners were actively discouraged from using these. Of the professionals that advice was sought from, only 30% were veterinarians, and 16% were pet shippers, highlighting opportunities for these two groups of professionals, in particular, to better support and advise pet owners before their dog's air travel and improve welfare outcomes for canine travellers. This survey also highlights improvements that can be made at airports, such as better provision for toileting areas and dedicated airport lounges for pets.

Supplementary Materials: The following supporting information can be downloaded at: https://www.mdpi.com/article/10.3390/ani13193093/s1, PDF of Survey Monkey© Questionnaire.

Author Contributions: Conceptualisation, K.J., J.L., T.D. and K.S.; methodology, K.J., J.L. and T.D.; software, J.L.; validation, K.J., J.L., T.D. and K.S.; formal analysis, J.L.; investigation, K.J. and J.L.; resources, K.J., J.L. and T.D.; data curation, K.J. and J.L.; writing—original draft preparation, K.J.; writing—review and editing, K.J., J.L., T.D. and K.S.; supervision, J.L., T.D. and K.S.; project administration, K.J.; visualisation, K.J. All authors have read and agreed to the published version of the manuscript.

Funding: This research received no external funding.

Institutional Review Board Statement: Ethical review and approval were waived for this study due to it being a retrospective survey not directly involving research subjects.

Informed Consent Statement: Informed consent was obtained from all subjects involved in the study.

Data Availability Statement: Not applicable.

Conflicts of Interest: The authors declare no conflict of interest.

References

1. U.S. Department of Transportation. Pets. Available online: www.transportation.gov/tags/pets (accessed on 20 August 2023).
2. Condor Ferries. *Pet Travel Statistics 2020–2021*. Available online: www.condorferries.co.uk/pet-travel-statistics (accessed on 20 August 2023).
3. Bergeron, R.; Scott, S.L.; Emond, J.-P.; Mercier, F.; Cook, N.J.; Schaefer, A.L. Physiology and behavior of dogs during air transport. *Can. J. Vet. Res.* **2002**, *66*, 211. [PubMed]
4. Leadon, D.P.; Mullins, E. Relationship between kennel size and stress in greyhounds transported short distances by air. *Vet. Rec.* **1991**, *129*, 70–73. [CrossRef] [PubMed]
5. Stewart, M.; Foster, T.M.; Waas, J.R. The effects of air transport on the behaviour and heart rate of horses. *Appl. Anim. Behav. Sci.* **2003**, *80*, 143–160. [CrossRef]
6. Snyder, R.J.; Perdue, B.M.; Powell, D.M.; Forthman, D.L.; Bloomsmith, M.A.; Maple, T.L. Behavioral and Hormonal Consequences of Transporting Giant Pandas from China to the United States. *J. Appl. Anim. Welf. Sci.* **2012**, *15*, 1–20. [CrossRef]
7. U.S. Department of Transportation Consumer Report 2022. Available online: www.transportation.gov/individuals/aviation-consumer-protection/air-travel-consumer-reports-2022 (accessed on 20 August 2023).
8. U.S. Department of Transportation Consumer Report 2019. Available online: www.transportation.gov/airconsumer/air-travel-consumer-reports-2019 (accessed on 20 August 2023).
9. Tiira, K.; Lohi, H. Reliability and Validity of a Questionnaire Survey in Canine Anxiety Research. *Appl. Anim. Behav. Sci.* **2014**, *155*, 82–92. [CrossRef]
10. Salonen, M.; Mikkola, S.; Hakanen, E.; Sulkama, S.; Puurunen, J.; Lohi, H. Reliability and validity of a dog personality and unwanted behavior survey. *Animals* **2021**, *11*, 1234. [CrossRef]
11. Cull, W.L.; O'Connor, K.G.; Sharp, S.; Tang, S.-F.S. Response rates and response bias for 50 surveys of pediatricians. *Health Serv. Res.* **2005**, *40*, 213–226. [CrossRef]

12. Wensley, S.P. Animal welfare and the human-animal bond: Considerations for veterinary faculty, students, and practitioners. *J. Vet. Med. Educ.* **2008**, *35*, 532–539. [CrossRef]
13. Russow, L.M. Ethical implications of the human-animal bond in the laboratory. *ILAR J.* **2002**, *43*, 33–37. [CrossRef]
14. Centers for Disease Control and Prevention. Division of Global Migration and Quarantine. Perspective from the Field: Illegal Puppy Imports Uncovered a JFK Airport. Available online: www.cdc.gov/ncezid/dgmq/feature-stories/operation-dog-catcher.html#:~:text=Since%20Operation%20Dog%20Catcher\T1\textquoterights%20start,it%20harder%20to%20breathe%20normally (accessed on 20 August 2023).
15. Packer, R.M.A.; O'Neill, D.G.; Fletcher, F.; Farnworth, M.J. Great expectations, inconvenient truths, and the paradoxes of the dog-owner relationship for owners of brachycephalic dogs. *PLoS ONE* **2019**, *14*, e0219918. [CrossRef]
16. Research staff, Texas A&M University, commissioned by IPATA—International Pet and Animal Transportation Association, Air Transportation of Brachycephalic Dog Breeds, 2018. (Unpublished work). (Unpublished work). 2018.
17. Tiffany, S.; Parr, J.M.; Templeman, J.; Shoveller, A.K.; Manjos, R.; Yu, A.; Verbrugghe, A. Assessment of dog owners' knowledge relating to the diagnosis and treatment of canine food allergies. *Can. Vet. J.* **2019**, *60*, 268–274. [PubMed]
18. Kogan, L.R.; Schoenfeld-Tacher, R. The Internet and Pet Health Information: Perceptions and Behaviors of Pet Owners and Veterinarians. *Internet J. Vet. Med.* **2009**, *8*, 1–19. [CrossRef]
19. Lunardo, R.; Jaud, D.; Jaspers, E. Engagement in vice food and beverage consumption: The role of perceived lack of control. *Psychol. Mark.* **2022**, *39*, 2221–2239. [CrossRef]
20. NPR—National Public Radio, USA, Dog Dies in Overhead Bin on United Flight; Airline Apologizes. Available online: https://www.npr.org/sections/thetwo-way/2018/03/14/593479827/dog-dies-in-overhead-bin-on-united-flight-airline-apologizes (accessed on 20 August 2023).
21. DH (Daily Hive) News, Canada, Owner Seeks Damages after Dog Dies on Flight to Vancouver. Available online: www.dailyhive.com/vancouver/dog-dies-flight-china-vancouver (accessed on 20 August 2023).
22. Jahn, K.; DePorter, T. Feline Stress Management During Air Travel: A Multimodal Approach. *J. Feline Med. Surg.* **2023**, *25*, 1098612X221145521. [CrossRef]
23. Tateo, A.; Zappaterra, M.; Covella, A.; Padalino, B. Factors influencing stress and fear-related behaviour of cats during veterinary examinations. *Ital. J. Anim. Sci.* **2021**, *20*, 46–58. [CrossRef]
24. Tennyson, A. Air transport of sedated pets may be fatal. *J. Am. Vet. Med. Assoc.* **1995**, *207*, 684.
25. IATA (International Air Transport Association) LAR (Live Animal Regulations) 2022. Available online: www.iata.org/en/publications/store/live-animals-regulations/ (accessed on 20 August 2023).
26. Piotti, P.; Uccheddu, S.; Alliani, M.; Mariti, C.; Nuti, V.; Ogi, A.; Pierantoni, L.; Gazzano, A. Management of specific fears and anxiety in the behavioral medicine of companion animals: Punctual use of psychoactive medications. *Dog Behav.* **2019**, *5*, 23–30.
27. Seibert, L.; Crowell-Davis, S. Antipsychotics. In *Veterinary Psychopharmacology*, 2nd ed.; Crowell-Davis, S., Murray, T.F., De Souza Dantas, L.M., Eds, Eds.; Wiley-Blackwell: Hoboken, NJ, USA, 2019; pp. 201–215.
28. Crowell-Davis, S.L.; Landsberg, G.M. Pharmacology and pheromone therapy. In *BSAVA Manual of Canine and Feline Behavioural Medicine*, 2nd ed.; Horwitz, D.F., Mills, D.S., Eds.; British Small Animal Veterinary Association: Gloucester, UK, 2018; pp. 245–258.
29. Beata, C.; Beaumont-Graff, E.; Diaz, C.; Marion, M.; Massal, N.; Marlois, N.; Muller, G.; Lefranc, C. Effects of alpha-casozepine (Zylkene) versus selegiline hydrochloride (Selgian, Anipryl) on anxiety disorders in dogs. *J. Vet. Behav. Clin. Appl. Res.* **2007**, *2*, 175–183. [CrossRef]
30. Michelazzi, M.; Berteselli, G.V.; Talamonti, Z.; Cannas, S.; Scaglia, E.; Minero, M.; Palestrini, C. Efficacy of L-Theanine in the treatment of noise phobias in dogs: Preliminary results. *Veterinaria* **2015**, *29*, 53–59.
31. Pageat, P.; Gaultier, E. Current research in canine and feline pheromones. *Vet. Clin. North Am. Small Anim. Pract.* **2003**, *33*, 187–211. [CrossRef]
32. McPeake, K.; Affenzeller, N.; Mills, D. Noise sensitivities in dogs: A new licensed treatment option. *Vet. Rec.* **2017**, *180*, 353–355. [CrossRef] [PubMed]
33. Ley, J.; Kerr, K.; Seksel, K. Results on the use of dog appeasement phreromone (DAP) collars in a selection of Australian dogs with anxiety disorders. *J. Vet. Behav.* **2010**, *5*, 45–46. [CrossRef]
34. Amat, M.; Le Brech, S.; Camps, T.; Manteca, X. Separation-Related Problems in Dogs. *Adv. Small Anim. Care* **2020**, *1*, 1–8. [CrossRef]
35. Mills, D.; Karagiannis, C.; Zulch, H. Stress-its effects on health and behavior: A guide for practitioners. *Vet. Clin. North Am. Small Anim. Pract.* **2014**, *44*, 525–541. [CrossRef] [PubMed]

animals

MDPI

Article

Innovative e-Learning Training Modules to Improve Animal Welfare during Transport and Slaughter of Pigs: A Pretest–Posttest Study to Pre-Evaluate the General Didactical Concept

Rudi Isbrandt [1], Nina Langkabel [1], Marcus G. Doherr [2], Sebastian Haase [3] and Diana Meemken [1,*]

1 Working Group Meat Hygiene, Institute of Food Safety and Food Hygiene, School of Veterinary Medicine, Freie Universität Berlin, Königsweg 67, 14163 Berlin, Germany; r.isbrandt@fu-berlin.de (R.I.); nina.langkabel@fu-berlin.de (N.L.)

2 Institute for Veterinary Epidemiology and Biostatistics, School of Veterinary Medicine, Freie Universität Berlin, Königsweg 67, 14163 Berlin, Germany; marcus.doherr@fu-berlin.de

3 School Pedagogy and School Improvement Research, Department of Education and Psychology, Freie Universität Berlin, Habelschwerdter Allee 45, 14195 Berlin, Germany; sebastian.haase@fu-berlin.de

* Correspondence: diana.meemken@fu-berlin.de

Simple Summary: Legal regulations at European and national level create the basis for the protection of animals. A certificate of competence is obligatory when animals intended for human consumption are handled or slaughtered. Within the project "eSchulTS2", e-learning training courses for employees of transport companies and abattoirs were developed to the improve animal welfare of pigs during transport and slaughter. This study was to assess the impact on the respective knowledge of employee groups with different educational level and background at one pig abattoir in Germany and to pre-evaluate the innovative and low-barrier didactical concept of the e-learning training courses. By using questions of knowledge, it was shown that more questions of knowledge were answered correctly after conducting the courses. Together with interactions shown by further statistical methods, this pre-evaluation showed that with the underlying didactical concept, an increase in knowledge was achieved.

Abstract: In addition to the information on the possession of a certificate of competence, there are no concrete obligations for repetitive training for personnel handling live animals at transport and slaughter. Deficiencies in the animal-welfare-friendly handling of pigs are known. The developed pilot modules "Handling of pigs" and "Electrical stunning" were tested in a pretest–posttest study in German and Romanian using questions of knowledge before and after the implementation of the modules. In this study, 45 and 46 datasets of participants could be analyzed. The mean percentages of correctly answered questions in the posttest increased by 5.6% in the module "Handling of pigs" and by 10.6% in the module "Electrical stunning". A significant interaction was found for the language match and trend categories in the module "Handling of pigs". No Romanian native speaker had a positive trend in this module. For both modules separately, participant education level significantly interacted with the language match and the presence or absence of a certificate of competence. Comparing the percentages of the correct given answers, significant interactions in the subgroups were more common in the module "Electrical stunning". One question in "Electrical stunning" was correctly answered significantly more often in the posttest. Because of the positive mean trends of knowledge within this pre-evaluation, we assume the didactical concept was suitable for our target groups. Holders of a certificate of competence also gave more correct answers in the post-test. This underlines the importance of repetitive training. Differences in the trends of knowledge gain seem to be topic and experience related.

Keywords: education; online training; slaughterhouse; abattoir; pig; animal well-being; didactic; knowledge test

Citation: Isbrandt, R.; Langkabel, N.; Doherr, M.G.; Haase, S.; Meemken, D. Innovative e-Learning Training Modules to Improve Animal Welfare during Transport and Slaughter of Pigs: A Pretest–Posttest Study to Pre-Evaluate the General Didactical Concept. *Animals* **2023**, *13*, 3593. https://doi.org/10.3390/ani13233593

Academic Editor: Elbert Lambooij

Received: 21 October 2023
Revised: 15 November 2023
Accepted: 19 November 2023
Published: 21 November 2023

1. Introduction

1.1. Animal Welfare during Transport and Slaughter

Welfare of livestock intended for slaughter is in the interest of consumers, who are increasingly concerned about animal welfare in intensified animal husbandry systems [1]. Despite the increasing numbers of animals that abattoir personnel have to care for, positive human–animal contact is important, and therefore, knowledge of how to correctly handle animals is required [2]. The social and political importance of animal welfare is reflected in European and German legislation dealing with the transport and slaughter of animals intended for human consumption. The German Animal Welfare Act explicitly states that no one may inflict pain, suffering, or harm on an animal without reasonable cause [3]. The necessity for authorization of transporters and the establishment of a certificate of competence for personnel transporting animals is defined in Regulation (EC) No 1/2005 on the protection of animals during transport and related operations [4]. Further topics and more specific information, such as recognition and withdrawal of the certificate of competence, are addressed in the German national law on the transportation of animals [5]. The certificate of competence, which personnel working at the abattoir with live animals or slaughtering them also have to hold, and the conditions under which it is issued are addressed in Regulation (EC) No 1099/2009 on the protection of animals at the time of killing [6]. The German national law on the protection of animals during slaughter specifies the competences needed, the content of teaching units, and the conduct of examinations to receive this certificate. A certificate of competence is valid for an unlimited period, but can be withdrawn if the requirements laid down in Regulation (EC) No 1099/2009 are violated several times [7]. The need for repeated training regarding animal welfare at the day of slaughter is not further addressed in European or German national law. However, in some German districts, repetitive training courses on animal welfare are implemented [8]. The German official Working Group for Consumer Protection recommends that training is regularly conducted in order to sustain the level of knowledge of employees working with animals [9]. The responsibility to conduct training repeatedly can lie with the animal welfare officers of larger abattoirs, since they have to guarantee that all employees know and understand the standard operating procedures [10].

1.2. Preliminary Work and Approach to Identify Suitable Training Content

The project "Development of target group-specific e-learning modules to improve animal welfare during transport and slaughter of cattle and pigs" (Acronym: eSchulTS2) was mainly implemented by members of the Freie Universität Berlin, School of Veterinary Medicine: Institute of Food Safety and Food Hygiene, Working Group Meat Hygiene, Institute of Animal Welfare, Animal Behavior and Laboratory Animal Science and Institute for Veterinary Epidemiology and Biostatistics. The core team was supported by the Department of Education and Psychology, Further Education and Educational Management of Freie Universität Berlin as well as by an industry partner.

The bases for preparing the training courses for pigs were a questionnaire on the status quo of animal welfare training at German and Austrian abattoirs [11] and the results of a systematic literature review to find out the "Impact of procedures and human–animal interactions during transport and slaughter of pigs" [12]. After conducting a Delphi-type expert elicitation in which experts rated the potential impacts, the topic areas for the final e-learning training modules were set [13]. Initially, the two training modules, "Handling of pigs" and "Electrical stunning", were designed to assess the didactical and technical concept of the e-learning modules with the potential to refine and finalize the concept. At the end of the project, a total of seven e-learning training modules in the following courses will be provided online: Handling of pigs; Fitness for transport; Stunning procedure (including Electrical-stunning; Stunning with carbon dioxide; Check of stunning effectiveness; Post-stun with captive bolt); and Bleeding. All e-learning modules will be available in multiple languages and will be free of charge for all interested users.

1.3. Didactical Concept for the e-Learning Training Courses

In the final e-learning courses, each participant can choose between the languages German, Romanian, Polish, Hungarian, Bulgarian, and English. The didactical concept is based on clear structures and a specific color scheme for each course, which ensures easy orientation. The minimum use of text components and use of simple language helps the participants to focus on the content of the modules. These techniques are how content and visual overload, especially for low-qualified participants, is avoided. A table of contents is always visible and also has the function of a progress bar (Figure 1).

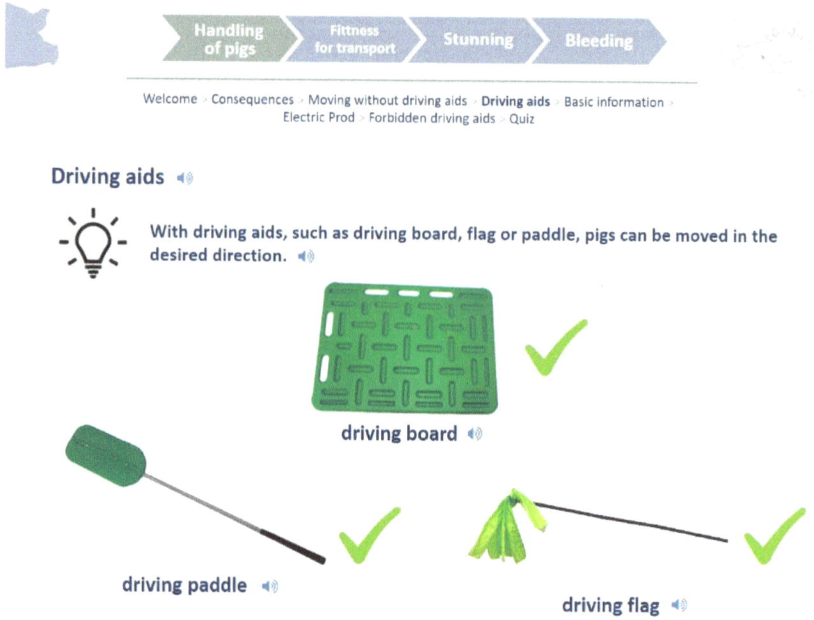

Figure 1. Screenshot from website "tet.folio", module "Handling of pigs", topic "Driving aids"; translated for publication.

The basis for communication of information is combinations of pictures and symbols. Videos with spoken text (voiceover) serve as the central knowledge transfer instrument in order to also include illiterates. The voiceover is available in the different selectable languages, which, apart from English, have been proven to be the most common mother languages of employees in abattoirs in German-speaking countries [11]. Selectable information boxes provide knowledge beyond the basic training content that is especially relevant for animal welfare officers, official veterinarians, or other interested participants. It is possible to repeat individual parts of the modules and videos or the entire modules at any time. An estimated working time of not more than 15 min for each module is intended to maintain concentration and to prevent the fatigue of participants. At the end of each module, participants can assess their own learning success by completing a quiz. A certificate of participation can be printed on request. Our aim was to pre-evaluate the underlying didactical concept of two e-learning modules with a pretest–posttest study design before elaborating on the other modules. This pre-evaluation was carried out by identifying short-term changes in knowledge at the level of individual participants as well as within groups of participants with similar demographic characteristics. In addition, we wanted to identify potential influences on different knowledge trends among participant groups, at the single-question level and between the two modules. The experience gained during this pre-evaluation process can also be incorporated into the intended final evaluation of all modules. The evaluation will follow the same approach and will be complemented by an additional posttest after a longer period of time to analyze the long-term gain of knowledge. This gives the opportunity to eliminate weaknesses and to adopt strengths of the methodology in the future.

To the best of the authors' knowledge, there are no comparable training materials available in Europe to improve animal welfare that were specifically developed for employees of pig transport and abattoir companies that have a limited education background and that lack (German) language proficiency.

2. Materials and Methods

2.1. Method Used and Participants Included

We used the method of pretest–posttest design [14] to evaluate the two e-learning training modules "Handling of pigs" and "Electrical stunning". In this design, questions to test knowledge were asked at two timepoints, directly before (pretest) and directly after (posttest) completion of the respective modules, to measure the short-term increase of knowledge.

The questions were designed especially for knowledge testing within the frame of developing and evaluating the concept of our e-learning training modules; as such, they will not be included in the final e-learning platform for all participants. At the time of this evaluation, the training modules were available only in German and Romanian. Participants with other mother tongues carried out the training in German. It was previously established by the animal welfare officer that they speak and understand German sufficiently well to very well. Six easy understandable questions were asked for the module "Handling of pigs" and five questions were asked for the module "Electrical stunning" (Supplementary Material File S1). For each question, three answer options (the correct answer and two distractors) were provided. There was the possibility to select between none, one, two, and all answers, but the given answers were rated as correct only when the correct answer was chosen. Participants were not told that only one answer option was correct. The same questions were asked both directly before conducting and directly after completing the module. The order of questions and the respective answer options were randomized for every participant and timepoint. After answering, the participants were not told if they had chosen the correct answer, in order not to influence the posttest.

The e-learning training courses were tested and pre-evaluated in one of the industry partner's pig abattoirs, where a computer pool room was available. Therefore, all participants were employed by the industry partner. Stunning was performed with carbon dioxide. Necessary emergency slaughters were performed after manual electrical stunning, whereas re-stunning was also performed with the captive bolt. The e-learning training courses and tests were conducted in September and October of 2022. It was not necessary for the participants to have specific prior knowledge or a certificate of competence, which also means that employees not handling live animals were allowed to take part. The group of employees without a certificate of competence should represent newcomers in this field currently undergoing training to receive a certificate of competence in the near future. The pretests and posttests were implemented in the e-learning platform "tet.folio" of Freie Universität Berlin, and made available online. Individual login details were generated that allowed us to link the responses from the pretest and posttest assessments for each person. The on-site implementation of the tests and supervision was within the responsibility of the local animal welfare officer of the abattoir. Participants, either in groups of up to ten persons in a computer pool room or individually in an office room, were given a standardized introduction by the animal welfare officer. The animal welfare officer was in the room the entire time and made sure that the participants did not communicate with each other. Furthermore, the officer was not allowed to help the participants if they had technical questions or if they had questions about the module content. This should simulate the aimed final e-learning situation in order to show whether the platform is intuitive to use.

However, questions and issues were noted and shared by the animal welfare officer with the project team for further improvements of the modules and platform settings. The participants' demographic data (country of origin, education, working position at the abattoir, and information regarding certificate of competence) were linked pseudonymously to the login details by the animal welfare officer and made available only to the project team for the purpose of analysis. The identity of the participants was not disclosed to the project team, and the individual participant's performance was not shared with the animal welfare officer or the employer.

2.2. Statistical Analysis

Responses were collected in "tet.folio" and downloaded as an MS Excel file (MS Office LTSC Professional Plus 2021, MS Excel Version 2108). The linked demographic data were received in a separate Excel file and merged with the response information through the internal identifier (login-ID). For the various subgroup analyses, participants were categorized regarding possession of a certificate of competence for transporting/handling/slaughtering animals (present or absent), and the training effect was compared between these two groups. In addition, the association between language match (module in mother tongue available or not; estimated on the basis of the records of the animal welfare officer regarding the participants' countries of origin) and training success was analyzed. Educational background was categorized based on the personal educational level (education level): (A) secondary school education, (B) secondary school education and professional training, (C) tertiary education. In another dimension, categories were defined by participants with specialization (education specialization): (A) secondary school education, (B) secondary school education and professional training in the field or tertiary degree in the field (of slaughter, animal husbandry, veterinary medicine, etc.), and (C) secondary school education and professional training in another field or tertiary degree in another field of competence. Another group of analyses was used to assess the change in averages of the individual participant's proportions of correct answers (given in percent) between the two time points. Here, only those questions were included for which an answer was actually provided.

For the last set of analyses, each participant's trend of changing knowledge (answer status correct or not before and after training) was defined and, therefore, we determined whether each participant (+) improved, (=) remained the same or (−) worsened. We defined this clustering of trends into three types, as trend-category-3. However, because of the small number of participants and to ensure better statistical results, we reduced this to a trend-category-2. Hence, we also used two categories "same or worse" (combination of the categories "=" and "−" from variable trend-category-3) and "better" (+) in our analysis.

Descriptive statistics were generated in MS Excel. The average percentages of correct answers were calculated overall and for participants in defined subgroups. Further analyses were carried out with IBM SPSS Statistics version 28 for windows. In addition to the description of frequencies (for categorical data), we tested the independent variables with the Fisher–Freeman–Halton exact test via cross tables to identify possible influences of the demographical subgroups (regarding certificate of competence, education, and language match) on the trend categories. In addition, we crossed the subgroups among themselves, also with the Fisher–Freeman–Halton exact test.

By using the Wilcoxon signed rank test for dependent repeated samples, the values (over all participants) of the average percentages of correct answers given at the two timepoints (before and after training) were analyzed overall and within each of the subgroups (certificate of competence, mother tongue, education level, and education specialization).

Cross-tables and the McNemar test statistic were used to assess the influence of training results at the single-question level (right or wrong answer given) in the pretest and posttest.

The knowledge testing was approved by the ethics committee of Freie Universität Berlin under ZEA Nr. 2022-016.

3. Results

3.1. Participants

In total, datasets of 45 participants in the module "Handling of pigs" were included for statistical analysis. The module "Electrical stunning" was performed by 46 participants. Demographical data regarding education or educational level were missing for two participants in the module "Handling of pigs" and for an additional participant in the module "Electrical stunning" (n = 3). As the questions were not compulsory, not all participants selected an answer to every question.

3.2. Overall Knowledge Gain

After calculating the average percentage of correct answers for the pretest and posttest assessment over all participants, it turned out that the percentage of correctly answered questions was higher in the posttest phase in both modules. The percentage of correct answers increased from pretest to posttest by 5.6% in the module "Handling of pigs" and by 10.6% in the module "Electrical stunning" (Figure 2).

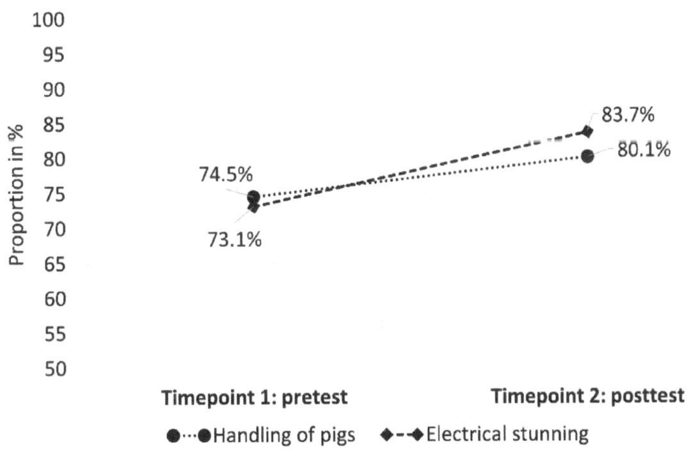

Figure 2. Average percentage of correct answers given in the pretest and posttest.

In the module "Handling of pigs", 7 out of 45 participants (15.6%) worsen, 26 (57.8%) had the same, and 12 (26.7%) had better results in the posttest compared to the pretest (Table 1). In the module "Electrical stunning", 4 out of 46 participants (8.7%) achieved worse results, 19 (41.3%) the same, and 23 (50.0%) better results in the posttest than in the pretest (Table 2).

3.3. Subgroup Analysis: Certificate of Competence, Language Match, Education

In the following, only some general results from the subgroup analysis are presented. The overview is shown in Tables 1–3.

About three quarters of the participants currently hold a certificate of competence. Around half of the participants could choose their mother tongue German in the modules, while the choice of Romanian as mother tongue was possible for about 30% of the participants (Tables 1 and 2). For both modules separately, the group of participants for whom the level of education was known was the same.

When examining the presence or absence of a certificate of competence in combination with the knowledge trends (trend-category-3 and trend-category-2), no statistically significant associations were detected for each of the modules. In the module "Handling of pigs", 29.4% of participants with a current certificate of competence had better results in the posttest, i.e., they answered more questions correctly) than in the pretest. Within the module "Electrical stunning", it was noticeable that no participant without a certificate of competence obtained worse results in the posttest compared to the pretest, and the majority obtained better results (63.6%).

Table 1. Results of demographical data, trend-category-3, and means of correct answers in the module "Handling of pigs".

	n/N %	Trend-Category-3: Individual Knowledge Trends	Mean of Correct Answers in Pretest in %	Mean of Correct Answers in Posttest in %	Mean of Trend (Change from Before to After) in %
all participants	45/45 100.0%	− 15.6% (7/45) = 57.8% (26/45) + 26.7% (12/45)	74.5	80.1	+5.6
Certificate of competence					
yes, hold a certificate of competence	34/45 75.6%	− 17.6% (6/34) = 52.9% (18/34) + 29.4% (10/34)	73.9	79.7	+5.8
no certificate of competence	11/45 24.4%	− 9.1% (1/11) = 72.7% (8/11) + 18.2% (2/11)	79.5	84.6	+5.1
Selection of mother-tongue					
German possible	21/45 46.7%	− 14.3% (3/21) = 52.4% (11/21) + 33.3% (7/21)	74.5	80.1	+5.6
Romanian possible	13/45 28.9%	− 30.8% (4/13) = 69.2% (9/13) + 0.0% (0/13)	74.1	79.0	+5.0
not possible	11/45 24.4%	− 0.0% (0/11) = 54.5% (6/11) + 45.5% (5/11)	71.0	80.6	+9.6
Education-level					
secondary school education	5/43 11.6%	− 0.0% (0/5) = 40.0% (2/5) + 60.0% (3/5)	73.8	79.6	+5.8
secondary school education and professional training	28/43 65.1%	− 21.4% (6/28) = 57.1% (16/28) + 21.4% (6/28)	74.5	80.1	+5.6
tertiary education	10/43 23.3%	− 0.0% (0/10) = 70.0% (7/10) + 30.0% (3/10)	75.6	82.8	+7.2
Education-specialization					
secondary school education	5/43 11.6%	− 0.0% (0/5) = 40.0% (2/5) + 60.0% (3/5)	73.8	79.6	+5.8
secondary school education and professional training in the field or tertiary degree in the field	13/43 30.2%	− 23.1% (3/13) = 61.5% (8/13) + 15.4% (2/13)	73.9	79.7	+5.8
secondary school education and professional training in another field or tertiary degree in another field	25/43 58.1%	− 12.0% (3/25) = 60.0% (15/25) + 28.0% (7/25)	74.3	79.7	+5.4

"−": worse/negative trend; "=": same/no trend; "+": better/positive trend.

In the e-learning module "Handling of pigs", the possibility to choose one's own mother tongue significantly interacted with both trend-category-3 (p = 0.031, Fisher–Freeman–Halton exact test) and trend-category-2 (p = 0.016; Fisher–Freeman–Halton exact test). No significant interaction was found in the module "Electrical stunning". Concerning the selection of the mother tongue in the "Handling of pigs" module, no participant who could chose Romanian as the mother tongue had a positive knowledge trend (Tables 1 and 2). In the "Electrical stunning" module, 28.6% of the participants who chose Romanian as their mother tongue achieved a positive knowledge trend. This percentage was around half of the percentage of those who chose German as their mother tongue (61.9%) and those who were not able to choose their mother tongue (54.5%).

Table 2. Results of demographical data, trend-category-3, and means of correct answers in the module "Electrical stunning".

	n/N %	Trend-Category-3: Individual Knowledge Trends	Mean of Correct Answers in Pretest in %	Mean of Correct Answers in Posttest in %	Mean of Trend (Change from Before to After) in %
all participants	46/46 100.0%	− 8.7% (4/46) = 41.3% (19/46) + 50.0% (23/46)	73.1	83.7	+10.6
Certificate of competence					
yes, hold a certificate of competence	35/46 76.1%	− 11.4% (4/35) = 42.9% (15/35) + 45.7% (16/35)	74.0	83.3	+9.3
no certificate of competence	11/46 23.9%	− 0.0% (0/11) = 36.4% (4/11) + 63.6% (7/11)	76.4	88.5	+12.1
Selection of mother-tongue					
German possible	21/46 45.7%	− 4.8% (1/21) = 33.3% (7/21) + 61.9% (13/21)	73.1	83.7	+10.6
Romanian possible	14/46 30.4%	− 14.3% (2/14) = 57.1% (8/14) + 28.6% (4/14)	73.9	83.4	+9.5
not possible	11/46 23.9%	− 9.1% (1/11) = 36.4% (4/11) + 54.5% (6/11)	78.8	86.7	+7.9
Education-level					
secondary school education	5/43 11.6%	− 0.0% (0/5) = 40.0% (2/5) + 60.0% (3/5)	76.6	85.8	+9.2
secondary school education and professional training	28/43 65.1%	− 10.7% (3/28) = 50.0% (14/28) + 39.3% (11/28)	73.1	83.7	+10.6
tertiary education	10/43 23.3%	− 0.0% (0/10) = 30.0% (3/10) + 70.0% (7/10)	81.2	90.0	+8.8
Education-specialization					
secondary school education	5/43 11.6%	− 0.0% (0/5) = 40.0% (2/5) + 60.0% (3/5)	76.6	85.8	+9.2
secondary school education and professional training in the field or tertiary degree in the field	13/43 30.2%	− 0.0% (0/13) = 76.9% (10/13) + 23.1% (3/13)	74.0	83.3	+9.3
secondary school education and professional training in another field or tertiary degree in another field	25/43 58.1%	− 12.0% (3/25) = 28.0% (7/25) + 60.0% (15/25)	74.6	84.8	+10.2

"−": worse/negative trend; "=": same/no trend; "+": better/positive trend.

Regarding education, the mean trends were similar within the groups "education-level" and "education-specialization" for the modules "Handling of pigs" and "Electrical stunning". In both modules, the choice of their own mother tongue as the module language significantly interacted with the category education level (p = 0.042; Fisher–Freeman–Halton exact test). Participants who chose Romanian as their mother tongue had nearly all secondary school education and completed professional training (90.9%; 10/11); the other participant had secondary school education only (9.1%; 1/11). None of the Romanian speakers had a tertiary degree. When choosing German as mother tongue was possible, one participant had secondary school education only (4.8%; 1/21), whereas twelve participants had secondary school education and completed professional training (57.1%; 12/21) and eight had a tertiary degree (38.1%; 8/21). A slightly different distribution of the educational levels was seen for participants who could not choose their mother tongue (graduation only: 27.3% (3/11); graduation and professional training: 54.5% (6/11); graduation and studies: 18.2% (2/11)). The three participants with only secondary school graduation were from Russia (n = 1) and Turkey (n = 2). Participants with additional professional training came from Greece (n = 3), Poland (n = 2), and Bulgaria (n = 1) and previously worked as carpenters, tilers, electricians, locksmiths, or butchers. Participants with a tertiary degree came from Greece (n = 1) and Poland (n = 1) with former jobs as teachers.

Table 3. Comparing the means of correct answers given in pretest and posttest with the Wilcoxon signed rank test.

	Module "Handling of Pigs"			Module "Electrical Stunning"		
	n/N %	Test Statistic Value	*p*-Value	n/N %	Test Statistic Value	*p*-Value
all participants	45/45 100.0%	372.0	0.533	46/46 100.0%	322.0	0.001 *
Certificate of competence						
yes, hold a certificate of competence	34/45 75.6%	252.0	0.683	35/46 76.1%	165.0	0.024 *
no certificate of competence	11/45 24.4%	13.0	0.596	11/46 23.9%	28.0	0.018 *
Selection of mother-tongue						
German possible	21/45 46.7%	81.0	0.226	21/46 45.7%	104.0	0.001 *
Romanian possible	13/45 28.9%	0.0	0.002 *	14/46 30.4%	12.0	0.750
not possible	11/45 24.4%	40.0	0.036 *	11/46 23.9%	23.5	0.105
Education-level						
secondary school education	5/43 11.6%	9.0	0.144	5/43 11.6%	6.0	0.102
secondary school education and professional training	28/43 65.1%	148.5	0.483	28/43 65.1%	82.0	0.063
tertiary education	10/43 23.3%	13.5	0.102	10/43 23.3%	28.0	0.016 *
Education-specialization						
secondary school education	5/43 11.6%	9.0	0.144	5/43 11.6%	6.0	0.102
secondary school education and professional training in the field or tertiary degree in the field	13/43 30.2%	26.5	0.554	13/43 30.2%	6.0	0.102
secondary school education and professional training in another field or tertiary degree in another field	25/43 58.1%	120.5	0.556	25/43 58.1%	148.0	0.006 *

* Significant interaction found by comparing the percentage of correct given answers in the pretest to the percentage of correct given answers in the posttest; alpha level of significance = 0.05.

For both modules, the presence or absence of a certificate of competence significantly interacted with the category education level ($p = 0.015$; Fisher–Freeman–Halton exact test). All participants with only secondary school graduation had a certificate of competence (100.0%; 5/5). From the 28 participants with secondary school education and professional training, the majority held a certificate of competence (82.1%; 23/28) and five did not (17.9%). The distribution was different among participants with a tertiary education, where the certificate of competence was held by four participants (40.0%; 4/10), but was not held by six (60.0%; 6/10), respectively. A significant interaction was also found for the presence or absence of the certificate of competence and the language match in both modules separately ("Handling of pigs" $p = 0.02$; Fisher–Freeman–Halton exact test/"Electrical stunning" $p = 0.001$; Fisher–Freeman–Halton exact test). Of the participants without a certificate of competence, ten were German and one originated from Poland.

Participants with a certificate were equally distributed. Eleven of these participants were German, thirteen and fourteen of these participants in the modules "Handling of pigs" and "Electrical stunning", respectively, were Romanian, whereas ten participants without certificates originated from other countries.

Crossing education subgroups and trend categories from the two modules, a significant interaction was found only in the "Electrical stunning" module for education specialization and trend-category-3 ($p = 0.05$; Fisher–Freeman–Halton exact test). Only participants with a secondary school education and professional training or tertiary degree from another field showed worse results in the posttest (12.0%), although more in this group had the same (28.0%) or better results in the posttest (60.0%). Participants with a secondary school education and professional training or tertiary degree in the field had the same results (76,9%) or obtained better results (23.1%). Participants with a secondary school education only had the same results (40.0%) or a positive trend (60.0%) in the posttest.

When comparing the percentages of correct given answers in the pretest and posttest, significant associations were found for both modules (Table 3).

3.4. Analysis on Single-Question Level

At the single-question level in the "Handling of pigs" module, a great variety regarding trend-category-3 was seen for question No. 3 (Figure 3). In this module, question No. 3 produced the greatest positive and negative trends. Also, at the single-question level, participants produced better results (i.e., have given more correct answers) in the "Electrical stunning" module than in the other module.

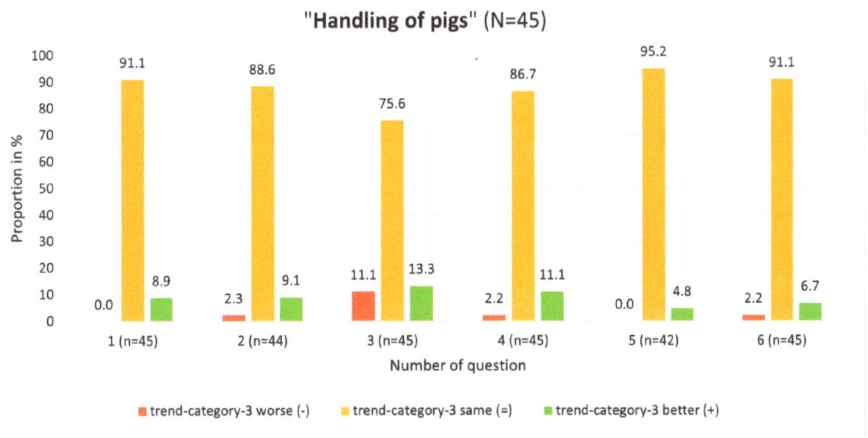

Figure 3. Results on single-question level in relation to trend-category-3 in the module "Handling of pigs".

In the "Electrical stunning" module, participants' knowledge was worse in the posttest compared to the pretest in only two questions. The participants achieved worse results in four questions in the "Handling of pigs" module (Figures 3 and 4). Comparing the pretest and posttest results for question No. 3 ("How many seconds must the minimum current be held so that safe stunning is achieved in fattening pigs?") in the "Electrical stunning" module, a strong significant interaction was found ($p < 0.001$; McNemar test). On answering this question, 24.4% of the participants produced a positive knowledge trend (so gave wrong answers in the pretest, but correctly answered in the posttest), which was the highest among all questions (Figure 4).

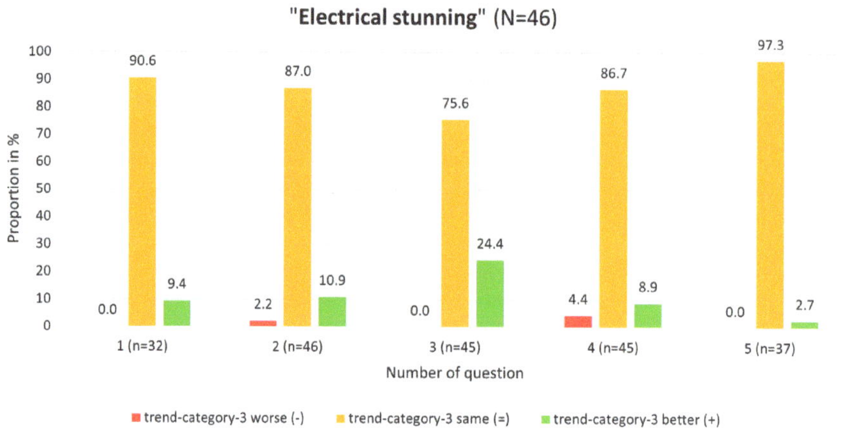

Figure 4. Results on single-question level in relation to trend-category-3 in the module "Electrical stunning".

4. Discussion

4.1. Study Design and General Limitations

In pretest–posttest studies, a group of participants perform the same test before and after an educational intervention [14]. This study type is often used in the evaluation of medical and management training [15,16]. A general criticism of one-group pretest–posttest designs is that this study type does not have a comparative group [17]. In order to include the highest possible total number of participants for our study, we used the single-group pretest–posttest design. In addition, it was difficult to reach a larger number of participants for this pre-evaluation due to the specific study topic and the generally difficult access to abattoir employees. However, the division into subgroups, each with a resultant relatively small participant number, hampered our statistical analysis and descriptive statistics, because spikes could result from each subgroup's small number of participants. Afterwards, the subgroups were compared with each other. Further interpretations have been refrained from in order to avoid overinterpretation of the underlying data. In hindsight, we should have set all questions as obligatory, thereby making participants answer all of them. This could have shown more deficits in knowledge and, therefore, could have impacted the results. The small number of knowledge questions in the pre- and posttests reflected only a part of the information imparted in the modules. In order to minimize the dropout rate and increase motivation, we decided that the module tests would both contain five to six questions. We consider this number of questions sufficient for a general statement on the didactical concept used in this pre-evaluation study.

4.2. Overall Knowledge Gain

The increased mean percentages of correct answers at the posttest showed that the participants gained knowledge during the e-learning lectures in both modules, at least in the short term. Therefore, we suppose that the didactical concept used was suitable for our target groups. As shown by the frequencies of participants worsening or improving in the posttest compared to the pretest, there were no differences between the modules. The module "Handling of pigs" reflects people's day-to-day work with pigs and does not have as many questionable facts as the "Electrical stunning" module. In our opinion, however, the questions in the modules were about equally difficult. Although there were some more complex questions on the use of the electrical prod in the module "Handling of pigs", the correct answers in general were communicated well in the module, in written words, within pictures and in the videos with voiceovers. This could be due to the fact that employees were already sensitized to the use of this driving aid after gaining a certificate of competence or taking part in other trainings. A reason for the higher positive learning trend in the module "Electrical stunning" may be the fact that in the abattoir where the study took place, stunning with carbon dioxide is used as the standard method. That means that electrical stunning is not part of the day-to-day work and is only used for emergency stunning or killing. Nevertheless, facts about this stunning device should be known by the employees who perform stunning. Reasons for the higher positive knowledge trend resulting from the "Electrical stunning" module could be that in general, more knowledge was imparted, and more animal welfare-related facts are necessary when performing electrical stunning.

It is known from other studies that most facts are forgotten in the first hours after memorization [18] and also, that repetition strengthens the retrieval of information [19]. Moreover, practice and repetition promote storage in the long-term memory [20]. These facts can be reasons for the positive learning trend found after knowledge testing directly after conducting the modules. Another possibility to be mentioned is that the increase in correct answered questions could be by chance. For the final evaluation of the modules with regard to long-term memorization, it will be interesting to ask the same questions again in a follow-up test after a few days or weeks. For our aim of a pre-evaluation of the general didactical concept, though, the positive trend and overall failures were more of interest and the improvements will be implemented in the pre-final versions of the modules. Furthermore, analyses of the data will focus on participants with the same learning trend. Care must be taken when interpreting the data as they are paired data. The influence of the individual question level is not considered further. The authors wanted to give a general overview

4.3. Subgroup Analysis: Certificate of Competence, Language, Education

Since it was not our aim to test the e-learning training modules only with persons who held a certificate of competence, we also enrolled participants without one. Because of legal requirements, all employees that have contact with living animals at the abattoir must hold the certificate before they are allowed to transport, move, stun, or kill animals, or check the stunning effectiveness [4,6]. This is reflected by the high number of participants with a certificate of competence. This fact may be the reason for a relatively high knowledge level in the pretest and resulting knowledge trends with a consequently flatter increase in the "Handling of pigs" module. For subsequent evaluations, participants with a certificate of competence should be asked whether they are currently working with live animals or not. This could provide new findings with regard to the subgroups and the interpretation of the didactical concept.

In this study, the countries of origin of the participants who were not from Germany or Romania were in line with the results of our former study in which we asked for the countries of origin of the employees of 29 slaughter companies in Germany and Austria [11]. Surprisingly, participants who could choose Romanian as their mother tongue had worse results than the other language subgroups. The reason for this effect remains unclear. The existing career changing in migration workers [21] might be one speculative reason for the worse results we detected in Romanian-speaking participants, as semi-trained and untrained employees have fewer basic skills [22]. It could be that the Romanian employees in our study represent more a group of career changers, and the results are not related to the country of origin per se. A comparison with another group of career changers to date has not been possible, because the modules have only been available in German and Romanian. In addition, a high impact of individual results on the group result for this language group could have occurred because of the relatively small number of Romanian native speakers who participated.

All participants had at least completed a secondary school education. This is in accordance with the results of our previous survey, in which over 30% of the participants estimated that 80–100% of the employees have a secondary school education [11]. We suppose that participants with a secondary school education more likely work in animal welfare sensitive areas, because all of them had a certificate of competence. In contrast, fewer participants with a tertiary education had a certificate of competence and could work in other areas within the abattoir. A reason for this could be that the slaughter industry has a lack of skilled workers [21], and in Germany, many people working in that industry are career changers originating from other European countries [23]. This is supported by the fact that teachers from other countries were also employed in the animal welfare sensitive areas of the abattoir where the study took place.

By comparing the percentages of the correct given answers in the pretest and posttest, a greater number of significant interactions were found in the module "Electrical stunning". This finding supports the fact that the trends in knowledge change were influenced by and related to the individual module topic.

4.4. Analysis on Single-Question Level

Based on the result for question No. 3 "Which statement about moving pigs is correct?" in the module "Handling of pigs" (Supplementary Material File S1), it can be seen that the presentation of information in the module videos and the question in the knowledge test can influence the result. To correctly answer this question, the participants had to mentally transform a negative statement from a video into a positive answer from the question. In addition, the relatively long answers, which are generally not recommended [24], could have had an impact, too. The wrong answer, "In a calm environment, pigs run slower in the desired direction because they explore everything curiously," was given often. For us, this answer was not surprising as we have observed that people who move pigs often interact a lot with them. In our experience, employees often think that pigs have to be stressed that they start to move or move in the desired direction. As a consequence, this fact should be addressed more during hands-on training in the future so that employees understand a calm environment as positive for the animals and the workflow. The question, "Which statement about moving pigs is correct?", could be judged as a "best answer" question and not a clear "right or wrong" question [24], which could have made the question more difficult. The way the questions were asked and answers were provided in relation to what was taught was shown to have an influence. This will be changed accordingly in the updated version of the modules in the questions for the final evaluation. The high percentage of participants improving in question No. 3 at the posttest also affected the mean trend in the module "Electrical stunning" because of the few questions asked.

5. Conclusions

In both e-learning training modules, "Handling of pigs" and "Electrical stunning", a knowledge gain was shown. The mean knowledge trend differed between the modules. The knowledge trend seems to be topical and educational and therefore experience related. No clear associations were identified between knowledge trends and whether or not participants held a certificate of competence for animal welfare sensitive areas. The influence of the possibility to choose training in one's own mother tongue is not clear. We speculate that the factor of career changing by Romanian speaking participants might be a reason for their worse results compared to the other participants. The fact that an increase in knowledge was also seen among participants with a certificate of competence underlines the importance of repetitive training courses in animal welfare topics. The design of the questions in relation to the knowledge taught in the modules can have an influence on the given answers. Because of the positive overall knowledge trend, we interpret the general didactical concept with intuitive color schemes, simple language, little text, voiceover videos, and easily understandable animations as suitable for the included target groups.

Supplementary Materials: The following supporting information can be downloaded at: https://www.mdpi.com/article/10.3390/ani13233593/s1, File S1: Questions of knowledge for pre- and posttest in the modules "Handling of pigs" and "Electrical stunning".

Author Contributions: Conceptualization, R.I., D.M. and N.L.; methodology, R.I., D.M. and N.L.; software, S.H.; validation, R.I., D.M., N.L. and M.G.D.; formal analysis, R.I., D.M., N.L. and M.G.D.; investigation, R.I.; resources, all; data curation, R.I., M.G.D. and D.M.; writing—original draft preparation, R.I.; writing—review and editing, all; visualization, R.I.; supervision, D.M. and N.L.; project administration, D.M. and N.L.; funding acquisition, D.M., M.G.D. and N.L. All authors have read and agreed to the published version of the manuscript.

Funding: This research was conducted in the framework of the joint research project "eSchulTS^2" (Development of target group-specific e-learning modules to improve animal welfare during transport and slaughter of cattle and pigs), which was funded by the German Ministry of Food and Agriculture (grant no. 2817806A18). The publication of this article was funded by Freie Universität Berlin.

Institutional Review Board Statement: The knowledge testing was approved by the ethics committee of Freie Universität Berlin under ZEA Nr. 2022-016 on 15 September 2022.

Informed Consent Statement: All participants were informed about the anonymity and voluntary nature of the knowledge tests and modules before carrying them out.

Data Availability Statement: The datasets that support the findings of this study, with the exception of the published Figures, Tables, and Supplementary Material, are not openly available due to reasons of sensitivity, but are available from the corresponding author upon reasonable request.

Acknowledgments: The authors thank the entire project team of "eSchulTS^2" for the constant and enriching thematic exchange and cooperation that has taken place, with special thanks to Nora Baer (Institute of Food Safety and Food Hygiene, Working Group Meat Hygiene, School of Veterinary Medicine, Freie Universität Berlin, Germany) for final editing of the videos and maintenance of the module platform, Vanessa Menzel (Animal Welfare Officer, Tönnies ApS & Co. KG) for the supervision of the training modules and Harm Kuper and Lisa Buchwald (both Department of Education and Psychology, Further Education and Educational Management, Freie Universität Berlin, Germany) for the discussions regarding the didactical concept.

Conflicts of Interest: The authors declare no conflict of interest. The funders had no role in the design of the study; in the collection, analyses, or interpretation of data; in the writing of the manuscript; or in the decision to publish the results.

References

1. Alonso, M.E.; González-Montaña, J.R.; Lomillos, J.M. Consumers' Concerns and Perceptions of Farm Animal Welfare. *Animals* **2020**, *10*, 385. [CrossRef] [PubMed]
2. Boivin, X.; Lensink, J.; Tallet, C.; Veissier, I. Stockmanship and Farm Animal Welfare. *Anim. Welf.* **2003**, *12*, 479–492. [CrossRef]
3. TierSchG. Tierschutzgesetz in der Fassung der Bekanntmachung vom 18. Mai 2006 (BGBl. I S. 1206, 1313), das zuletzt durch Artikel 2 Absatz 20 des Gesetzes vom 20. Dezember 2022 (BGBl. I S. 2752) geändert worden ist. 2006. Available online: https://www.gesetze-im-internet.de/tierschg/BJNR012770972.html (accessed on 17 April 2023).
4. European Commission. Council Regulation (EC) No 1/2005 of 22 December 2004 on the Protection of Animals during Transport and Related Operations and Amending Directives 64/432/EEC and 93/119/EC and Regulation (EC) No 1255/97. 2005. Available online: https://eur-lex.europa.eu/legal-content/EN/TXT/PDF/?uri=CELEX:32005R0001 (accessed on 24 April 2023).
5. TierSchTrV. Verordnung zum Schutz von Tieren beim Transport und zur Durchführung der Verordnung (EG) Nr. 1/2005 des Rates (Tierschutztransportverordnung-TierSchTrV) vom 11. Februar 2009 (BGBl. I S. 375), die zuletzt durch Artikel 9 Absatz 14 des Gesetzes vom 3. Dezember 2015 (BGBl. I S. 2178) geändert worden ist. 2009. Available online: https://www.gesetze-im-internet.de/tierschtrv_2009/ (accessed on 17 April 2023).
6. European Commission. Council Regulation (EC) No 1099/2009 of 24 September 2009 on the Protection of Animals at the Time of Killing. 2009. Available online: https://eur-lex.europa.eu/legal-content/EN/TXT/PDF/?uri=CELEX:32009R1099&from=DE (accessed on 17 April 2023).
7. TierSchlV. Verordnung zum Schutz von Tieren im Zusammenhang mit der Schlachtung oder Tötung und zur Durchführung der Verordnung (EG) Nr. 1099/2009 des Rates (Tierschutz-Schlachtverordnung-TierSchlV). 2012. Available online: https://www.gesetze-im-internet.de/tierschlv_2013/ (accessed on 17 April 2023).
8. Frisch, C.; Merl, K. Tierschutzkontrollen in handwerklichen Schlachtbetrieben. *RFL* **2019**, *71*, 338–341.
9. Anonymous. Handbuch Tierschutzüberwachung bei der Schlachtung und Tötung. AG Tierschutz der Länderarbeitsgemeinschaft Verbraucherschutz (LAV). 2019. Available online: https://www.openagrar.de/servlets/MCRFileNodeServlet/openagrar_derivate_00037082/Handbuch-Tierschutzueberwachung-Schlachten-2019-12-Aktualisierung-2021.pdf (accessed on 17 April 2023).

10. European Commission. The Animal Welfare Officer in the European Union. 2012. Available online: https://op.europa.eu/de/publication-detail/-/publication/bb4e4dba-e9ce-4e39-9d51-e3271f069744/language-en (accessed on 25 April 2023).
11. Isbrandt, R.; Meemken, D.; Langkabel, N. Animal welfare training at cattle and pig slaughterhouses-Results of an online survey in German-speaking countries. *Berl. Münch. Tierärztl. Wochenschr.* **2022**, *135*, 1–16.
12. Isbrandt, R.; Wiegard, M.; Meemken, D.; Langkabel, N. Impact of Procedures and Human-Animal Interactions during Transport and Slaughter on Animal Welfare of Pigs: A Systematic Literature Review. *Animals* **2022**, *12*, 3391. [CrossRef] [PubMed]
13. Duckwitz, V.; Eichler, F.; Isbrandt, R.; Nicolaisen, S.; Langkabel, N.; Wiegard, M.; Meemken, D.; Thöne-Reineke, C.; Doherr, M. Resultate einer Expert:innenbefragung zur Tierschutzrelevanz einzelner Prozessschritte und deren Verbesserungspotenzial durch Schulungen bei Transport und Schlachtung von Rind und Schwein. In Proceedings of the 22. Fachtagung für Fleisch-und Geflügelfleischhygiene, Berlin, Germany, 1–2 March 2022.
14. Marsden, E.; Torgerson, C.J. Single group, pre- and post-test research designs: Some methodological concerns. *Oxf. Rev. Educ.* **2012**, *38*, 583–616. [CrossRef]
15. Borg Sapiano, A.; Sammut, R.; Trapani, J. The effectiveness of virtual simulation in improving student nurses' knowledge and performance during patient deterioration: A pre and post test design. *Nurse Educ. Today* **2018**, *62*, 128–133. [CrossRef] [PubMed]
16. Kuraoka, Y. Effect of an experiential learning-based programme to foster competence among nurse managers. *J. Nurs. Manag.* **2018**, *26*, 1015–1023. [CrossRef] [PubMed]
17. Knapp, T.R. Why Is the One-Group Pretest-Posttest Design Still Used? *Clin. Nurs. Res.* **2016**, *25*, 467–472. [CrossRef] [PubMed]
18. Hofmann, E.; Löhle, M. *Erfolgreich Lernen: Effiziente Lern-und Arbeitsstrategien für Schule, Studium und Beruf*, 3rd ed.; Hogrefe Verlag GmbH & Co. KG: Göttingen, Germany, 2016; pp. 155–156.
19. Zhan, L.; Guo, D.; Chen, G.; Yang, J. Effects of Repetition Learning on Associative Recognition Over Time: Role of the Hippocampus and Prefrontal Cortex. *Front. Hum Neurosci.* **2018**, *12*, 277. [CrossRef] [PubMed]
20. Payk, T.R. Aufmerksamkeit. Lernen. Gedächtnis. In *Psychopathologie: Vom Symptom zur Diagnose*, 5th ed.; Springer: Berlin/Heidelberg, Germany, 2021; pp. 243–265.
21. Wagner, B.; Hassel, A. Posting, subcontracting and low-wage employment in the German meat industry. *Transfer* **2016**, *22*, 163–178. [CrossRef]
22. Flake, R.; Malin, L.; Middendorf, L.; Seyda, S. Lebenssituation und Potenziale An- und Ungelernter. In *Berufsbildung für Geringqualifizierte—Barrieren und Erträge*; Bundesinstitut für Berufsbildung: Bonn, Germany, 2017; p. 13ff.
23. Bosch, G.; Hüttenhoff, F.; Weinkopf, C. Fleischwirtschaft. In *Kontrolle von Mindestlöhnen*; Bosch, G., Hüttenhoff, F., Weinkopf, C., Eds.; Springer: Wiesbaden, Germany, 2019; pp. 191–236.
24. Smolle, J. *Medizinische MC-Fragen: Rasch und Einfach Erstellen*, 2nd ed.; Walter de Gruyter GmbH Co. KG: Berlin, Germany; New York, NY, USA, 2010.

animals

MDPI

Article

Effects of Electrolyte Multivitamins and Neomycin on Immunity and Intestinal Barrier Function in Transported Lambs

Cui Xia [1,2,†], Chunhui Duan [1,†], Conghui Chen [1], Xinyu Yang [2], Yingjie Zhang [1], Yueqin Liu [1,*,‡] and Yuzhong Ma [2,*,‡]

1 College of Animal Science and Technology, Agricultural University of Hebei, Baoding 071001, China; xiacui950114@163.com (C.X.); duanchh211@126.com (C.D.); zhangyingjie66@126.com (Y.Z.)
2 College of Veterinary Medicine, Agricultural University of Hebei, Baoding 071001, China
* Correspondence: liuyueqin66@126.com (Y.L.); dkma@hebau.edu.cn (Y.M.)
† These authors contributed equally to this work and should be regarded as co-first authors.
‡ These authors also contributed equally to this work.

Simple Summary: Transport stress damages the body health and reduces the immunity of animals. Currently, drugs such as vitamins and antibiotics, etc., are used to alleviate transport stress. In this experiment, lambs were fed diets with electrolytic multivitamin and neomycin, respectively. The weight, stress hormones and immune indicators of sera were examined. At the same time, the villus height, crypt depth and the ratio of villus height to crypt depth (V/C) were measured. Furthermore, the mRNA expressions of Occludin and MUC1, and the protein expression of Occludin in the jejunal mucosa, as well as the mRNA expressions of ZO-1 and Occludin and the protein expression of Occludin in the colonic mucosa were examined. Otherwise, the mRNA expressions of TRAF6, TLR4, MyD88 and NF-kB, and the protein expressions of TLR4 and NF-kB p65, as well as the mRNA expressions of TRAF6, TLR4 and NF-kB in the colon were measured. Adding 375 mg/d/lamb electrolytic multivitamin in the basal diet from 2 d before transportation to 7 d after transportation could potentially improve the immunity and intestinal barrier function. This provides a reference for the application of anti-stress additives to alleviate transport stress of lambs.

Abstract: Animals experience stress when they are transported. In this experiment, sixty 4-month-old lambs were randomly divided into three groups: CG (basal diet), EG (basal diet + 375 mg/d/lamb electrolytic multivitamin) and NG (basal diet + 200 mg/d/lamb neomycin). The transportation day was recorded as the 0th day. Blood, liver, spleen, jejunum and colon were collected on the 0th, 7th and 14th day. The results were as follows: In EG and NG groups, the lamb weights ($p < 0.01$), IgA and IgG ($p < 0.05$) increased significantly. The concentrations of ACTH, E, COR, IL-1β, IL-6 and IFN-γ decreased significantly ($p < 0.01$). The content of colonic propionate increased significantly ($p < 0.05$). The villus height and V/C increased, and crypt depth decreased significantly ($p < 0.01$). The mRNA expressions of Occludin and MUC1, and the protein expression of Occludin in the jejunal mucosa, the mRNA expressions of ZO-1 and Occludin, and the protein expression in the colonic mucosa increased significantly ($p < 0.01$). The mRNA expression of TRAF6 and the protein expression of TLR4 in the jejunum decreased significantly ($p < 0.05$), as well as the mRNA expressions of TLR4, MyD88 and NF-kB, and the protein expression of NF-kB p65 and the mRNA expressions of TRAF6, TLR4 and NF-kB in the colon ($p < 0.01$). In conclusion, an electrolytic multivitamin could potentially improve the immunity and intestinal barrier function, and when it was added with 375 mg/d in the basal diet for each lamb from 2 d before transportation to 7 d after transportation, it had a better effect than neomycin.

Keywords: lambs; transport stress; electrolytic multivitamin; neomycin; immune function; intestinal health

Citation: Xia, C.; Duan, C.; Chen, C.; Yang, X.; Zhang, Y.; Liu, Y.; Ma, Y. Effects of Electrolyte Multivitamins and Neomycin on Immunity and Intestinal Barrier Function in Transported Lambs. *Animals* **2024**, *14*, 177. https://doi.org/10.3390/ani14020177

Academic Editor: Elbert Lambooij

Received: 24 November 2023
Revised: 30 December 2023
Accepted: 2 January 2024
Published: 5 January 2024

1. Introduction

Transport stress is an important factor in the production of stress in animals, and it inhibits the immune system, causes changes in body metabolism [1], and even causes serious damage to animal tissues and organs [2]. It also leads to weight loss and the suppression of the immune system in animals, as well as disrupting the stability of the gastrointestinal ecosystem of animals, causing changes in the bacterial community, releasing endotoxins, and weakening disease resistance, thereby increasing the risk of gastrointestinal infections [3]. It also affects animal welfare. Wei [4] found that transport stress could induce intestinal oxidative stress in pigs, destroy the antioxidant defense system, damage the intestinal barrier function, increase the risk of bacterial translocation, and lead to the incidence rate increasing. Therefore, the question of how to take effective measures to prevent and alleviate transport stress is crucial to improving animal immune function, reducing the incidence rate and improving animal welfare. So, the prevention and treatment of transport stress has become the focus of research. Currently, the drugs used to prevent and treat transport stress mainly include vitamins and antibiotics. Other researchers [5–7] found that adding a certain amount of vitamin C and electrolyte to the diet could alleviate anorexia, indigestion and diarrhea, and decrease fat and immunity and intestinal inflammatory reactions caused by stress in animals. It could effectively alleviate stress. In clinical practice, neomycin has good preventive and therapeutic effects on various animal diseases caused by bacteria. Transport stress often leads to a decrease in the immunity of animals, which can lead to a series of diseases after infection by pathogenic microorganisms. Therefore, the prevention and treatment of transport stress is extremely important. At present, the effect and mechanism of electrolytic multivitamin and neomycin in the immunity and intestinal barrier function of transported lambs are unclear.

Therefore, this experiment aimed to investigate the effects of different treatments on the stress hormone levels in sera, immune function and intestinal health of lambs by adding an electrolytic multivitamin before and after transportation, and adding neomycin after transportation. This experiment can provide a reference for the selection and application of anti-stress additives. It is meaningful for the rational utilization of animal resources and the improvement of animal welfare.

2. Materials and Methods

2.1. Experimental Design

Sixty male 4-month-old Hu lambs with similar weight were randomly divided into a control group (CG), electrolytic multivitamin (EG) and neomycin group (NG), respectively, with 20 for each group. The day of transportation was recorded as the 0th day. The lambs in the CG were fed with a basal diet, the lambs in EG had 375 mg/d electrolytic multivitamin (Anhui Lingmu Biotechnology Co., Ltd., Fuyang, China) added to the basal diet for each lamb from 2 d before transportation to 7 d after transportation, and the lambs in the NG were treated with neomycin (Henan Weilong Veterinary Medicine Co., Ltd., Kaifeng, China) 200 mg/d per lamb after transportation from 0 d to 7 d. The basic feed was provided by the sheep farm, and the lambs were fed twice a day at 7:00 and 17:00 to ensure that all lambs had leftover feed in their feed tanks every day. The composition and nutritional levels of the feed are shown in Table 1. On the 0th, 7th, and 14th days after transportation, five lambs were randomly selected from each group and weighed. Then, 5 mL jugular vein blood was collected for separating sera; it was placed in a sterile tube and frozen at −20 °C for further analysis.

Before feeding, first, an electrolytic multivitamin and neomycin were fully mixed with a small amount of feed, respectively. This ensured that the additives were fully consumed by each lamb. And then lambs could freely feed on basal diets. The lambs in each group were raised in their groups. The main components and contents of electrolytic multivitamins are shown in Table 2.

Table 1. Composition and nutrient levels of the basal diet (dry matter basis).

Ingredients	Content %	Nutritional Level	Content
Cracked Corn	55.00	Neutral Detergent Fibers	33.33
Soybean meal	20.00	Crude Protein	18.06
Peanut seeding	12.50	Acidic Detergent Fibers	14.49
Peanut meal	9.00	Metabolic Energy (MJ/kg) [2]	12.50
Premix [1]	2.50	Ca	0.75
NaCl	0.50	NaCl	0.50
Baking soda	0.50	P	0.31
Total	100.00		

[1] Each kg premix contained the following: VA 15356 IU, VD 4300 IU, VE 50 mg, Fe 88.70 mg, Zn 70.90 mg, Mn 51.80 mg, Cu 13.75 mg, Se 0.23 mg, I 1.50 mg, Co 0.49 mg. [2] Metabolic energy was a calculated value, while the others were measured values.

Table 2. The main components and contents of electrolytic multivitamins.

Components	Contents
Vitamin C	≥2000 IU/kg
Vitamin B2	≥750 mg/kg
Vitamin A	30,000–5,000,000 IU/kg
Vitamin D3	75,000–20,000,000 IU/kg
Vitamin E	≥500 IU/kg
Vitamin B1	≥500 mg/kg
Water	≤10%
Folate	30 mg/kg
Taurine	20,000 mg/kg
Zn	1000 mg/kg
Mn	1000 mg/kg
Fe	1000 mg/kg
Cu	600 mg/kg

After blood collection, lambs were anesthetized with a dose of 0.2 mg/kg of xylazine (Hebei Mingeng Biotechnology Co., Ltd., Shijiazhuang, China) and then exsanguinated, and tissues were collected. The livers and spleens were collected and weighed for organ indices. The jejunum and colon tissues were taken out, and the contents were collected in sterile tubes and frozen in liquid nitrogen for volatile fatty acid (VFA) determination. And the middle part of the jejunum and colon was collected, and then the contents rinsed with saline and fixed in 4% formaldehyde solution for morphological examination. The remaining jejunum and colon samples were cut into two parts. One part of the jejunum and colon samples was stored in liquid nitrogen for later use. The other part of the jejunum and colon samples was used to collect the jejunal and colonic mucosa, which was scraped carefully with a glass slide and then stored in liquid nitrogen for later analysis.

2.2. Feeding and Management

This experiment was conducted from 1 June 2021 to 30 June 2021 at Zhangjiakou Lanhai Livestock Breeding Co., Ltd. (40°37′ N 115°03′ E and ~1300–1600 m above sea level, Zhangjiakou, China). The enclosure and breeding tools were thoroughly disinfected during the experiment period. Sixty male 4-month-old Hu sheep with similar weights were selected, and all experimental lambs were in three pens in the same building, with one pen for each group. The experimental lambs were transported for 8 h from 09:00 to 17:00 on June 18. They were transported locally in one truck, with a density of 1.7 m^2/lamb. The speed of transportation was between 30 and 80 km/h. Driving sections included country roads and highways. On the day of transportation, the outdoor temperature was 18–29 °C and the relative humidity was 64%. There were no compartments inside the carriage, and the lambs were not tied down. During transportation, food and water was not provided and animals were allowed to drink within 12 h following transport.

2.3. Determination of Immune Function

2.3.1. Determination of Immune Organ Index

The livers and spleens were weighed and the weights recorded. The organ index was calculated as follows:

$$\text{Immune organ index} = \text{Immune organ weight (g)/Live weight (kg)}$$

Note: live weight refers to the weight of the lamb without feeding and drinking for 8 h.

2.3.2. Assay of Sera Hormones and Immune Indices

Sera samples were thawed and mixed prior to testing. Concentrations of sera adrenocorticotropic hormone (ACTH), epinephrine (E), cortisol (COR), immunoglobulin A (IgA), immunoglobulin G (IgG), immunoglobulin M (IgM), and cytokines interleukin-1β (IL-1β), interleukin-2 (IL-2), interleukin-4 (IL-4), interleukin-6 (IL-6), interleukin-12 (IL-12), and interferon-γ (IFN-γ) were measured using enzyme-linked immunosorbent assay kits (Nanjing Jiancheng Bioengineering Institute, Nanjing, China) according to the standard procedures described by the manufacturer.

2.4. Determination of Intestinal Health

2.4.1. Examination of Intestinal Morphology

Methods for determining intestinal morphology were performed according to Ma [8]. Remove the jejunal and colonic tissue samples fixed with 4% formaldehyde solution for more than 24 h, and then ethanol with increasing gradient concentrations is used for dehydration, xylene for cleared, and paraffin for permeation. The tissue samples were embedded with KD-BM III (Zhejiang Jinhua Kedi Instrument Equipment Co., Ltd., Jinhua, China). Several cross-sections were cut to a thickness of 5 μm and mounted on adherent slides, and an XDS-1B inverted biological microscope (Chongqing Guangdian Corporation, Chongqing, China) was used to observe the tissue morphology in the field of view at different magnifications. Three clear, intuitive visions of each tissue were selected for observation. Image J software v. 1 was used to measure the villus width, villus height, crypt depth, and the villus height-to-crypt depth ratio (V/C), and each field of view measured 2 data points per indicator.

2.4.2. Examination of VFA

Assays for determining VFA were performed according to Zhang [9]. Samples of jejunal and colonic contents were thawed and thoroughly mixed immediately before testing. Concentrations of acetate, propionate, butyrate, and isovalerate were determined with the Daojin Gas Chromatograph System GC2010, as described in the instructions. Weigh 0.5 g of the jejunal and colonic contents into a centrifuge tube. Add 1.5 mL of ultrapure water, vortex-oscillate for 3–5 min, and centrifuge at 5000× g for 10 min. Take 1 mL of supernatant and transfer it to a centrifuge tube, then add 0.2 mL 25% metaphosphate solution, cover tightly, and shake well. And place it in an ice water bath for 30 min, and centrifuge at 10,000× g for 10 min. Take the supernatant and use a gas chromatograph to determine the concentration of VFA in the jejunal and colonic contents. The injection volume was 2 μL. The chromatographic column was an HP-INNOWAX (19091N-133) capillary column (30 m × 0.25 mm × 0.25 μm). The detector temperature was 220 °C. The column temperature was programmed to rise for 1 min at 80 °C, followed by a rise of 15 °C/min to 170 °C and was then maintained for 1.5 min. The carrier gas was high-purity nitrogen.

2.4.3. Quantitative Real-Time PCR for Gene Expression Analysis

Total RNA was isolated from frozen jejunal and colonic mucosa and tissue (50 mg) with Trizol reagent (TransGen Biotech, Beijing, China). The concentration of total ribonucleic acid (RNA) in the samples was evaluated using a spectrophotometer (NanoDrop-2000, Thermo Fisher Scientific, Waltham, MA, USA) at 260 nm and 280 nm, respectively. Ratios of absorption (260:280 nm) between 1.8 and 2.0 for all samples were accepted as "pure" for RNA. RNA (1 μg) was used to generate copy DNA (cDNA) using the Reverse Transcription Kit according to the manufacturer's instructions (Transgen Biotech, Beijing, China). Primer sequences for ZO-1, Occludin, MUC 1, MUC 2, TLR4, TRAF6, NF-kB, and MyD88 in jejunal and colonic mucosa and tissue were designed using the GenBank database from the National Center for Biotechnology Information and primer 5 design software. Quantitative real-time PCR was performed with SYBR Green Premix Es Taq (Takara, Beijing, China) using a StepOnePlus real-time PCR system (Applied Biosystems (Thermo-Fisher Scientific, Shanghai, China)) on 96-well plates with 20 μL of total reaction volume of 10 μL SYBR Green Premix, 1 μL cDNA, 0.4 μL of forward and 0.4 μL of reverse primers, and 8.2 μL of double-distilled water. Each reaction was run in duplicate. The PCR cycling protocol included one cycle of pre-incubation at 95 °C for 2 min, 40 cycles of denaturation at 95 °C for 5 s, and annealing at 60 °C for 30 s; each cycle was increased by 0.5 °C at 65–95 °C. β-actin was used as an internal control in this study. The average expression of the target genes relative to β-actin was determined using the $2^{-\Delta\Delta Ct}$ method, as described by Livak [10]. Primers for reverse transcription quantitative real-time PCR were synthesized by the Beijing Genomics Institution (BGI Genomics Co., Ltd., Beijing, China; Table 3).

Table 3. Gene primer information.

Primer		Sequence (5′→3′)	Length (bp)	Annealing Temperature (°C)	Accession Number
ZO-1	Forward	CATCACGCCAGCATACAA	177	52	XM_042235171
	Reverse	GCAGACTTCAGGAGGGTTT			
Occludin	Forward	CCAGCGTTGTAAGGTCAG	218	51	>XM_042233706.1
	Reverse	ATCCACGTAGAGTCCAGTAG			
MUC1	Forward	CTCCCTCACAGGACCCATCA	126	56	>XM_027976040.2
	Reverse	CGTTCCCACTCCCGTTTC			
MUC2	Forward	TCACCTGCCCTGACTTTGA	144	57	>DQ152978.1
	Reverse	TGTGCGAAATCTCCCTCGT			
TRAF6	Forward	TTCAACAGTTAGAGGGTCG	330	53	XM_015101111.3
	Reverse	GGGATATGTAGTTGGCACA			
TLR4	Forward	AAGGTGATTGTCGTGGTG	178	57	>NM_001135930.1
	Reverse	TGTTCAGAAGGCGATAGA			
MyD88	Forward	AGTACAAGCCAATGAAGAAAGA	100	58	>DQ152979.1
	Reverse	AGGCGAGTCCAGAACCAG			
NF-kB	Forward	CAACTGCTCTACCTCCTG	290	58	>XM_027966801.2
	Reverse	CTTCACTGTTACGGGTTC			
β-actin	Forward	CCCTGGAGAAGAGCTACGAG	98	58	NM-102179831
	Reverse	CAGGAAGGAAGGCTGGAAGA			

2.4.4. Western Blot Measured Protein Expression Levels

The protein was extracted from the jejunal and colonic mucosa and tissue using a protein extraction kit (Solarbio Biotech, Beijing, China) according to the manufacturer's instructions. The protein concentration was determined as described above. Western blot analysis was carried out as previously described [11]. In brief, electrophoresis was performed with Sodium Dodecyl Sulfate-Polyacrylamide Gel Electrophoresis, followed by membrane transfer, blocking, the incubation of primary antibodies, membrane washing with tris buffered saline (TBS), the incubation of secondary antibodies, and film washing with TBS. Next, the membrane was automatically exposed and a picture was taken; its grayscale values were analyzed using Image J software v. 1 [12].

2.5. Statistical Analyses

All data were analyzed using single factor ANOVA in SPSS 21.0 statistical software [13] for the differences of indicators when using different treatments at the same time and the same treatment at different times. Duncan's method for multiple comparisons was used. The General Linear Model (GLM) was used to analyze the effects of treatment and time. Effects $p \leq 0.05$ were considered different. Values were expressed as "Mean $\pm$ standard error" (M $\pm$ SE).

3. Results

3.1. Weight and Immune Organ Index

From Table 4, the weight in EG was significantly higher than that in CG on day 7 ($p < 0.01$). The weights in NG and EG were significantly higher than that in CG on day 14 ($p < 0.01$). The weight of lambs gradually increased along with increasing time ($p < 0.01$).

Table 4. Effects of different treatments on weight and immune organ index of lambs.

Items	Time	Groups			p-Value		
		CG	NG	EG	Time	Treatment	Time × Treatment
Weight/kg	0 d	19.88 ± 0.16^{B}	20.25 ± 0.35^{C}	20.88 ± 0.38^{B}	<0.010	<0.010	0.679
	7 d	20.55 ± 0.40^{Bb}	21.64 ± 0.40^{Bab}	21.86 ± 0.25^{Ba}			
	14 d	22.13 ± 0.36^{Ab}	23.48 ± 0.23^{Aa}	23.62 ± 0.41^{Aa}			
Liver index (g/kg)	0 d	17.47 ± 1.12	18.95 ± 1.39	17.31 ± 1.60	0.562	0.814	0.977
	7 d	18.22 ± 1.84	19.07 ± 1.97	19.21 ± 1.52			
	14 d	19.16 ± 2.14	19.37 ± 1.30	19.43 ± 1.44			
Spleen index (g/kg)	0 d	0.88 ± 0.06	0.94 ± 0.07	0.83 ± 0.09	0.720	0.875	0.928
	7 d	0.89 ± 0.09	0.88 ± 0.10	0.89 ± 0.07			
	14 d	0.87 ± 0.10	0.82 ± 0.05	0.83 ± 0.07			

Note: ^{ab} Different lowercase letters superscripted in the same row indicate a significant difference between different treatments ($p < 0.05$). ^{ABC} The superscript values in the same column with different uppercase letters indicate significant difference at different times ($p < 0.05$). The same letter or no letter indicates no significant difference ($p > 0.05$). There is no meaning between uppercase and lowercase letters.

3.2. Stress Hormones, Inflammatory Factors, and Immunoglobulins in the Sera

As shown in Table 5, on day 0, the concentrations of ACTH and COR in EG were significantly lower than those in CG and NG ($p < 0.01$). The concentration of E in CG was significantly higher than those in NG and EG ($p < 0.01$), and the concentration of E in NG was significantly higher than that in EG ($p < 0.01$). On day 7, the concentration of ACTH in EG was significantly lower than that in CG ($p < 0.01$). The concentration of COR in EG was significantly lower than those in CG and NG ($p < 0.01$). The concentrations of E in EG and NG were significantly lower than that in CG ($p < 0.05$). On day 14, the concentration of E in EG was significantly lower than those in CG and NG ($p < 0.01$). The concentration of COR in EG was significantly lower than that in CG ($p < 0.01$). The concentration of IgG in EG was significantly higher than those in EG and CG ($p < 0.05$). The concentrations of IgA in EG and NG were significantly higher than that in CG ($p < 0.05$). On day 0 and 7, the concentrations of IL-1β and IL-6 in EG were significantly lower than those in CG ($p < 0.01$). On day 7 and 14, the concentration of IFN-γ in EG was significantly lower than that in CG ($p < 0.01$). With the extension of feeding time, the concentrations of ACTH and E decreased significantly ($p < 0.05$), and the concentrations of COR, IL-1β, IL-2, IL-6, IL-12, and IFN-γ decreased significantly ($p < 0.01$). Otherwise, the concentrations of IL-4, IgA, IgG, and IgM increased significantly ($p < 0.01$).

Table 5. Effects of different treatments of transport stress on stress hormones and immune indices in the sera of lambs.

Items		Time	Groups			p-Value		
			CG	NG	EG	Time	Treatment	Time × Treatment
Stress hormones	ACTH (ng/L)	0 d	56.91 ± 1.12 ^{Aa}	53.91 ± 3.51 ^{a}	46.33 ± 2.59 ^{b}			
		7 d	53.56 ± 1.38 ^{ABa}	49.71 ± 2.75 ^{ab}	45.04 ± 1.32 ^{b}	0.027	<0.010	0.943
		14 d	50.53 ± 1.98 ^{B}	47.41 ± 2.84	43.27 ± 2.46			
	E (ng/mL)	0 d	1.03 ± 0.02 ^{Aa}	0.84 ± 0.02 ^{b}	0.75 ± 0.03 ^{c}			
		7 d	0.91 ± 0.03 ^{Ba}	0.80 ± 0.04 ^{b}	0.73 ± 0.03 ^{b}	0.031	<0.010	0.278
		14 d	0.87 ± 0.05 ^{Ba}	0.83 ± 0.02 ^{a}	0.70 ± 0.05 ^{b}			
	COR (ng/mL)	0 d	47.07 ± 1.95 ^{a}	44.25 ± 1.39 ^{Aa}	39.15 ± 1.75 ^{b}			
		7 d	45.85 ± 1.27 ^{a}	43.89 ± 1.29 ^{ABa}	38.08 ± 0.90 ^{b}	<0.010	<0.010	0.979
		14 d	42.75 ± 2.47 ^{a}	39.47 ± 1.40 ^{Bab}	35.40 ± 1.34 ^{b}			
Inflammatory factors	IL-1β (ng/L)	0 d	28.41 ± 1.40 ^{a}	25.84 ± 1.19 ^{Aab}	23.87 ± 1.04 ^{Ab}			
		7 d	24.04 ± 1.79 ^{a}	23.01 ± 1.50 ^{Bab}	20.81 ± 0.78 ^{Bb}	<0.010	<0.010	0.925
		14 d	22.94 ± 1.89	21.60 ± 0.90 ^{B}	19.82 ± 1.05 ^{B}			
	IL-2 (ng/L)	0 d	36.42 ± 1.42 ^{A}	34.47 ± 2.40	33.30 ± 2.39 ^{A}			
		7 d	30.13 ± 1.64 ^{B}	31.79 ± 1.36	29.42 ± 1.32 ^{AB}	<0.010	0.361	0.935
		14 d	28.60 ± 2.66 ^{B}	27.43 ± 0.98	27.16 ± 1.35 ^{B}			
	IL-4 (ng/L)	0 d	48.88 ± 1.63 ^{B}	49.99 ± 4.34	48.27 ± 1.49			
		7 d	53.47 ± 2.17 ^{AB}	55.54 ± 3.04	54.80 ± 3.26	<0.010	0.726	0.997
		14 d	57.28 ± 2.61 ^{A}	57.64 ± 1.17	57.31 ± 3.11			
	IL-6 (ng/L)	0 d	60.26 ± 1.75 ^{Aa}	56.99 ± 1.98 ^{Aab}	51.51 ± 2.11 ^{b}			
		7 d	56.03 ± 2.00 ^{ABa}	53.93 ± 2.35 ^{ABab}	47.31 ± 2.17 ^{b}	<0.010	<0.010	0.876
		14 d	50.06 ± 2.60 ^{B}	50.17 ± 1.83 ^{B}	46.05 ± 1.15			
	IL-12 (ng/L)	0 d	160.67 ± 4.21	159.46 ± 3.86 ^{A}	158.59 ± 3.27			
		7 d	154.22 ± 2.13	153.36 ± 1.77 ^{AB}	151.15 ± 3.03	<0.010	0.416	0.958
		14 d	155.77 ± 3.77	148.78 ± 1.96 ^{B}	148.47 ± 3.82			
	IFN-γ (ng/L)	0 d	69.12 ± 2.67	64.58 ± 2.97	61.32 ± 2.20 ^{A}			
		7 d	65.14 ± 3.44 ^{a}	60.32 ± 1.59 ^{ab}	55.92 ± 2.92 ^{ABb}	<0.010	<0.010	0.994
		14 d	60.53 ± 1.30 ^{a}	57.69 ± 4.36 ^{ab}	51.66 ± 1.82 ^{Bb}			
Immunoglobulins	IgA (mg/mL)	0 d	1.09 ± 0.03	0.97 ± 0.06 ^{B}	1.09 ± 0.09 ^{B}			
		7 d	1.12 ± 0.16	1.33 ± 0.09 ^{AB}	1.42 ± 0.03 ^{A}	<0.010	0.033	0.105
		14 d	1.31 ± 0.05 ^{b}	1.53 ± 0.06 ^{Aa}	1.52 ± 0.09 ^{Aa}			
	IgG (mg/mL)	0 d	11.29 ± 0.07	11.39 ± 0.62 ^{B}	10.49 ± 0.43 ^{B}			
		7 d	11.85 ± 0.71	12.30 ± 0.52 ^{A}	13.14 ± 0.46 ^{A}	<0.010	0.015	0.357
		14 d	12.19 ± 0.49 ^{b}	12.63 ± 0.30 ^{Ab}	13.80 ± 0.31 ^{Aa}			
	IgM (mg/mL)	0 d	1.92 ± 0.09 ^{B}	2.04 ± 0.10 ^{B}	2.05 ± 0.09 ^{B}			
		7 d	2.17 ± 0.10 ^{AB}	2.22 ± 0.14 ^{AB}	2.22 ± 0.13 ^{B}	<0.010	0.138	0.602
		14 d	2.31 ± 0.06 ^{A}	2.54 ± 0.07 ^{A}	2.65 ± 0.06 ^{A}			

Note: ^{abc} Different lowercase letters superscripted in the same row indicate significant differences between different treatments ($p < 0.05$). ^{AB} The superscript values in the same column with different uppercase letters indicate significant differences at different times ($p < 0.05$). The same letter or no letter indicates no significant difference ($p > 0.05$). There is no meaning between uppercase and lowercase letters.

3.3. VFA in the Jejunum and Colon

The contents of propionate in the colon of EG and NG were significantly higher than that in CG on day 7 ($p < 0.05$). The contents of acetate, propionate, butyrate, and isovalerate in the jejunum and colon of lambs gradually increased along with increasing time ($p < 0.01$).

3.4. Morphology of Intestinal Tissues

As shown in Table 6, Figures 1 and 2, in the jejunum, on day 7, the crypt depth in EG was significantly lower than that in CG ($p < 0.01$), and the V/C was significantly higher than that in CG ($p < 0.01$). On day 14, the villus height in EG was significantly higher than that in CG ($p < 0.01$). The crypt depths in EG and NG were significantly lower than that in CG ($p < 0.01$), and the V/C was significantly higher than that in the CG ($p < 0.01$). In the colon, on day 7, the villus height and V/C in EG and NG were significantly higher than those in CG ($p < 0.01$), and the crypt depth was significantly lower than that in CG ($p < 0.01$). The V/C in NG was significantly higher than that in CG on day 14 ($p < 0.01$). Otherwise, there was a tendency for interaction between treatment and time f for V/C in the colon. With the extension of time, the villus width in the jejunum increased significantly ($p < 0.05$). Otherwise, the villus width in the colon, villus height, and V/C in the jejunum and colon increased significantly ($p < 0.01$). The crypt depth in the jejunum and colon decreased significantly ($p < 0.01$).

Table 6. Effects of different treatments of transport stress on the morphology of the jejunum and colon of lambs.

Items		Time	Groups			*p*-Value		
			CG	NG	EG	Time	Treatment	Time × Treatment
Jejunum	Villus width (μm)	0 d	131.29 ± 3.84 ^{B}	133.26 ± 4.84	136.75 ± 6.90	0.019	0.713	0.996
		7 d	136.65 ± 4.00 ^{AB}	141.57 ± 4.74	140.37 ± 5.40			
		14 d	147.02 ± 6.32 ^{A}	148.36 ± 7.59	150.39 ± 10.36			
	Villus height (μm)	0 d	531.95 ± 11.19	546.68 ± 8.61	561.45 ± 10.46 ^{B}	<0.010	<0.010	0.796
		7 d	540.74 ± 14.59	565.68 ± 15.22	581.36 ± 13.40 ^{AB}			
		14 d	548.04 ± 13.27 ^{b}	582.95 ± 12.16 ^{ab}	609.21 ± 13.56 ^{Aa}			
	Crypt depth (μm)	0 d	171.77 ± 8.88	166.81 ± 3.89 ^{A}	154.29 ± 6.47 ^{A}	<0.010	<0.010	0.840
		7 d	165.48 ± 3.67 ^{a}	154.05 ± 4.38 ^{Bab}	144.77 ± 7.56 ^{ABb}			
		14 d	158.82 ± 2.62 ^{a}	140.85 ± 4.09 ^{Cb}	132.58 ± 5.79 ^{Bb}			
	V/C [1]	0 d	3.17 ± 0.22	3.29 ± 0.31 ^{C}	3.69 ± 0.19 ^{B}	<0.010	<0.010	0.547
		7 d	3.27 ± 0.07 ^{b}	3.69 ± 0.13 ^{Bab}	4.11 ± 0.27 ^{ABa}			
		14 d	3.46 ± 0.10 ^{b}	4.16 ± 0.13 ^{Aa}	4.64 ± 0.20 ^{Aa}			
Colon	Villus width (μm)	0 d	72.61 ± 3.63	77.40 ± 3.71	70.75 ± 3.00 ^{B}	<0.010	0.051	0.298
		7 d	75.23 ± 2.82	86.60 ± 2.81	81.42 ± 2.38 ^{A}			
		14 d	81.06 ± 2.32	82.79 ± 2.78	85.00 ± 2.41 ^{A}			
	Villus height (μm)	0 d	188.86 ± 6.27	208.09 ± 11.95 ^{B}	196.80 ± 7.60 ^{B}	<0.010	<0.010	0.394
		7 d	199.50 ± 4.35 ^{b}	223.63 ± 4.91 ^{Ba}	236.10 ± 10.90 ^{Aa}			
		14 d	222.31 ± 27.50	254.39 ± 20.09 ^{A}	245.90 ± 34.12 ^{A}			
	Crypt depth (μm)	0 d	51.59 ± 2.00	45.68 ± 1.16 ^{A}	47.14 ± 1.55 ^{A}	<0.010	<0.010	0.367
		7 d	49.06 ± 2.74 ^{a}	42.22 ± 1.21 ^{Bb}	39.84 ± 1.13 ^{Bb}			
		14 d	46.49 ± 1.93	41.21 ± 0.93 ^{B}	43.51 ± 1.17 ^{AB}			
	V/C	0 d	3.72 ± 0.28	4.54 ± 0.22 ^{C}	4.23 ± 0.27 ^{B}	<0.010	<0.010	0.071
		7 d	4.14 ± 0.24 ^{b}	5.32 ± 0.16 ^{Ba}	5.93 ± 0.57 ^{Aa}			
		14 d	4.87 ± 0.41 ^{b}	6.28 ± 0.24 ^{Aa}	5.62 ± 0.32 ^{Aab}			

[1] V/C = the ratio of villus height to crypt depth. Note: ^{ab} different lowercase letters superscripted in the same row indicate significant differences between different treatments ($p < 0.05$). ^{ABC} The superscript values in the same column with different uppercase letters indicate significant differences at different times ($p < 0.05$). The same letter or no letter indicates no significant difference ($p > 0.05$). There is no meaning between uppercase and lowercase letters.

3.5. Expression Levels of Important Genes in the Jejunal and Colonic Mucosa

As shown in Table 7 and Figure 3, in the jejunal mucosa, the mRNA expression of MUC1 on day 7 and the protein expression of Occludin on day 14 in EG was significantly higher than those in NG and CG ($p < 0.01$). The mRNA expression of Occludin on days 7 and 14 and the mRNA expression of MUC1 on day 14 in EG were significantly higher than that in CG ($p < 0.01$). In the colonic mucosa, the mRNA expressions of ZO-1 and Occludin in EG were significantly higher than those in CG on days 0 and 7 ($p < 0.01$). The mRNA expression of Occludin in EG was significantly higher than that in CG on day 14 ($p < 0.01$). The protein expression of Occludin in EG was significantly higher than those in CG and NG on days 0, 7, and 14 ($p < 0.01$). The mRNA expressions of ZO-1, Occludin, MUC1, and MUC2 and the protein expression of Occludin and MUC2 in the jejunal and colonic mucosa increased along with increasing time ($p < 0.01$).

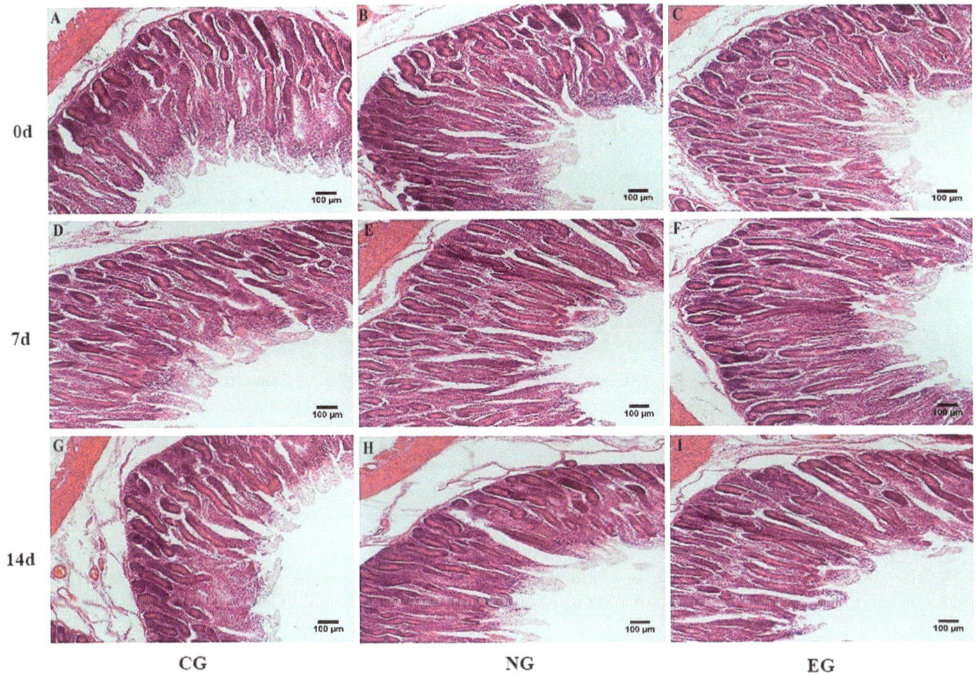

Figure 1. Effects of different treatments of transport stress on the jejunal morphology of lambs. (**A**) Jejunum of CG on day 0, (**B**) jejunum of NG on day 0, (**C**) jejunum of EG on day 0, (**D**) jejunum of control group on day 7, (**E**) jejunum of NG on day 7, (**F**) jejunum of EG on day 7, (**G**) jejunum of CG on day 14, (**H**) jejunum of NG on day 14, (**I**) Jejunum of EG on day 14.

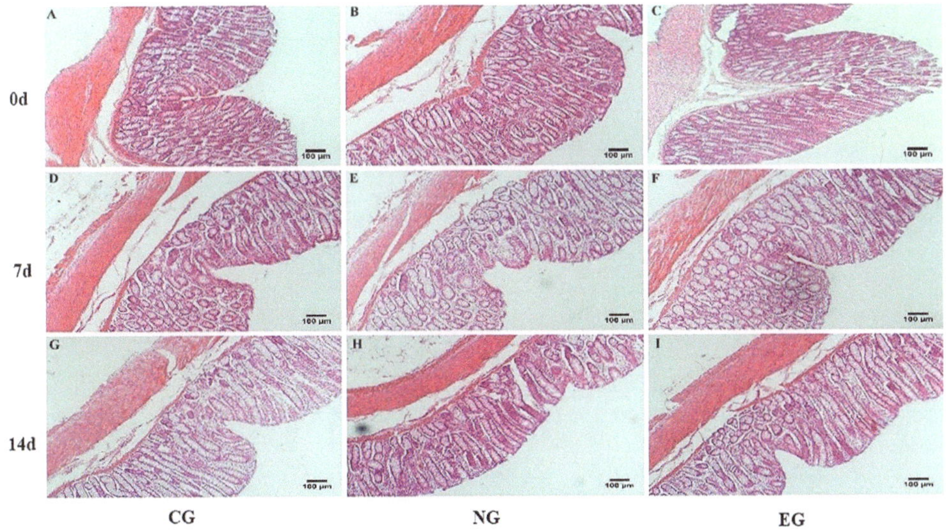

Figure 2. Effects of different treatments of transport stress on the colonic morphology of lambs. (**A**) Colon of CG on day 0, (**B**) colon of NG on day 0, (**C**) colon of EG on day 0, (**D**) colon of control group on day 7, (**E**) colon of NG on day 7, (**F**) colon of EG on day 7, (**G**) colon of CG on day 14, (**H**) colon of NG on day 14, (**I**) colon of EG on day 14.

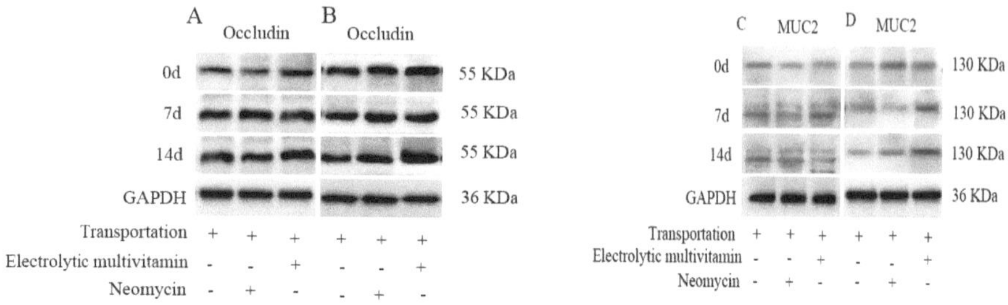

Figure 3. Effects of different treatments of transport stress on the protein expression of relevant genes in the jejunal and colonic mucosa. (**A**) The effect of different times and treatments on the protein expression of Occludin in the jejunal mucosa. (**B**) The effect of different times and treatments on the protein expression of Occludin in the colonic mucosa. (**C**) The effect of different times and treatments on the protein expression of MUC2 in the jejunal mucosa. (**D**) The effect of different times and treatments on the protein expression of MUC2 in the colonic mucosa. "+" represents lambs that were transported and treated with the addition of electrolytic multidimensional or neomycin, and "-" represents lambs that did not have the treatment addition. The original images are included in the Supplementary Materials.

Table 7. Effects of different treatments of transport stress on the expression levels of important genes in the jejunal and colonic mucosa.

Items			Time	Groups			*p*-Value		
				CG	NG	EG	Time	Treatment	Time × Treatment
Jejunal mucosa	mRNA expression	ZO-1	0 d	1.14 ± 0.13 ^{B}	1.30 ± 0.16 ^{C}	1.46 ± 0.13 ^{B}	<0.010	0.170	0.971
			7 d	1.83 ± 0.20 ^{B}	1.95 ± 0.23 ^{B}	2.01 ± 0.25 ^{B}			
			14 d	2.41 ± 0.25 ^{A}	2.60 ± 0.22 ^{A}	2.88 ± 0.27 ^{A}			
		Occludin	0 d	1.46 ± 0.12 ^{B}	1.57 ± 0.19 ^{B}	1.89 ± 0.20 ^{B}	<0.010	<0.010	0.835
			7 d	1.82 ± 0.16 ^{ABb}	2.31 ± 0.18 ^{Aab}	2.56 ± 0.20 ^{Aa}			
			14 d	2.20 ± 0.21 ^{Ab}	2.61 ± 0.20 ^{Aab}	2.95 ± 0.21 ^{Aa}			
		MUC1	0 d	1.41 ± 0.15 ^{B}	1.63 ± 0.15 ^{B}	1.86 ± 0.20 ^{B}	<0.010	<0.010	0.979
			7 d	1.76 ± 0.11 ^{ABb}	1.95 ± 0.14 ^{ABb}	2.34 ± 0.16 ^{Aa}			
			14 d	2.01 ± 0.16 ^{Ab}	2.21 ± 0.16 ^{Aab}	2.60 ± 0.12 ^{Aa}			
		MUC2	0 d	1.54 ± 0.14	1.63 ± 0.13	1.80 ± 0.15	0.106	0.220	0.986
			7 d	1.70 ± 0.12	1.75 ± 0.13	1.83 ± 0.11			
			14 d	1.79 ± 0.10	1.89 ± 0.10	1.94 ± 0.15			
	Protein expression	Occludin	0 d	0.64 ± 0.01 ^{Bb}	0.84 ± 0.04 ^{Ba}	0.91 ± 0.05 ^{Ba}	<0.010	<0.010	0.273
			7 d	0.86 ± 0.11 ^{B}	0.93 ± 0.05 ^{AB}	1.02 ± 0.06 ^{B}			
			14 d	1.09 ± 0.01 ^{Ab}	1.04 ± 0.04 ^{Ab}	1.30 ± 0.08 ^{Aa}			
		MUC2	0 d	0.09 ± 0.004 ^{B}	0.10 ± 0.02 ^{B}	0.11 ± 0.02 ^{B}	<0.010	0.508	0.154
			7 d	0.12 ± 0.01 ^{AB}	0.12 ± 0.01 ^{AB}	0.13 ± 0.02			
			14 d	0.12 ± 0.01 ^{A}	0.13 ± 0.01 ^{A}	0.13 ± 0.01			
		ZO-1	0 d	1.30 ± 0.14 ^{Bb}	1.60 ± 0.12 ^{Bab}	1.92 ± 0.21 ^{Ba}	<0.010	<0.010	0.989
			7 d	1.65 ± 0.16 ^{Bb}	1.94 ± 0.18 ^{Bab}	2.32 ± 0.18 ^{ABa}			
			14 d	2.25 ± 0.22 ^{A}	2.53 ± 0.15 ^{A}	2.75 ± 0.18 ^{A}			

Table 7. *Cont.*

			Time	Groups			p-Value		
Items				CG	NG	EG	Time	Treatment	Time × Treatment
Colonic mucosa	mRNA expression	Occludin	0 d 7 d 14 d	1.79 ± 0.21^{Bb} 2.06 ± 0.19^{ABb} 2.44 ± 0.17^{Ab}	2.02 ± 0.19^{Bab} 2.43 ± 0.17^{ABab} 2.90 ± 0.17^{Aab}	2.42 ± 0.19^{Ba} 2.83 ± 0.19^{ABa} 3.25 ± 0.16^{Aa}	<0.010	<0.010	0.979
		MUC1	0 d 7 d 14 d	1.49 ± 0.14 1.60 ± 0.11 1.77 ± 0.11	1.60 ± 0.12 1.74 ± 0.10 1.89 ± 0.11	1.65 ± 0.10^{B} 1.77 ± 0.10^{AB} 1.98 ± 0.10^{A}	<0.010	0.136	0.999
		MUC2	0 d 7 d 14 d	1.51 ± 0.16 1.78 ± 0.19 1.80 ± 0.18	1.78 ± 0.19 1.80 ± 0.15 1.86 ± 0.22	1.89 ± 0.18 1.82 ± 0.23 1.92 ± 0.14	0.679	0.497	0.915
	Protein expression	Occludin	0 d 7 d 14 d	0.62 ± 0.05^{Bb} 0.84 ± 0.07^{Ab} 0.96 ± 0.03^{Ab}	0.76 ± 0.05^{Bb} 0.93 ± 0.05^{Bb} 1.10 ± 0.04^{Ab}	0.89 ± 0.04^{Ca} 1.16 ± 0.05^{Ba} 1.46 ± 0.07^{Aa}	<0.010	<0.010	0.207
		MUC2	0 d 7 d 14 d	0.09 ± 0.01^{B} 0.13 ± 0.02^{AB} 0.27 ± 0.04^{A}	0.10 ± 0.01^{B} 0.17 ± 0.01^{B} 0.30 ± 0.04^{A}	0.11 ± 0.03^{B} 0.21 ± 0.02^{AB} 0.35 ± 0.06^{A}	<0.010	0.077	0.784

Note: ^{ab} Different lowercase letters superscripted in the same row indicate significant differences between different treatments ($p < 0.05$). ^{ABC} The superscript values in the same column with different uppercase letters indicate significant differences at different times ($p < 0.05$). The same letter or no letter indicates no significant difference ($p > 0.05$). There is no meaning between uppercase and lowercase letters.

3.6. Expression Levels of Important Genes and Proteins in the Jejunal and Colonic Tissues

From Table 8 and Figure 4, in the jejunum, the mRNA expressions of TLR4 and MyD88 in EG were significantly lower than those in CG on day 0 ($p < 0.01$). The mRNA expressions of TLR4, MyD88, and NF-kB and the protein expression of NF-kB p65 in EG were significantly lower than those in CG on day 7 ($p < 0.01$). On day 14, the mRNA expression of TRAF6 and the protein expression NF-kB p65 in NG and EG were significantly lower than those in CG ($p < 0.05$). The mRNA expressions of MyD88 and NF-kB in EG were significantly lower than those in CG ($p < 0.01$). In the colon, the mRNA expression of TRAF6 in EG was significantly lower than that in CG on day 7 ($p < 0.01$). The mRNA expressions of TLR4 and NF-kB in EG were significantly lower than those in CG on day 7 and 14 ($p < 0.01$). With the extension of time, the mRNA expressions of MyD88 in the jejunum, the mRNA expressions of TRAF6, TLR4, and NF-kB and the protein expressions of TLR4, NF-kB p65, and MyD88 in the jejunum and colon decreased significantly ($p < 0.01$).

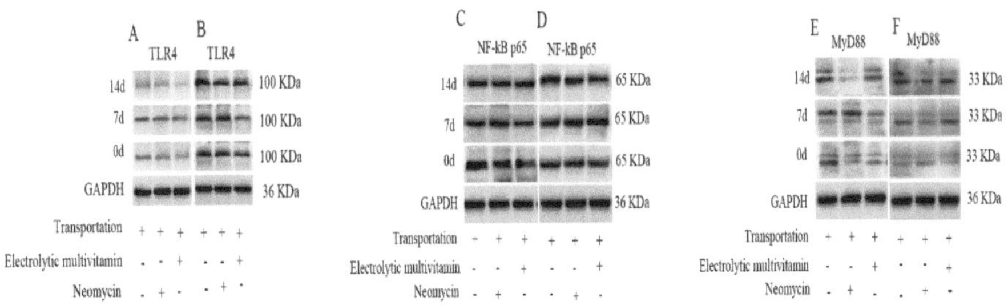

Figure 4. Effects of different treatments of transport stress on the expression levels of relevant genes in the jejunum and colon. (**A**) The effect of different times and treatments on the protein expression of TLR4 in the jejunum. (**B**) The effect of different times and treatments on the protein expression of TLR4 in the colon. (**C**) The effect of different times and treatments on the protein expression of NF-kB p65 in the jejunum. (**D**) The effect of different times and treatments on the protein expression of NF-kB p65 in the colon. (**E**) The effect of different times and treatments on the protein expression

of MyD88 in the jejunum. (**F**) The effect of different times and treatments on the protein expression of MyD88 in the colon. "+" represents lambs that were transported and were treated with the addition of electrolytic multidimensional or neomycin, and "-" represents lambs that did not have the treatment addition. The original images are included in the Supplementary Materials.

Table 8. Effects of different treatments of transport stress on the expression levels of relevant genes in the jejunum and colon.

Items			Time	Groups			*p*-Value		
				CG	NG	EG	Time	Treatment	Time × Treatment
Jejunum	mRNA expression	TRAF6	0 d 7 d 14 d	3.01 ± 0.15 ^{A} 2.65 ± 0.20 ^{AB} 2.26 ± 0.15 ^{Ba}	2.82 ± 0.16 ^{A} 2.50 ± 0.18 ^{A} 1.82 ± 0.13 ^{Bb}	2.70 ± 0.19 ^{A} 2.34 ± 0.31 ^{A} 1.52 ± 0.13 ^{Bb}	<0.010	0.011	0.765
		TLR4	0 d 7 d 14 d	2.81 ± 0.15 ^{Aa} 2.43 ± 0.13 ^{ABa} 2.10 ± 0.19 ^{B}	2.51 ± 0.16 ^{Aab} 2.17 ± 0.11 ^{ABab} 1.86 ± 0.16 ^{B}	2.33 ± 0.16 ^{Ab} 1.90 ± 0.12 ^{ABb} 1.67 ± 0.21 ^{B}	<0.010	<0.010	0.998
		MyD88	0 d 7 d 14 d	3.03 ± 0.15 ^{Aa} 2.71 ± 0.12 ^{ABa} 2.33 ± 0.17 ^{Ba}	2.74 ± 0.15 ^{Aab} 2.41 ± 0.15 ^{ABab} 2.07 ± 0.13 ^{Bab}	2.36 ± 0.23 ^{Ab} 2.13 ± 0.17 ^{ABb} 1.69 ± 0.16 ^{Bb}	<0.010	<0.010	0.997
		NF-kB	0 d 7 d 14 d	2.92 ± 0.16 ^{A} 2.54 ± 0.18 ^{ABa} 2.06 ± 0.18 ^{Ba}	2.50 ± 0.22 2.15 ± 0.17 ^{ab} 1.97 ± 0.18 ^{ab}	2.56 ± 0.12 ^{A} 1.83 ± 0.13 ^{Bb} 1.54 ± 0.16 ^{Bb}	<0.010	<0.010	0.556
	Protein expression	TLR4	0 d 7 d 14 d	0.53 ± 0.04 ^{Aa} 0.37 ± 0.04 ^{B} 0.27 ± 0.05 ^{B}	0.38 ± 0.06 ^{b} 0.29 ± 0.04 0.19 ± 0.09	0.45 ± 0.01 ^{Aab} 0.26 ± 0.06 ^{B} 0.14 ± 0.01 ^{B}	<0.010	0.028	0.774
		NF-kB p65	0 d 7 d 14 d	0.94 ± 0.05 ^{A} 0.72 ± 0.07 ^{Ba} 0.67 ± 0.04 ^{Ba}	0.89 ± 0.02 ^{A} 0.64 ± 0.05 ^{Bab} 0.49 ± 0.05 ^{Cb}	0.74 ± 0.08 ^{A} 0.50 ± 0.07 ^{Bb} 0.38 ± 0.05 ^{Bb}	<0.010	<0.010	0.799
		MyD88	0 d 7 d 14 d	0.21 ± 0.02 ^{A} 0.17 ± 0.03 ^{A} 0.09 ± 0.005 ^{B}	0.15 ± 0.04 ^{A} 0.14 ± 0.04 ^{AB} 0.04 ± 0.01 ^{B}	0.20 ± 0.01 ^{A} 0.13 ± 0.02 ^{B} 0.06 ± 0.004 ^{C}	<0.010	0.052	0.662
		TRAF6	0 d 7 d 14 d	2.66 ± 0.12 ^{A} 2.49 ± 0.14 ^{ABa} 2.13 ± 0.17 ^{B}	2.41 ± 0.17 2.23 ± 0.13 ^{ab} 1.95 ± 0.18	2.28 ± 0.15 ^{A} 1.96 ± 0.12 ^{ABb} 1.72 ± 0.13 ^{B}	<0.010	<0.010	0.981
Colon	mRNA expression	TLR4	0 d 7 d 14 d	3.07 ± 0.13 ^{Aa} 2.65 ± 0.19 ^{ABa} 2.22 ± 0.21 ^{Ba}	2.60 ± 0.17 ^{Ab} 2.29 ± 0.20 ^{ABab} 1.82 ± 0.19 ^{Bab}	2.37 ± 0.17 ^{Ab} 1.95 ± 0.17 ^{ABb} 1.54 ± 0.14 ^{Bb}	<0.010	<0.010	0.998
		MyD88	0 d 7 d 14 d	2.86 ± 0.14 2.76 ± 0.17 2.61 ± 0.14	2.71 ± 0.10 2.61 ± 0.12 2.52 ± 0.22	2.74 ± 0.15 2.64 ± 0.18 2.59 ± 0.34	0.432	0.706	0.999
		NF-kB	0 d 7 d 14 d	2.91 ± 0.17 ^{A} 2.58 ± 0.12 ^{ABa} 2.18 ± 0.19 ^{Ba}	2.71 ± 0.15 ^{A} 2.34 ± 0.18 ^{Aab} 1.82 ± 0.10 ^{Bab}	2.51 ± 0.10 ^{A} 1.93 ± 0.13 ^{Bb} 1.55 ± 0.20 ^{Bb}	<0.010	<0.01	0.903
	Protein expression	TLR4	0 d 7 d 14 d	0.20 ± 0.01 ^{A} 0.12 ± 0.02 ^{B} 0.09 ± 0.003 ^{B}	0.18 ± 0.05 ^{A} 0.10 ± 0.02 ^{AB} 0.05 ± 0.01 ^{B}	0.20 ± 0.01 ^{A} 0.10 ± 0.01 ^{B} 0.02 ± 0.003 ^{C}	<0.010	0.104	0.505
		NF-kB p65	0 d 7 d 14 d	1.07 ± 0.10 ^{A} 0.95 ± 0.05 ^{AB} 0.73 ± 0.07 ^{B}	0.98 ± 0.11 0.84 ± 0.08 0.77 ± 0.04	0.94 ± 0.10 0.78 ± 0.09 0.62 ± 0.09	<0.010	0.184	0.916
		MyD88	0 d 7 d 14 d	0.38 ± 0.05 ^{A} 0.24 ± 0.04 ^{AB} 0.18 ± 0.05 ^{B}	0.30 ± 0.05 ^{A} 0.19 ± 0.04 ^{AB} 0.11 ± 0.02 ^{B}	0.33 ± 0.03 ^{A} 0.22 ± 0.05 ^{AB} 0.15 ± 0.04 ^{B}	<0.010	0.193	0.996

Note: ^{ab} Different lowercase letters superscripted in the same row indicate significant differences between different treatments ($p < 0.05$). ^{ABC} The superscript values in the same column with different uppercase letters indicate significant differences at different times ($p < 0.05$). The same letter or no letter indicates no significant difference ($p > 0.05$). There is no meaning between uppercase and lowercase letters.

4. Discussion

Animals are affected by various stressors during transportation: their catabolism increases, nutrient consumption accelerates, and immunity decreases. Gou et al. [14] found a positive correlation between transportation time and weight loss. Furthermore, Hongjian et al. [15] found that vitamin D could significantly improve animal production and alleviate stress symptoms. In this experiment, the weight of lambs in the EG and NG was significantly higher than that in the CG. On day 7, the weight in the EG increased compared to the CG. This indicated that electrolytic multivitamin and neomycin could effectively increase the weight of lambs and alleviate transport stress, and the electrolytic multivitamin had a better effect than neomycin. The possible reason was that the supplementation of vitamins could improve the immunity of lambs [16]. In addition, with the extension of time, the weight of lambs gradually increased within 0–14 days, and transport stress was gradually alleviated.

The relative weight of immune organs represents the development status of immune organs [17], which affects the immune function of the body. It was found that the liver weight of calves decreased significantly with 11 h transportation, which was due to an increase in vitamin consumption by stress [18]. The electrolytic multivitamin had a better effect on increasing immune organ indices than neomycin, and the possible reason was that the energy consumption increased after lambs were transported, which increased demand for vitamins in lambs; the electrolytic multivitamin could supplement vitamins to the body, so the immune organ indices increased [19,20].

Under stress, the function of the hypothalamic pituitary adrenal axis (HPA axis) in animals enhances, leading to an increase in the synthesis and secretion of related hormones such as ACTH, COR, and E [21]. Zeng et al. [22] found that COR levels in serum reduced significantly and the anti-stress of cows was improved by adding vitamins to the diet. In this experiment, the concentrations of ACTH, COR, and E in the EG were the lowest, and the concentration of E in the NG was significantly lower than that in the CG. This indicated that the electrolytic multivitamin and neomycin could alleviate transport stress effectively, and the effect of the electrolytic multivitamin was better than neomycin. With the extension of time, the concentrations of stress hormones such as ACTH, E, and COR in the sera of lambs decreased gradually, and transport stress alleviated gradually.

Inflammatory factors are important mediators in the neuroendocrine and immune networks. They participate in and regulate the physiological functions of the body, and play an important role in regulating immune and inflammatory responses [23]. It was found that various stress factors could induce the secretion of inflammatory factors such as IL-1β, IL-6, and TNF-α, which could lead to inflammatory reactions in the body [24]. Shojadoost et al. [19] found that adding an appropriate amount of vitamins to the diet could reduce the expressions of intestinal inflammatory factors, and it helped to keep a balance of intestinal inflammatory factors and improve the immune function of body. In this experiment, after adding the electrolytic multivitamin, the concentrations of IL-1β, IL-6, and IFN-γ in EG decreased significantly, which indicated that the electrolytic multivitamin could reduce the inflammatory reaction, which was helpful to alleviate transport stress. In addition, with the extension of time, the concentrations of IL-2, IL-6, IL-12, and IFN-γ in sera decreased gradually, and the concentration of IL-4 increased gradually. This indicated that there was still a serious inflammatory reaction in the lamb within 7 days after transportation, and the inflammatory reaction weakened and the stress alleviated gradually on day 14.

Immunoglobulins such as IgA, IgM, and IgG are the most common indicators for evaluating humoral immune function. Amaral et al. [25] found that stress could lead to the decrease of immunoglobulin in serum and the increase of the incidence rate in animals. Liu et al. [26] found that adding vitamin E to the diet could increase the content of immunoglobulin in the sera of chicken. In this experiment, the concentrations of IgA in the EG and NG were significantly higher than that in the CG. IgA plays an important role in intestinal mucosal immune response [27]. This indicated that the electrolytic multivitamin and neomycin could block the penetration of pathogenic microorganisms into the intestinal mucosa and exert immune regulatory effects. The concentration of IgA was the highest in EG, which indicated that the electrolytic multivitamin had a better effect than neomycin. It was also found that the concentration of IgG in the EG was significantly higher than those in NG and CG. This indicated that the electrolytic multivitamin could enhance the immunity of lambs effectively. The electrolytic multivitamin and neomycin had no effect on the concentration of IgM in sera; the specific reasons need further research. Stanger et al. [28] found that the immune cells returned to normal levels after the adult cattle landed for 6 days after they had been transported for 72 h. In this experiment, the concentrations of IgA, IgG, and IgM increased within 0–14 days, the transportation stress of lambs was gradually relieved, and the immune function of lambs enhanced gradually.

The VFA in intestinal contents is the degraded product of carbohydrates by intestinal microorganisms [29], which can regulate intestinal pH, inhibit the proliferation of harmful bacteria [30], protect the intestinal mucosal barrier [31], and provide energy for the body and intestinal cells [32]. In this experiment, the content of colonic propionate in the EG and NG increased significantly, and the possible reasons were related to the types of microorganisms in the intestine [33], the addition levels of the electrolytic multivitamin and neomycin, and the pH of the intestine. The addition of the electrolytic multivitamin and neomycin had no effect on the contents of acetate, butyrate, isovalerate in the jejunum and colon, as well as the propionate in the jejunum. It is possible that the majority of VFA was absorbed in the rumen, so the change in the intestine was not significant. With the extension of time, the contents of acetate, propionate, butyrate, and isovalerate in the jejunum and colon increased gradually. The digestive function of lambs enhanced gradually, and the stress response alleviated gradually.

The intestine is an important site for digestion, absorption, and nutrient metabolism. It can easily become a target organ for transport stress [34], which results in a series of digestive system diseases. Yu et al. [35] found that there was significant damage to the intestine during heat stress in pigs, including damage to the top of the intestinal villi, reductions in villi height, and a deepening of the crypt depth. Xiao et al. [36] found that vitamins could promote intestinal development, maintain the integrity of intestinal barriers, and improve the digestion and absorption of nutrients. In this experiment, the crypt depth of the jejunum in the EG decreased, and the V/C and villus height increased. Moreover, the crypt depth of the jejunum in the NG decreased, and the V/C increased. This indicated that the electrolytic multivitamin and neomycin could promote jejunal development, and the electrolytic multivitamin had more effects than neomycin. In the colon, the villus height and V/C of EG and NG increased, and the crypt depth in NG decreased. This indicated that the electrolytic multivitamin and neomycin could maintain the colonic morphological integrity of the lambs. The electrolytic multivitamin could improve the morphology in the jejunum and colon of lambs, and it improved the nutrient utilization efficiency and growth and development of lambs. This was consistent with the result showing that the electrolytic multivitamin could promote the weight of lambs.

Tight junction proteins play an important role in maintaining intestinal homeostasis [37], and the decrease of their expression levels results in an increase in diarrhea rate [38]. The mucous layer is the first line of defense against foreign pathogens, and the decrease in the content of mucin results in intestinal barrier dysfunction [39]. Carol et al. [40] found that vitamin D could increase the expressions of tight junction proteins such as Claudin, Occludin, and ZO-1 in intestinal mucosal epithelial cells, which protected the intestinal mucosal barrier. In this experiment, the mRNA expressions of Occludin and MUC1 of the jejunal mucosa in EG, as well as the protein expression of Occludin in EG, increased. This indicated that the electrolytic multivitamin could increase the mRNA expressions of Occludin and MUC1 and protein expressions of Occludin in the jejunal mucosa of lambs; it potentially enhanced tight junctions. In the colonic mucosa, the mRNA expressions of ZO-1 and Occludin in the EG, as well as the protein expression of Occludin in the EG, increased. This indicated that the electrolytic multivitamin had a potential effect on improving the barrier function of colonic mucosa. With the extension of time, the mRNA expressions of

ZO-1, Occludin, MUC1, and MUC2, and the protein expressions of Occludin and MUC2 in jejunal and colonic mucosa, increased gradually, the tight junction recovered gradually, and the barrier function of the jejunal and colonic mucosa was enhanced. The electrolytic multivitamin had a better effect on protecting the intestinal digestive function, which is consistent with the change in the intestinal morphology of lambs. Therefore, it was speculated that the electrolytic multivitamin could maintain and improve physical and chemical barrier functions of the intestine by affecting the morphology of intestinal tissue, intestinal tight junction proteins, and mucins.

The TLRS/MyD88/NF-kB signaling pathway is an important pathway in the occurrence and development of inflammation, and it participates in biological processes such as immune regulation [41]. Yang et al. [24] found that the sustained activation of the TLR4/MyD88/NF-kB signaling pathway led to excessive immune inflammatory responses and ultimately damaged tissue. The expression levels of related signaling molecules in the TLR4/MyD88/NF-kB signaling pathway changed if an inflammatory response occurred. The electrolytic multivitamin downregulated the mRNA expressions of TRAF6, TLR4, MyD88, and NF-kB, and the protein expression of NF-kB p65 in the jejunum, and neomycin also downregulated the mRNA expression of TRAF6 and the protein expression of NF-kB p65. Meanwhile, the electrolytic multivitamin downregulated the mRNA expressions of TRAF6, TLR4, and NF-kB in the colon. This indicated that the electrolytic multivitamin could alleviate inflammation in the colon effectively. The mRNA expression levels of TRAF6, TLR4, MyD88, and NF-kB, and the protein expressions of TLR4, MyD88, and NF-kB p65 in the jejunum and colon, decreased along with increasing time, and the stress response of the jejunum and colon alleviated gradually. There was a consistent trend of change in the expression levels of TRAF6, TLR4, NF-kB, and MyD88 in the jejunum and colon, and the concentrations of IL-1β, IL-6, and IFN-γ in the sera of lambs. The electrolytic multivitamin was more beneficial than neomycin in reducing the inflammatory reaction. It is speculated that the electrolytic multivitamin may inhibit the expression levels of important genes in the TLR4/MyD88/NF-kB signaling pathway in the jejunum and colon of lambs, thereby inhibiting the production of inflammatory factors.

5. Conclusions

Dietary supplementation with a 375 mg/d/lamb electrolytic multivitamin had a negative effect on the immunity and intestinal barrier function of transported lambs. It has great potential to improve the effect of transport stress on immunity and intestinal barrier function.

Supplementary Materials: The following supporting information can be downloaded at: https://www.mdpi.com/article/10.3390/ani14020177/s1.

Author Contributions: Conceptualization, C.X. and C.D.; Formal analysis, C.C., X.Y. and Y.Z.; Investigation, C.X., C.D., C.C., X.Y. and Y.Z.; Writing—original draft, C.X. and C.D.; Writing—review and editing, C.X.; Project administration, Y.Z., Y.L. and Y.M.; Funding acquisition, Y.Z., Y.L. and Y.M. All authors have read and agreed to the published version of the manuscript.

Funding: This research was supported by the National Sheep Industry Technology System for Fluff (CARS39-23) and 2022 Basic Scientific Research Project of the Agricultural University of Hebei (KY2022029).

Institutional Review Board Statement: The experimental animal protocol of this study was approved by the Ethics Committee of Laboratory Animals of Hebei Agricultural University; the ethics committee name was the Ethics Committee of Laboratory Animals of Hebei Agricultural University. The approval code was 2020044, and the approval date was 26 December 2020. The relevant methods used in this test were implemented in accordance with relevant guidelines and regulations.

Informed Consent Statement: Not applicable.

Data Availability Statement: The data that support the findings of this study are available from the corresponding author at any time upon reasonable request. The original sequences we used for the primer sequence design can be found in GenBank (https://www.ncbi.nlm.nih.gov/genbank/, accessed on 18 November 2021.) under the accession numbers XM_042235171, >XM_042233706.1, >XM_027976040.2, >DQ152978.1, NM-102179831, XM_015101111.3, >NM_001135930.1, >DQ152979.1, and >XM_027966801.2.

Acknowledgments: We thank the Animal Science and Technology Laboratory of Hebei Agricultural University for providing technical support.

Conflicts of Interest: The authors declare no conflicts of interest.

References

1. Elisa, W.; Sébastien, G.; Martin, K.; Christina, W.; Marion, S.; Armin, T.; Brown, J. Effects of Transport Duration and Environmental Conditions in Winter or Summer on the Concentrations of Insulin-Like Growth Factors and Insulin-Like Growth Factor-Binding Proteins in the Plasma of Market-Weight Pigs. *Front. Endocrinol.* **2018**, *9*, 36.
2. Yulong, H.; Zhan, S.; Yisha, Z.; Xueyi, W.; Zeheng, G.; Lihua, H.; Dai, H. Transport stress induces pig jejunum tissue oxidative damage and results in autophagy/mitophagy activation. *J. Anim. Physiol. Anim. Nutr.* **2019**, *103*, 1921–1929.
3. Hogan, J.P.; Petherick, J.C.; Phillips, C.J.C. The physiological and metabolic impacts on sheep and cattle of feed and water deprivation before and during transport. *Nutr. Res. Rev.* **2007**, *20*, 17–28. [CrossRef] [PubMed]
4. Wei, H.K.; Xue, H.X.; Zhou, Z.X.; Peng, J. A carvacrol-thymol blend decreased intestinal oxidative stress and influenced selected microbes without changing the messenger RNA levels of tight junction proteins in jejunal mucosa of weaning piglets. *Animal* **2017**, *11*, 193–201. [CrossRef] [PubMed]
5. Gupta, D.; Ashutosh, M.; Kashyap, G.; Punetha, M.; Patel, B.; Ashutosh; Para, I. Physiological response to transportation stress at different flocking densities in hot humid and winter seasons in goats and assessment of pretreatment of Vitamin C, Electrolyte and Jaggery. *Indian J. Anim. Res.* **2019**, *53*, 299–306. [CrossRef]
6. Barry, P.; Wang, J.G. Yeast models for amyloid disease. *Essays Biochem.* **2014**, *56*, 85–97.
7. Pabst, O. New concepts in the generation and functions of IgA. *Nat. Rev. Immunol.* **2012**, *12*, 821–832. [CrossRef]
8. Ma, J.; Shah, A.M.; Wang, Z.S.; Hu, R.; Zou, H.W.; Wang, X.Y.; Kong, X.Y. Dietary supplementation with glutamine improves gastrointestinal barrier function and promotes compensatory growth of growth-retarded yaks. *Animal* **2021**, *15*, 100108. [CrossRef]
9. Zhang, R.; Zhang, W.B.; Bi, Y.L.; Tu, Y.; Ma, T.; Dong, L.F.; Diao, Q.Y. Sanguinarine and resveratrol affected rumen fermentation parameters and bacterial community in calves. *Anim. Feed Sci. Technol.* **2019**, *251*, 64–75. [CrossRef]
10. Livak, K.J.; Schmittgen, T.D. Analysis of relative gene expression data using real-time quantitative PCR and the 2(-Delta Delta C(T)) Method. *Methods* **2001**, *25*, 402–408. [CrossRef]
11. Xiaoling, C.; Dahui, L.; Zhiqing, H.; Gang, J.; Hua, Z.; Guangmang, L. Quercetin regulates skeletal muscle fiber type switching via adiponectin signaling. *Food Funct.* **2021**, *12*, 2693–2702.
12. Shengyue, S.; Lianyu, Y.; Wenyan, Y. Effects of Acanthopanax Saponin B on Expression of Hsp70 and Stress Hormone mRNA and Protein in Hypothalamus of Cold-stressed Mice. *J. Jilin Agric. Univ.* **2022**, *44*, 365–371. (In Chinese)
13. Kangli, W.; Haoran, Z.; Qianjie, H.; Junhong, L.; Guangyong, C.; Guangtian, C.; Yang, C. Effects of astragalus and ginseng polysaccharides on growth performance, immune function and intestinal barrier in weaned piglets challenged with lipopolysaccharide. *J. Anim. Physiol. Anim. Nutr.* **2020**, *104*, 1096–1105.
14. Gou, Z.; Abouelezz, K.F.M.; Fan, Q.; Li, L.; Lin, X.; Wang, Y.; Ma, X. Physiological effects of transport duration on stress biomarkers and meat quality of medium-growing Yellow broiler chickens. *Animal* **2021**, *15*, 10079. [CrossRef] [PubMed]
15. Hongjian, X.; Quanyu, Z.; Lihua, W.; Chengrui, Z.; Yang, L.; Yonggen, Z. Effects of 25-Hydroxyvitamin D3 and Oral Calcium Bolus on Lactation Performance, Ca Homeostasis, and Health of Multiparous Dairy Cows. *Animals* **2021**, *11*, 1576.
16. Hansen, L.D.E. Pre-transit vitamin C injection improves post-transit performance of beef steers. *J. Anim. Sci.* **2020**, *14*, 2083–2090.
17. Yibing, W.; Yuanyuan, W.; Baikui, W.; Xiaoqiang, M.; Shouqun, J.; Weifen, L. Protocatechuic acid improved growth performance, meat quality, and intestinal health of Chinese yellow-feathered broilers. *Poult. Sci.* **2019**, *98*, 3138–3149.
18. Fernandez, X.; Monin, G.; Culioli, J.; Legrand, I.; Quilichini, Y. Effect of duration of feed withdrawal and transportation time on muscle characteristics and quality in Friesian-Holstein calves. *J. Anim. Sci.* **1996**, *74*, 1576–1583. [CrossRef]
19. Bahram, S.; Alexander, Y.; Mohammadali, A.; Raveendra, K.; Jake, A.; Nitish, B.; Sharif, S. Effects of vitamins A, D, E and C on the chicken immune system: A Review. *Poult. Sci.* **2021**, *100*, 100930.
20. Farhad, G.; AliAhmadAlaw, Q.; Marina, S.; Anatolievich, M.A.; Ivan, G.; Alireza, S.; Gorlov, I.; Seidavi, A. Effects of in ovo feeding of vitamin E or vitamin C on egg hatchability, performance, carcass traits and immunity in broiler chickens. *Anim. Biotechnol.* **2023**, *34*, 456–461.
21. He, X.; Lu, Z.; Ma, B.; Zhang, L.; Li, J.; Jiang, Y.; Gao, F. Chronic heat stress alters hypothalamus integrity, the serum indexes and attenuates expressions of hypothalamic appetite genes in broilers. *J. Therm. Biol.* **2019**, *81*, 110–117. [CrossRef] [PubMed]
22. Zeng, H.; Xi, Y.; Li, Y.; Wang, Z.; Zhang, L.; Han, Z. Analysis of Astragalus Polysaccharide Intervention in Heat-Stressed Dairy Cows' Serum Metabolomics. *Animals* **2020**, *10*, 574. [CrossRef]
23. Yi, Y.X.; Hua, W.L.; Farrar, W.L. A Role for PPARgamma in the Regulation of Cytokines in Immune Cells and Cancer. *PPAR Res.* **2008**, *2008*, 961753.

24. Yang, Z.M. The effect of transport stress on blood biochemical indicators and concentration of inorganic ion in Qingjiao Ma chicken. *Chin. J. Anim. Sci.* **2021**, *57*, 261–264.
25. Amaral, B.C.; Connor, E.E.; Tao, S.; Hayen, M.J.; Bubolz, J.W.; Dahl, G.E. Heat stress abatement during the dry period influences metabolic gene expression and improves immune status in the transition period of dairy cows. *J. Dairy Sci.* **2011**, *94*, 86–96. [CrossRef] [PubMed]
26. Liu, Y.J.; Zhao, L.H.; Mosenthin, R.; Zhang, J.Y.; Ji, C.; Ma, Q.G. Protective Effect of Vitamin E on laying performance, antioxidant capacity, and immunity in laying hens challenged with Salmonella Enteritidis. *Poult. Sci.* **2019**, *98*, 5847–5854. [CrossRef]
27. Chen, H.; Ao, R.G.L.; Wang, C.J.; Si, M.J.D.; Andrew, R.C.; Zhang, J.; Meng, L. Comparative study on immune and antioxidant functions of different breeds of cattle under heat stress. *J. China Agric. Univ.* **2019**, *24*, 72–77.
28. Stanger, K.J.; Ketheesan, N.; Parker, A.J.; Coleman, C.J.; Lazzaroni, S.M.; Fitzpatrick, L.A. The effect of transportation on the immune status of Bos indicus steers. *J. Anim. Sci.* **2005**, *83*, 2632–2636. [CrossRef]
29. Jiayu, M.; Jian, W.; Shad, M.; Shenfei, L.; Di, W.; Jie, G.; Piao, X. Supplementation of Mixed Organic Acids Improves Growth Performance, Meat Quality, Gut Morphology and Volatile Fatty Acids of Broiler Chicken. *Animals* **2021**, *11*, 3020.
30. Veronica, K.S.; Nagaraju, I.; Leslie, H.M.; Bonnie, V.; Samuel, B.J.; Darko, S.; Melgar, A.; Spitzer, A. Using Structural Equation Modeling to Understand Interactions Between Bacterial and Archaeal Populations and Volatile Fatty Acid Proportions in the Rumen. *Front. Microbiol.* **2021**, *12*, 611951.
31. Ma, J.; Piao, X.; Mahfuz, S.; Long, S.; Wang, J. The interaction among gut microbes, the intestinal barrier and short chain fatty acids. *Anim. Nutr.* **2022**, *9*, 159–174. [CrossRef] [PubMed]
32. Junhua, L.; Tingting, X.; Weiyun, Z.; Shengyong, M. High-grain feeding alters caecal bacterial microbiota composition and fermentation and results in caecal mucosal injury in goats. *Br. J. Nutr.* **2014**, *112*, 416–427.
33. Chen, J.; Yang, Z.; Dong, G. Niacin nutrition and rumen-protected niacin supplementation in dairy cows: An updated review. *Br. J. Nutr.* **2019**, *122*, 1103–1112. [CrossRef]
34. Gu, X.H.; Hao, Y.; Wang, X.L. Overexpression of heat shock protein 70 and its relationship to intestine under acute heat stress in broilers. 2. Intestinal oxidative stress. *Poult. Sci.* **2012**, *91*, 790–799. [CrossRef] [PubMed]
35. Yu, J.; Yin, P.; Liu, F.; Cheng, G.; Guo, K.; Lu, A.; Zhu, X.; Luan, W. Effect of heat stress on the porcine small intestine: A morphological and gene expression study. *Comp. Biochem. Physiol. Part A* **2010**, *156*, 119–128. [CrossRef] [PubMed]
36. Xiao, L.; Cui, T.; Liu, S.; Chen, B.; Wang, Y.; Yang, T.; Li, T.; Chen, J. Vitamin A supplementation improves the intestinal mucosal barrier and facilitates the expression of tight junction proteins in rats with diarrhea. *Nutrition* **2018**, *57*, 97–108. [CrossRef] [PubMed]
37. Mirjam, B.Z.; Punita, D.; Thomas, F.B. Tight junction proteins in gastrointestinal and liver disease. *Gut* **2019**, *68*, 547–561.
38. Nabuurs, M.J.; Essen, G.J.V.; Nabuurs, P.; Niewold, T.A.; Meulen, V.D.J. Thirty minutes transport causes small intestinal acidosis in pigs. *Res. Vet. Sci.* **2001**, *70*, 123–127. [CrossRef]
39. Desai, M.S.; Seekatz, A.M.; Koropatkin, N.M.; Kamada, N.; Hickey, C.A.; Wolter, M.; Kitamoto, S.; Muller, A. A Dietary Fiber-Deprived Gut Microbiota Degrades the Colonic Mucus Barrier and Enhances Pathogen Susceptibility. *Cell* **2016**, *167*, 1339–1353 e1321. [CrossRef]
40. Carol, L.; Ethan, L.; Sinobol, C.; Rachel, F.; Bo, L.; Haitao, Z.; Yamato, M. Protective effects of vitamin D against injury in intestinal epithelium. *Pediatr. Surg. Int.* **2019**, *35*, 1395–1401.
41. Xiao, Y.; Yan, H.; Diao, H.; Yu, B.; He, J.; Yu, J.; Zheng, P.; Mao, X. Early Gut Microbiota Intervention Suppresses DSS-Induced Inflammatory Responses by Deactivating TLR/NLR Signalling in Pigs. *Sci. Rep.* **2017**, *7*, 3224. [CrossRef] [PubMed]

animals

Article

Farm Animal Welfare during Transport and at the Slaughterhouse: Perceptions of Slaughterhouse Employees, Livestock Drivers, and Veterinarians

Maja Lipovšek [1], Andrej Kirbiš [1], Iztok Tomažič [2], Alenka Dovč [3] and Manja Križman [1,*]

[1] Institute for Food Safety, Feed and Environment, Veterinary Faculty, University of Ljubljana, 1000 Ljubljana, Slovenia; majalip11@gmail.com (M.L.); andrej.kirbis@vf.uni-lj.si (A.K.)
[2] Department of Biology, Biotechnical Faculty, University of Ljubljana, 1000 Ljubljana, Slovenia; iztok.tomazic@bf.uni-lj.si
[3] Clinic for Birds, Small Mammals and Reptiles, Veterinary Faculty, University of Ljubljana, 1000 Ljubljana, Slovenia; alenka.dovc@vf.uni-lj.si
* Correspondence: manja.krizman@vf.uni-lj.si; Tel.: +386-1-4779-185

Simple Summary: This study investigated the level of knowledge and the current situation with regard to the welfare of farm animals during transportation and in beef, pork, and poultry slaughterhouses. For this purpose, a questionnaire was developed to obtain data on respondents' understanding of their work, knowledge of legislation, training, and attitudes towards animal welfare. Slaughterhouse employees and professional animal livestock drivers participated in the study. Slaughterhouse employees showed more knowledge about animal welfare than livestock drivers, but both groups were not sufficiently familiar with animal welfare legislation and regulations. All respondents agreed that animals are sentient beings and almost all respondents were unfamiliar with the concept of biosecurity. This study found that the results of the veterinary experts' observations were generally lower than the results of the employees' and livestock drivers' self-assessments. Based on the research findings, it can be concluded that there is a need to improve the awareness and knowledge of slaughterhouse employees and livestock drivers regarding animal handling and welfare regulations. This would include providing better hands-on training, better knowledge of legislation, and raising awareness of the benefits of certain procedures and standards in slaughterhouses and during transportation.

Abstract: Animal welfare is a multidimensional concept that includes several physical and psychological parameters of the animal. The aim of this study was to assess animal welfare during transportation and in Slovenian beef, pork, and poultry slaughterhouses. A questionnaire was used for this study. Several parameters of animal welfare were rated on a 5-point scale, such as health status, animal behavior, lairage or transport vehicle conditions, and driver regulation compliance. The scale was also used for the second part of the study. This consisted of two studies: (1) self-assessment by slaughterhouse employees and livestock transport drivers and (2) animal welfare observational assessment performed by two veterinarians. The results were compared with each other. Ten large slaughterhouses and nine livestock drivers took part in the survey. The results showed that slaughterhouse employees knew more about animal welfare than livestock truck drivers, but both groups were not sufficiently familiar with animal welfare laws and regulations. This study found that the experts' assessments were generally lower than the self-assessments of employees and livestock drivers. Based on the research findings, it can be concluded that there is a need to improve the awareness and knowledge of slaughterhouse employees and livestock drivers regarding animal handling and animal welfare regulations.

Keywords: welfare; transport; livestock; slaughterhouse; legislation

Citation: Lipovšek, M.; Kirbiš, A.; Tomažič, I.; Dovč, A.; Križman, M. Farm Animal Welfare during Transport and at the Slaughterhouse: Perceptions of Slaughterhouse Employees, Livestock Drivers, and Veterinarians. *Animals* **2024**, *14*, 443. https://doi.org/10.3390/ani14030443

Academic Editor: Elbert Lambooij

Received: 4 December 2023
Revised: 24 January 2024
Accepted: 25 January 2024
Published: 29 January 2024

1. Introduction

Animal welfare means how an animal copes with the conditions in which it lives. An animal is in a good state of welfare if it is (scientifically proven) healthy, comfortable, well fed, safe, able to act out its innate behavior, and does not suffer from unpleasant conditions such as pain, fear, and distress [1]. Animal welfare requires disease prevention and veterinary treatment, appropriate housing, management, nutrition, humane treatment, and humane slaughter [2,3]. Various concepts of animal welfare have been developed over the last half decade [4].

Farm animals such as poultry, pigs, cows, sheep, goats, and horses are usually under stress due to high production and economic demands. In recent decades, we have gained a better overview of the parameters that influence animal welfare [5].

Protecting animal welfare during the slaughter process is about minimizing the pain, distress, and suffering of farm animals at the time of slaughter. Slaughterhouse staff must therefore apply various procedures. For example, they must regularly check that the animals show no signs of consciousness or sentience between the end of the stunning process and death [6]. Current European Union legislation on the welfare of animals at the time of killing requires that personnel working with live animals have a certificate of competence. Food business operators must ensure that certain slaughter operations, e.g., movement, stunning, and slaughter, are only carried out by persons holding such a certificate. The equipment used for slaughter and the design and construction of the slaughterhouse must also comply with the regulations [7]. In order to obtain the certificate of competence, personnel must complete a training course consisting of practical training on animal handling and correct slaughtering, basic training, and refresher training [8].

Recently, more data have been collected in slaughterhouses to monitor the welfare of animals on the farm, during transportation, and on the slaughterhouse premises [9]. The slaughterhouse can be a dangerous and psychologically stressful workplace [10]. Slaughterhouse employees and livestock drivers must comply with legal regulations and work quickly at the same time. This can be difficult for an employee if they do not have practical knowledge and skills. However, if the slaughterhouse does not have the appropriate facilities or technical equipment, it is even more difficult for slaughterhouse employees to move animals quickly and in accordance with best animal welfare practices.

The objective of our research project was to assess the knowledge of slaughterhouse employees and livestock drivers about animal welfare indicators. In Slovenia, there have been several animal welfare scandals in slaughterhouses in recent years. The aim of our study was therefore to find out whether the problem lies in the knowledge of animal welfare among people working with live animals and what knowledge can be improved. For this purpose, we used a questionnaire with (I) basic information, (II) general statements, and (III) statements about practical work with live animals. In addition, we assessed (IV) the perception of the participants' practical work on animal welfare based on an actual animal welfare situation in a slaughterhouse lairage.

2. Materials and Methods

Ten slaughterhouse employees and nine livestock truck drivers participated in this study (Table 1). Slovenia is divided into ten regions for official control in slaughterhouses (R1–R10 in Table 1) by the Slovenian Administration for Food Safety, Veterinary Sector and Plant Protection. For our study, we selected the largest slaughterhouse in Slovenia and one livestock driver from every selected slaughterhouse who transported animals for slaughter on that day. There are seven slaughterhouses for poultry meat in Slovenia, three small and four large ones. All four large slaughterhouses participated in our study. For red meat, there are 83 slaughterhouses in Slovenia, of which 58 (70%) are small (slaughtering less than 1000 livestock units per year) and 25 (30%) are large. We surveyed six large red meat slaughterhouses, which means that 24% of our sample consists of large slaughterhouses. However, these are the largest slaughterhouses in Slovenia, with the fastest slaughter lines and most cattle and pigs are slaughtered here. In Slovenian slaughterhouses, work is performed only during morning shifts. This means that the slaughterhouse staff usually consisted of two people who were suitable for the interview.

Table 1. Visits of slaughterhouses (n = 10) in regional offices in Slovenia, categories of animals approved for slaughter, stunning methods and slaughter line speed.

Slaughterhouse Employees/Livestock Drivers	Region for Official Control	Animal Species Approved for Slaughter/Animals Transported on the Truck	Stunning Method of the Slaughterhouse	Slaughter Line Speed
Slaughterhouse employees	R1	Turkeys	Electrical waterbath	1100 turkeys per hour
	R5	Bovine, caprine, ovine, horses	Penetrative captive bolt device and head-only electrical stunning	7 bovine animals per hour 20 sheep or goats per hour
	R3	Bovine, porcine, horses	Penetrative captive bolt device and carbon dioxide at high concentration	25 bovine animals per hour 100 pigs per hour
	R10	Chicken broilers	Carbon dioxide in two phases	8600 broilers per hour
	R1	Bovine, porcine	Penetrative captive bolt device and head-only electrical stunning	25 bovine animals per hour 30 saws per hour
	R2	Bovine, porcine	Penetrative captive bolt device and carbon dioxide at high concentration	30 bovine animals per hour 100 pigs per hour
	R3	Porcine	Carbon dioxide at high concentration	110 pigs per hour or 80 piglets
	R5	Chicken broilers	Electrical waterbath	4000 broilers per hour
	R9	Chicken broilers	Electrical waterbath	4000 broilers per hour
	R9	Porcine	Carbon dioxide at high concentration	100 pigs per hour
Livestock drivers	R1	Turkeys	NA	NA
	R5	Bovine (young bulls)	NA	NA
	R3	Bovine (milk cows)	NA	NA
	R10	Broilers	NA	NA
	R1	Bovine (young bulls)	NA	NA
	R3	Porcine	NA	NA
	R5	Chicken broilers	NA	NA
	R9	Chicken broilers	NA	NA
	R9	Porcine	NA	NA

NA = not applicable.

We first contacted the biggest slaughterhouses in Slovenia according to data from the Slovenian Administration for Food Safety, Veterinary Sector and Plant Protection. We contacted them by telephone to find out whether they wanted to participate in the animal welfare research project. If the slaughterhouse agreed, we arranged a date for the visit of veterinary experts. The slaughterhouse manager decided which slaughterhouse employee would take part in the project. However, the selected employee was asked by the interviewing veterinarian if they were willing to participate and, if so, they were presented with a consent form to sign. All slaughterhouse employees who participated in the study had a certificate of competence in accordance with Regulation (EC) 1099/2009.

To obtain the certificate of competence, the employee must complete a theoretical and practical part of the training. The first part is completed in a company accredited by the Slovenian Administration for Food Safety, Veterinary Sector and Plant Protection. The theoretical part covers the basics of animal welfare, animal welfare legislation, species-specific characteristics of animal movement, and methods of stunning and killing. After the theoretical part, employees must pass a written examination. The second, practical part takes place in the slaughterhouse where the employees work [11]. After the interview with the slaughterhouse employee, we asked the first livestock truck driver who brought the animals to the slaughterhouse that day whether he would like to participate in the research project. If he agreed, a consent form was presented to him for signature. In addition, all livestock truck drivers participating in the study were professional drivers for live animals and were trained in accordance with Regulation (EC) 1/2005. In one slaughterhouse R2, only local farmers were transporting livestock to the slaughterhouse on the day of the survey. As they do not require certification under Regulation (EC) 1/2005, they did not meet the criteria for the project.

The sample size of the red meat slaughterhouses is a limitation. However, 10 large slaughterhouses were contacted, 4 of which declined to participate for various reasons. The speed of the slaughter lines in the 15 remaining slaughterhouses is much slower than in the selected ones, which is why they were not included in the study.

2.1. Questionnaire with Protocol

For this study, an animal welfare questionnaire with protocol was created for slaughterhouse employees and livestock drivers (AWQ/SL and AWQ/D) (Supplementary Material S1). General data and participant demographics were collected first, followed by 29 items assessing participants' general beliefs about animal welfare: (1) the importance of information (seven items); (2) the influence of the public on animal welfare (five items); (3) animal welfare (five items); (4) the utilitarian/dominionistic view of farm animals (five items); and (5) the relationship between humans and animals (seven items). The third part of the questionnaire consisted of 25 questions about the participants' knowledge of animal welfare: (1) what animal welfare is, (2) water and feed requirements, (3) animal well being, and (4) health status. The protocol consisted of questions on the perceived importance of the participants and related observational assessment questions: (1) general status, (2) animal behavior, (3) lairage conditions, and (4) environmental conditions. The questionnaires on the importance of animal welfare at the slaughterhouse and during transport (self-assessment) were identical and were used to compare the assessment of the importance of animal welfare at slaughter and during transport. Observational assessment parameters were rated on a 5-point scale, while employee/driver perceived importance was rated on a 5-point Likert scale.

Two veterinarians familiar with slaughter and transportation assessed the welfare protocol. Both were previously trained by the project leader, a diplomate of the European College of Animal Welfare and Behavioral Medicine. The observation scoring points have the following meaning: (1) major deficiencies (immediate action required); (2) deficiencies warranting a warning; (3) minor deficiencies (advice required); (4) no deficiencies (compliance with standards); and (5) no deficiencies (above standards). Additional descriptions were provided for each observation point (Supplementary Material S1). The legal basis for establishing the point scale was Council Regulation (EC) No. 1/2005, Council Regulation (EC) No. 1099/2009, and the Animal Protection Act from the Official Gazette of the Republic of Slovenia, No. 38/13.

For example, the questions in the questionnaire designed to measure the importance of animal welfare for slaughterhouse employees/livestock drivers began with the question "In your opinion, how important is it that you move the animals quickly but in accordance with the law?" The scale indicates the degree of importance for slaughterhouse employees/livestock drivers: (1) not at all important, (2) not important, (3) undecided, (4) important, and (5) very important.

First, animal welfare was assessed using the protocol, and then slaughterhouse employees/livestock driver was asked about their views on welfare using a questionnaire. Slaughterhouses and livestock drivers were interviewed from 4 April to 14 September 2022.

2.2. Statistical Analysis

All raw data were first transferred to Microsoft Excel version 2312 build 16.0.17126.20132 and transformed for use in Statistical Package for the Social Sciences—SPSS (ver. 26). Mean values were calculated for each parameter of the questionnaire (general status, animal behavior, lairage/transport conditions, and environmental conditions) and these were compared between slaughterhouse employees and livestock drivers (Mann–Whitney test). In addition, the Wilcoxon test was applied to compare the observation results with the slaughterhouse employees/livestock drivers' perceived importance of animal welfare from the protocol and the questionnaire. Due to the small sample size, effect sizes were calculated to determine the strength of statistical differences using the formula $r = Z/\sqrt{N}$. Values < -0.2 or >0.2 were treated as significant [12].

3. Results

Ten slaughterhouse employees and nine livestock drivers took part in this study. The interview lasted an average of 55 to 75 min per participant, including observation by two experts of how the participants carry out their daily work in the field of animal welfare. The gender, age and education level of the participants are summarized in Table 2.

Table 2. Gender, age, and level of education of the participants.

Category	Slaughterhouse Employees	Livestock Drivers
Gender		
Male	7	9
Female	3	-
Age		
<40 years	2	2
>40 years	8	7
Education		
Primary school	1	1
High school	6	8
Professional degree	2	-
Unified Master's program	1	-

3.1. Animal Welfare Beliefs and Awareness Comparison between Slaughterhouse Employees and Livestock Drivers

We compared slaughterhouse employees' and livestock drivers' perceptions of animal welfare (Figure 1) and found that slaughterhouse employees generally expressed stronger beliefs than livestock drivers. Although all calculated *p*-values were above the statistical significance level (all $p > 0.05$), the calculated effect sizes indicated that there were no differences in the table for 12 of 29 items (see Supplementary Material S1). For four out of seven items assessing the importance of information (Figure 1, A section), slaughterhouse employees rated three statements more positively than livestock drivers. On the other hand, livestock drivers were more likely to think that animal welfare should be taught in schools than slaughterhouse employees (A_06). Several differences were also found in utilitarian/dominionistic attitudes towards farm animals. For example, three out of five items were rated by slaughterhouse employees as more in favor of animal welfare (Figure 1, D section). These participants do not see animals only as a source of income (D_01), do not believe that animals should be treated like machines (D_02), and do not think that sick animals should be killed by the farmers themselves (D_05). However, livestock drivers agree with the statement that sick animals should be euthanized by farmers. In the "animal welfare" category (Figure 1, C-section), two of the five items were rated differently by slaughterhouse employees and livestock drivers. Slaughterhouse employees were more likely to agree that animals should be housed appropriately (C_04) and that farm animals do not have a lower pain threshold (C_05). In the two remaining categories, public influence on animal welfare (Figure 1, B-section) and human–animal relations (Figure 1, E section), only one item each was rated differently by slaughterhouse employees and livestock drivers. Slaughterhouse employees were more likely to agree with the statement that the government should provide financial support to improve animal welfare in slaughterhouses (B_04) and were less likely to agree with the statement that animals only obey when they are afraid of their owners (E_04). Livestock drivers, on the other hand, were rather undecided on this parameter.

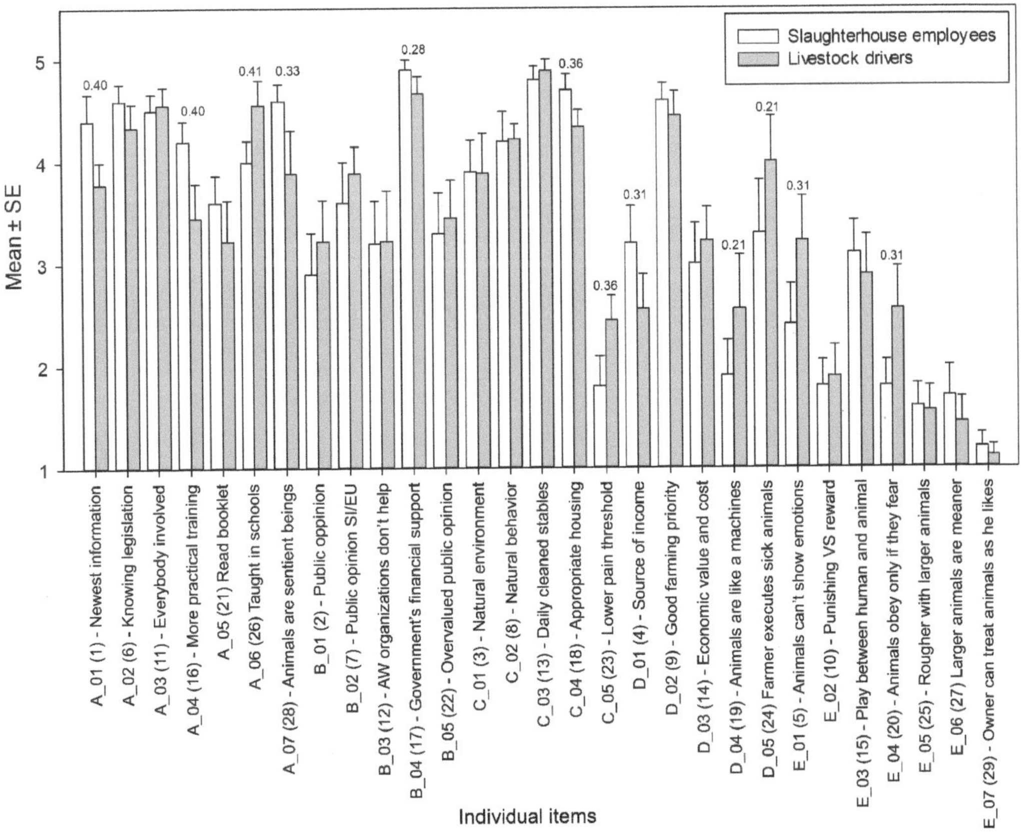

Figure 1. Comparison between slaughterhouse employees and livestock drivers in relation to animal welfare beliefs on the topics of knowledge of animal welfare laws (A), influence of public opinion on animal welfare (B), animal welfare (C), value of animals to humans (D) and relationship between humans and animals (E) (only effect size values > 0.20 are listed).

Next, we compared the knowledge of animal welfare between slaughterhouse employees and livestock drivers on four different topics (see Supplementary Material S1): what is animal welfare (A), water and feed requirements (B), animal welfare (C), and health status (D) (Figure 2). For three statements, the calculated p-values were statistically significant ($p < 0.05$). With regard to animal welfare (Figure 2, A section), slaughterhouse employees were undecided on the statement "due to domestication, farm animals have different needs than their ancestors living in the wild", while livestock drivers agreed with the statement (A_05). The statement that the welfare of the animal is taken care of when it is growing and in good physical condition was also rated higher by the livestock drivers (A_04). On the other hand, slaughterhouse employees rated the statement that we comply with animal welfare standards if the animal reproduces successfully higher (A_03). In section (B), water and feed requirements (Figure 2, B-section), all calculated values were above statistical significance. However, slaughterhouse employees were undecided on the statement (B_04) "farm animals do not need bedding in the slaughter pens". Livestock drivers were more likely to answer, "disagree with the statement". In the section on animal welfare (Figure 2, C section), there was a statistically significant difference in the statement "farm animals are sentient living beings", which slaughterhouse employees agreed with more strongly than livestock drivers (C_01). Slaughterhouse employees also rated the statements about mobile slaughterhouses and the advantages of slaughtering in regular slaughterhouses positively (C_05). Slaughterhouse employees tended to agree with both methods, while livestock drivers were undecided. In the last section (Figure 2, D-section) on health status, there was a statistically significant difference in the statement "the presence of disease does not affect animal welfare". Slaughterhouse employees disagreed with this statement, while livestock drivers were undecided (D_01). The most interesting finding was (D_06) that most slaughterhouse employees and livestock drivers were undecided about the statement "The term biosecurity means that the feed for the animals is sanitary". For the statement "The presence of diseases does not affect the meat products", both parties disagreed, but slaughterhouse employees agreed more than livestock drivers (D_07).

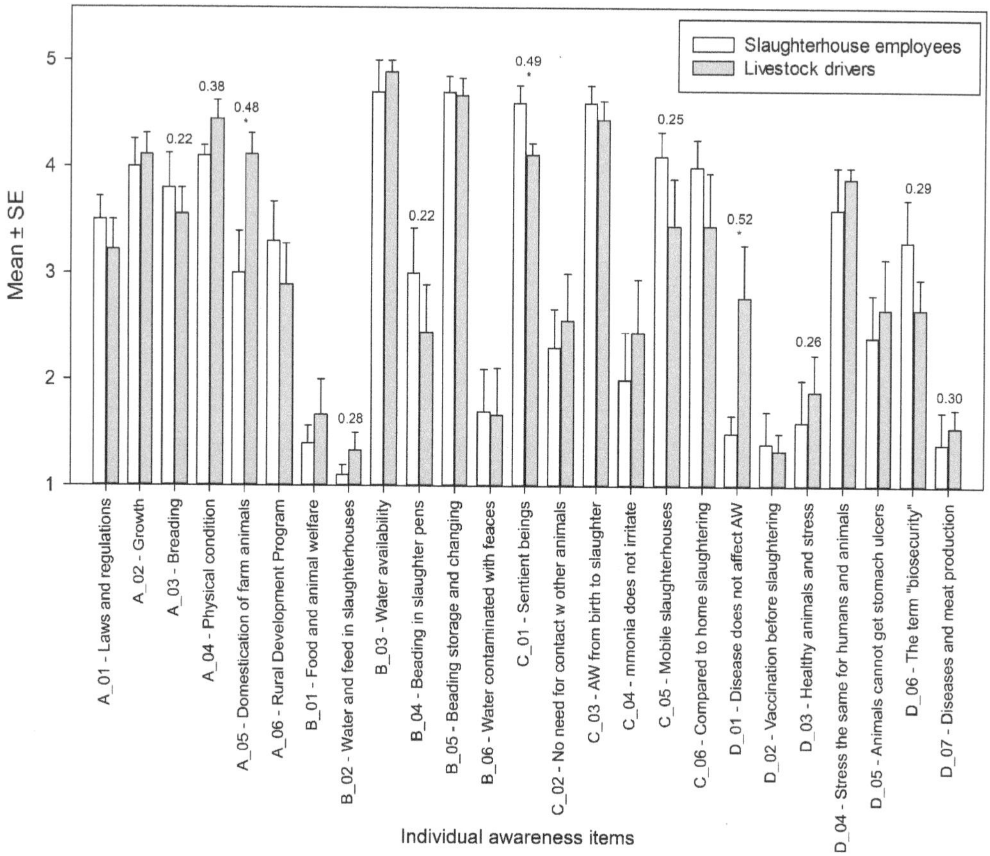

Figure 2. Comparison of animal welfare awareness between slaughterhouse employees and livestock drivers on the topics of animal welfare (A), water and feed requirements (B), the welfare of animals (C), and health status (D) (only effect size values > 0.20 are listed; * $p < 0.05$).

3.2. Comparison of Perceived and Actual Animal Welfare Standards in Slaughterhouse and on the Transport

3.2.1. Observational Assessment by the Veterinary Expert Compared to Self-Assessed Importance by the Slaughterhouse Employees

In the observational assessment by the veterinary experts, the highest average score for the slaughterhouse was awarded for the lairage conditions parameter and the lowest for the environmental conditions (Figure 3). The highest values for the self-assessed importance of animal welfare by the slaughterhouse employees were awarded for the two questions "In your opinion, how important is...", "...the number of animals per unit area", and "...that ventilation and heating are regulated/arranged in the lairage?" The comparison of the results of the veterinary expert's observational assessment with the slaughterhouse employees' self-assessment of the importance of animal welfare revealed statistically significant differences for five out of ten items (Wilcoxon test, all $p < 0.05$). Moreover, for all but one item, the calculated effect sizes were substantial and ranged from 0.42 to 0.84. In all these cases, the self-assessed importance of animal welfare was rated significantly higher than the score obtained from the veterinary expert's observational assessment. The most significant difference was clearest in the parameters for handling animals in the lairage (B3). This was followed by the parameters of stocking density in the pens (C1), the way employees enter the lairage (A1), the behavioral condition of the animals (B1 and B2), and the presence of thermometers and hygrometers in the lairage (D5).

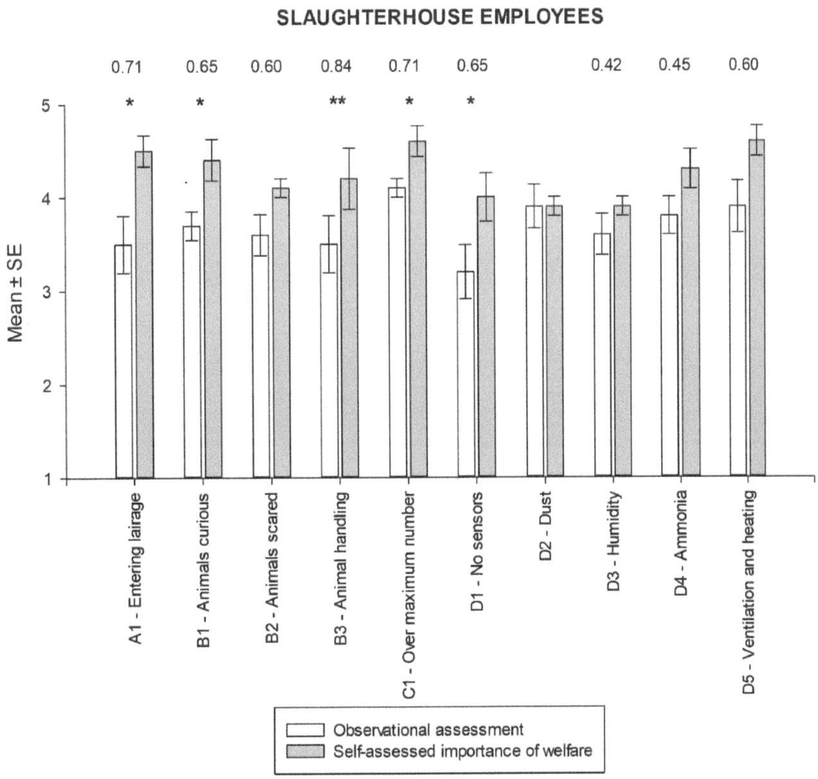

Figure 3. Comparison of perceived and actual animal welfare in the slaughterhouse on topics of General impression (A), Animal behavior (B), Lairage conditions (C), and Environmental conditions (D) (only effect size values > 0.20 are listed; * $p < 0.05$; ** $p < 0.01$).

3.2.2. Observational Assessment by the Veterinary Expert Compared to Self-Assessed Importance by the Livestock Drivers

In the observational assessment by the veterinary experts of transport, the highest average result was for the "compliance with regulations" factor and the lowest for the general condition and environment (Figure 4). The highest values for the self-assessed importance of animal welfare by the livestock drivers were obtained for the two questions "In your opinion, how important is it...", "...that you put the correct departure date and time on your driver statement", and "...that you have a driving license in accordance with Regulation (EC) 1/2005?". The comparison of the results of the observational assessment with the self-assessed importance of animal welfare revealed statistically significant differences for four out of thirteen items (Wilcoxon test, all $p < 0.05$). For nine of thirteen items, the calculated effect sizes were also considerable and ranged from 0.42 to 0.82. With the exception of two items, the self-assessed importance of animal welfare was significantly higher in all these cases than the score obtained from the observational assessment. However, for the parameter "compliance with regulations" in response to the question "How important do you think it is that you have not been fined for a violation of Regulation (EC) 1/2005?", the observation score was higher than the self-assessed importance (E6). This is due to the fact that one driver stated that it was not important whether he had received a penalty for the inappropriate transportation of live animals to the slaughterhouse. This is because he received a fine for not complying with the regulations for the transportation of live animals. However, apart from the fine, there were no consequences for him such as, for example, the ban on transporting live animals.

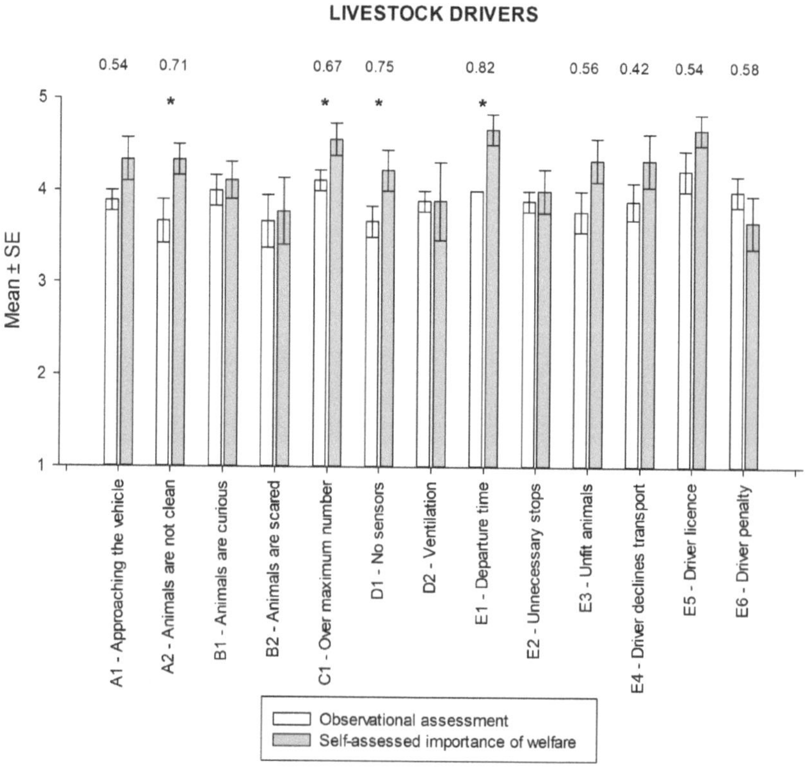

Figure 4. Comparison of perceived and actual animal welfare standards during transport on topics of General impression (A), Animal behavior (B), Transport conditions (C), Environmental conditions (D), Regulation compliance (E) (only effect size values > 0.20 are listed; * $p < 0.05$).

4. Discussion

Animal welfare is clearly a concept that can be studied scientifically, but our understanding of animal welfare, and even the science we do to assess and improve animal welfare, is influenced by value-based ideas about what is important or desirable for animals to live well [1].

Our comparison of the opinions of slaughterhouse employees and livestock drivers on animal welfare shows that there are significant differences in their perceptions. Slaughterhouse employees generally have a stronger opinion, while livestock drivers are more indecisive in their responses. A key finding was the statement about the need for practical training to ensure animal welfare. Slaughterhouse employees agree that more or better practical training is needed. In contrast, livestock drivers remained undecided, indicating a possible gap in their understanding and knowledge in this area. This highlights the importance of providing comprehensive and targeted training programs for both slaughterhouse employees and livestock drivers to improve their ability to effectively comply with animal welfare standards. European Union legislation on the protection of animals at the time of killing requires that persons working with live animals have a certificate of competence [7]. In Slovenia, however, only employees of slaughterhouses have a certificate of competence, but not livestock drivers. This does not comply with the regulation, as the drivers unload live animals from the vehicle at the slaughterhouse. According to the World Organization for Animal Health [2], people involved in unloading, moving, lairage, caring, restraining, stunning, slaughtering, and bleeding animals play an important role in their welfare. Their competence must be acquired through formal training and practical experience in the slaughterhouse. After the theoretical part, employees in Slovenia must pass a written examination. However, the degree of concern for animal welfare depends on various parameters, such as gender, age, education, place of residence, dietary habits, etc. [13–15]. Improving the welfare of farm animals is costly. The simplest approach is to do nothing, but doing nothing can come at a cost. These costs come in the form of risks. Public concern about farm animal welfare has been studied over a long period of time [16–19], and there is evidence that public concern is increasing [16,17]. From different perspectives, some of the costs are one-time costs associated with changing infrastructure and practices, others are ongoing operational costs, and others are costs to which all companies in an industry must indirectly contribute. All of these costs are likely to be important factors in deciding which improvements should be made [20].

In addition, slaughterhouse employees showed a higher level of understanding of animal welfare compared to livestock drivers. They did not agree with the statement that farm animals have no feelings or only act out of fear. Slaughterhouse employees strongly affirmed that animals are sentient beings and showed a deeper understanding of their needs and behaviors. On the other hand, livestock drivers tended to agree with the statement that farmers should kill sick animals themselves to avoid expensive treatment or euthanasia costs. This discrepancy indicates a potential risk in the decision making of livestock drivers when they encounter animals that are not fit for transportation. The views of animal drivers could be influenced by cost considerations that could affect their judgment about the appropriate actions to take when dealing with animals that are unfit for transport. According to Grandin [21], "the most important thing is to have an animal fit for transport". It is impossible to ensure the welfare of the animal during transportation if the animal is unfit for transport. However, the transportation of unfit animals is a common cause of non-compliance with animal health regulations [22]. The difference in knowledge and perspective between slaughterhouse employees and livestock drivers can be attributed to their different roles and responsibilities in the industry. Slaughterhouse employees who are directly involved in the day-to-day management and care of animals are likely to have more first-hand experience and knowledge of animal welfare. Livestock drivers, on the other hand, are primarily focused on transportation and may not have the same level of familiarity with animal welfare principles and practices. Encouraging collaboration and communication between slaughterhouse employees and livestock drivers can help create a shared commitment to animal welfare compliance throughout the supply chain. Where there are knowledge gaps in farm animal welfare, investment in research, development and knowledge enhancement is needed [20]. On the other hand, the welfare of farm animals can also have an impact on the quality of the end product. Extensive research has shown that meat quality improves when stress in cattle is reduced at slaughter [23–26]. Poor animal welfare during transportation and rearing can also lead to poorer product quality [27].

When comparing agreement with the statements on various topics relating to animal welfare, water and feed requirements, and health status, further differences emerged. The calculated p-values were statistically significant for three statements ($p < 0.05$). Firstly, the slaughterhouse employees were undecided on the statement that domesticated farm animals have different needs than their wild ancestors, while the livestock drivers agreed with this statement. This indicates a different understanding of the specific needs and characteristics of domesticated animals compared to their wild ancestors. Domestication is the process of adapting animals to live in close proximity to humans (domestic animals) or under human-made housing and husbandry conditions (livestock) [28]. Domestication can strongly influence the behavioral repertoire of an animal [29], i.e., selection pressure can change the behavior of animals during domestication [30]. There was also disagreement about the need for bedding in the slaughter pens, with slaughterhouse employees being undecided, indicating a lack of consensus among them, while livestock drivers tended to disagree. This suggests that they may have different views on the importance of bedding for animals in slaughter pens. If the animal is strongly motivated to perform behaviors that are no longer possible due to the lack of bedding, welfare problems may arise [31]. An example is the rooting behavior of pigs on concrete floors without bedding [32] while waiting overnight for slaughter. If the animal cannot perform desired behaviors because the required substrate is not available, this can lead to frustration [33]. Undecided responses in this case may indicate that further clarification or information on this issue is required. In the section on animal welfare, differences were found between opinions on mobile slaughterhouses and regular slaughterhouses. Slaughterhouse employees seemed more open-minded about both methods and were willing to consider alternative methods. One statement, for example, read: "Mobile slaughterhouses mean less suffering for the animals because the very stressful transportation for the animals is eliminated. Overall, the slaughterhouse employees agreed with this statement. However, the livestock drivers' responses were a mixture of possible answers, ranging from strongly disagree to strongly agree. This can be attributed to the fact that slaughterhouse employees are familiar with the different approaches and the potential advantages and disadvantages of each method. The indecisiveness of the livestock drivers, on the other hand, indicates a lack of experience or knowledge of the differences between these approaches. There are no mobile slaughterhouses in Slovenia. The main reason for this is that farms are located close to at least one slaughterhouse, but not further than 100 km away [34]. The second statistically significant statement was whether animals are sentient beings. Slaughterhouse employees agreed more strongly with this than livestock drivers. The third statistically significant statement was disagreement about the impact of disease on animal welfare. Slaughterhouse employees disagreed with the statement that diseases do not affect animal welfare. This response indicates that they recognize the negative impact of disease on animal welfare and the importance of disease prevention and management. In addition, slaughterhouse employees are taught about the concept of five freedoms as part of their theoretical training for the certificate of competence, and the third concept is freedom from pain, injury or disease [35]. In contrast, livestock drivers were undecided, indicating a possible lack of knowledge as they are not trained according to Regulation (EC) 1099/2009. In Slovenia, farmers who achieved more than the required animal welfare standards in 2021 were supported by the Decree on Animal Welfare Measures under the Rural Development Program in the Republic of Slovenia for the period 2014–2020 [36]. As already mentioned [2], the transportation and housing of farm animals also play a very important role in animal welfare. For slaughterhouses, the one-time costs associated with changing infrastructure and converting practices can be a major burden on the business [20]. Therefore, the next step could be to extend animal welfare legislation to financially support better transportation practices and modern slaughterhouse equipment. The most concerning finding was that both slaughterhouse employees and livestock drivers were mostly undecided about the statement "'biosecurity' means that the feed for the animals is sanitary". At a time when African Swine Fever (ASF) and Avian Influenza (AI) are threatening livestock populations, biosecurity is one of the most important preventative measures. It is very likely that trucks were a major factor in the spread of ASF in China [37], and trucks are also considered a potentially important risk for the spread of ASF in and around Europe [38]. Ssematimba [39] and other researchers investigated biosecurity measures in poultry farming and found that the main risk factors for the spread of AI are the movement of birds during thinning and restocking, most human movement when accessing the houses, and proximity to other poultry farms.

This study shows differences between the veterinarians' assessments and the self-assessment of animal welfare by slaughterhouse employees and livestock drivers. In general, the veterinarians gave lower average scores than the participants, with the exception of one item (dust in the pan/lairage). Slovenian slaughterhouse employees give high priority to all parameters of animal welfare. This is consistent with the results of other studies conducted worldwide, which show that people consider animal welfare and animal welfare laws to be important, want better animal welfare, and consider animal welfare to be an important social issue [14,16,17]. The fact that the highest self-assessed scores were obtained for the parameters of housing conditions and environmental conditions in the stables shows that Slovenian slaughterhouse employees attach great importance to these aspects of animal welfare. Lairage refers to the area where the animals are kept before slaughter and it is crucial that the conditions in this area are conducive to animal welfare. The high level of self-assessment indicates that the slaughterhouse employees interviewed in the study recognize the importance of appropriate housing and environmental conditions at this stage of the animals' journey. The high importance placed on lairage conditions and environmental conditions in the pens may be due to the recognition that the experiences animals have during this period can have a significant impact on their overall welfare and quality of life. Adequate ventilation, heating systems, and sufficient space are key factors that help to minimize stress, discomfort, and potential health problems for the animals [2,13–15]. The high self-assessment of these parameters also reflects the recognition of legal and ethical obligations related to animal welfare in the pre-slaughter phase. This indicates that slaughterhouse employees are aware of the importance of creating a favorable environment for animals in the pre-slaughter period.

The results in transportation also show a discrepancy between the experts' observations and the participants' self-assessments. In most cases, the experts' observations resulted in lower scores than the participants' self-assessments, indicating a possible difference in perception or awareness of animal welfare during transportation. However, for two parameters, the experts' observations largely agreed with the participants' self-assessments, particularly for environmental conditions, meaning that both experts and participants recognized the importance of adequate ventilation of the transport vehicle for animal welfare. However, a statistically significant difference ($p < 0.05$) was found for the parameter of the presence of thermometers and hydrometers. Livestock drivers indicated that temperature and humidity are important when transporting animals to the slaughterhouse. However, there are no procedures in place in case the temperature or humidity affects the welfare of the animals during transportation. The Dutch company Vion, for example, reduces the density of pigs on the vehicle when the outside temperature is 27 °C and prohibits the transportation of live animals when the temperature is 35 °C or higher [40]. One driver's response that he did not care if he received a penalty for improperly transporting live animals to the slaughterhouse may have influenced his lower overall self-assessment score. This suggests that livestock transport drivers may not be aware that improper driving can cause stress and discomfort to animals, which can lead to a lower level of animal welfare. Transporting live animals is a critical stage in the overall animal welfare process and it is essential to ensure that animals are handled and transported in a way that minimizes stress, discomfort, and the risk of injury. Penalties for improper transportation or handling of animals should enforce regulations and promote proper handling of animals during transport [2,41].

5. Conclusions

The main findings of our study show that there is a difference between the perception of animal welfare by slaughterhouse employees and livestock drivers. Slaughterhouse employees have a stronger opinion and understanding than livestock drivers. This is probably due to a lack of knowledge about animal welfare, as livestock drivers do not have a certificate of competence according to Regulation (EC) 1099/2009. The employees of Slovenian slaughterhouses and the livestock drivers give high priority to animal welfare in the lairage and on transport. However, the assessment of the veterinary experts was almost always lower than that of the slaughterhouse employees and livestock drivers, which shows that there is room for improvement. Overall, these findings suggest that both slaughterhouse employees and livestock drivers in Slovenia need to be better trained and made aware of animal welfare practices and regulations.

Improved training programs should address the identified knowledge gaps and promote a deeper understanding of animal welfare issues across the industry. It is understandable that improving the welfare of farm animals in slaughterhouses through better employee training and new equipment is costly. But there can also be high costs associated with doing nothing, e.g., fines from official veterinarians and the possible loss of special certificates, e.g., Hilton Foods, McDonald's, and market losses as a result. By improving the knowledge and perspective of all stakeholders, we can strive for better animal welfare outcomes in food production and transportation as well as better meat quality and food safety.

Supplementary Materials: The following supporting information can be downloaded at: https://www.mdpi.com/article/10.3390/ani14030443/s1, S1: Animal Welfare Questionnaire (AWQ/SL and AWQ/D).

Author Contributions: M.K., I.T. and A.D. conceived and designed the study. M.L., A.K. and M.K. conducted the study. I.T. analyzed the data. M.L. and M.K. wrote the manuscript. All authors have read and agreed to the published version of the manuscript.

Funding: This research was funded by the Target Research Project (CRP V4-2024) and Research Core Funding (P4-0092), both financed by the Slovenian Research Agency (ARRS) and the Ministry of Agriculture, Forestry and Food.

Institutional Review Board Statement: Slovenian legislation for behavioral research does not require ethical assessment and is not necessary for the implementation of research work. However, it defines some important requirements related to the protection of personal data. In short, firstly, it requires that participants provide the appropriate consent for the collection and use of their personal data; secondly, it requires that the data be adequately protected and used only for agreed-upon purposes; and, thirdly, it requires you to properly identify the data at the end of the survey, destroy information about who the individual data belongs to https://www.ff.uni-lj.si/fakulteta/komisija-za-etiko (accessed on 29 December 2023).

Informed Consent Statement: The evaluation of animal welfare in Slovenian slaughterhouses was carried out within the framework of the Slovenian Target Research Program (breeding of domestic animals by upgrading animal welfare in accordance with social requirements, No. V4-2024).

Data Availability Statement: The data presented in this study are available on request from the corresponding author. The data are not publicly available due to [protection of personal data].

Acknowledgments: We would like to thank the slaughterhouse personnel and livestock drivers of live animals for their cooperation in the project.

Conflicts of Interest: The authors declare no conflict of interest.

References

1. Fraser, D. Understanding animal welfare. *Acta Vet. Scand.* **2008**, *50*, 1–7. [CrossRef] [PubMed]
2. World Organization of animal Health. Introduction to the Recommendations for Animal Welfare. In *Terrestrial Animal Health Code;* World Organization for Animal Health: Paris, France, 2023; Available online: https://www.woah.org/fileadmin/Home/eng/Health_standards/tahc/2023/preface.pdf (accessed on 20 August 2023).
3. Carenzi, C.; Verga, M. Animal welfare: Review of the scientific concept and definition. *Ital. J. Anim. Sci.* **2009**, *8*, 21–30. [CrossRef]
4. Arndt, S.S.; Goerlich, V.C.; van der Staay, F.J. A dynamic concept of animal welfare: The role of appetitive and adverse internal and external factors and the animal's ability to adapt to them. *Front. Anim. Sci.* **2022**, *3*, 1–21. [CrossRef]
5. Temple, D.; Manteca, X. Animal welfare in extensive production systems is still an area of concern. *Front. Sustain. Food Syst.* **2020**, *4*, 545902. [CrossRef]
6. Nielsen, S.S.; Alvarez, J.; Bicout, D.J.; Calistri, P.; Depner, K.; Drewe, J.A.; Garin-Bastuji, B.; Gonzales Rojas, J.L.; Gortázar Schmidt, C.; Michel, V.; et al. Welfare of pigs at slaughter-Scientific Opinion on the welfare of pigs at slaughter. *EFSA J.* **2020**, *18*, 6148.
7. The Council of the European Union. Council Regulation (EC) No 1099/2009 of 24 September 2009 on the Protection of Animals at the Time of Killing. *Off. J. Eur. Union* **2009**, *303*, 1–30.
8. Regulation on the Implementation of the Regulation (EC) on the Protection of Animals during Slaughter. 2013. Available online: https://www.uradni-list.si/glasilo-uradni-list-rs/vsebina/2013-01-2716/uredba-o-izvajanju-uredbe-es-o-zasciti-zivali-pri-usmrtitvi (accessed on 23 August 2023).
9. Hernandez, R.O.; Romero, M.H.; Sanchez, J.A. Assessment of slaughterhouse-based measures as animal welfare indicators in fattening pigs. *Front. Anim. Sci.* **2023**, *4*, 1064933. [CrossRef]
10. Winders, D.J.; Abrell, E. Slaughterhouse Workers, Animals, and the Environment: The Need for a Rights-Centered Regulatory Framework in the United States That Recognizes Interconnected Interests. *Health Hum. Rights.* **2021**, *23*, 21–33. Available online: https://www.hhrjournal.org/wp-content/uploads/sites/2469/2021/12/abrell.pdf (accessed on 16 June 2023).
11. Republic of Slovenia. Decree implementing the Council Regulation (EC) on the protection of animals at the time of killing. In *Official Gazette of the Republic of Slovenia;* No. 5/17; Republic of Slovenia: Ljubljana, Slovenia, 2017.
12. Nadlučnik, E.; Golinar Oven, I.; Tomažič, I.; Plut, J.; Dovč, A.; Štukelj, M. Discrepancies between farmers' perceptions and actual animal welfare conditions on commercial pig farms. *Front. Vet. Sci.* **2022**, *29*, 1010791. [CrossRef]
13. Cornish, A.; Raubenheimer, D.; McGreevy, P. What We Know about the Public's Level of Concern for Farm Animal Welfare in Food Production in Developed Countries. *Animals* **2016**, *6*, 74. [CrossRef]
14. Sinclair, M.; Lee, N.Y.P.; Hötzel, M.J.; de Luna, M.C.T.; Sharma, A.; Idris, M.; Derkley, T.; Li, C.; Islam, M.A.; Iyasere, O.S.; et al. International perceptions of animals and the importance of their welfare. *Front. Animal Sci.* **2022**, *3*, 1–23. [CrossRef]
15. Hemsworth, P.H.; Coleman, G.J. Human-animal interactions and animal productivity and welfare. In *Human-Livestock- Interactions: The Stockperson and the Productivity and Welfare of Intensively Farmed Animals*, 2nd ed.; Hemsworth, P.H., Coleman, G.J., Eds.; Wallingford CABI: Wallingford, UK, 2011; pp. 47–83.
16. European-Commission. *Attitudes of EU Citizens towards Animal Welfare;* European Commission: Brussels, Belgium, 2007; Available online: https://www.politique-animaux.fr/fichiers/eurobarometer_-_attitudes_of_eu_citizens_towards_animal_welfare-_2007.pdf (accessed on 16 July 2023).
17. European-Commission. *Attitudes of Europeans towards Animal Welfare, Report;* European Commission: Brussels, Belgium, 2016; Available online: https://op.europa.eu/en/publication-detail/-/publication/9bc3a0b7-ec17-11e5-8a81-01aa75ed71a1/language-en (accessed on 27 July 2023).
18. Gracia, A. The determinants of the intention to purchase animal welfare-friendly meat products in Spain. *Anim. Welf.* **2013**, *22*, 255–265. [CrossRef]
19. Alonso, M.E.; González-Montaña, J.R.; Lomillos, J.M. Consumers' concerns and perceptions of farm animal welfare. *Animals* **2020**, *10*, 385. [CrossRef]
20. Fernandes, J.N.; Hemsworth, P.H.; Coleman, G.J.; Tilbrook, A.J. Costs and Benefits of Improving Farm Animal Welfare. *Agriculture* **2021**, *11*, 104. [CrossRef]
21. Grandin, T. Perspectives on transportation issues: The importance of having physically fit cattle and pigs. *J. Anim. Sci.* **2001**, *79*, 201–207. [CrossRef]
22. CFIA. Transporting Unfit or Compromised Animals. Available online: https://inspection.canada.ca/animal-health/terrestrial-animals/humane-transport/transporting-unfit-or-compromised-animals/eng/1582045810428/1582045810850 (accessed on 8 August 2023).
23. Grandin, T. The Effect of Stress on Livestock and Meat Quality Prior to and during Slaughter. *Int. J. Study Anim. Probl.* **1980**, *1*, 313–337. Available online: https://www.wellbeingintlstudiesrepository.org/acwp_faafp/20/ (accessed on 27 July 2023).
24. Grandin, T. Handling methods and facilities to reduce stress on cattle. *Vet. Clin. N. Am. Food Anim. Pract.* **1998**, *14*, 325–341. [CrossRef]
25. Grandin, T. Animal Welfare and Food Safety at the Slaughter Plant. In *Improving the Safety of Fresh Meat*, 1st ed.; Sofos, J.N., Ed.; Woodhead Publishing Limited: Cambridge, UK, 2005; pp. 244–255. Available online: https://ubblab.weebly.com/uploads/4/7/4/6/47469791/improving_the_safety_of_fresh_meat.pdf (accessed on 27 July 2023).
26. Grandin, T. Auditing animal welfare at slaughter plants. *Meat Sci.* **2010**, *86*, 56–65. [CrossRef] [PubMed]
27. Dawkins, M.S. Animal welfare and efficient farming: Is conflict inevitable? *Anim. Prod. Sci.* **2017**, *57*, 201–208. [CrossRef]

28. Price, E.O. Behavioral development in animals undergoing domestication. *Appl. Anim. Behav. Sci.* **1999**, *65*, 245–271. [CrossRef]
29. Rooney, N.; Bradshaw, J. Canine welfare science: An antidote to sentiment and myth. In *Domestic Dog Cognition and Behavior*, 1st ed.; Horowitz, A., Ed.; Springer: Heidelberg/Berlin, Germany, 2014; pp. 241–274.
30. Fraser, D.; Weary, D.M.; Pajor, E.A.; Milligan, B.N. A scientific conception of animal welfare that reflects ethical concerns. *Anim. Welfare* **1997**, *6*, 187–205. [CrossRef]
31. Miller, R.R.; Polack, C.W. Sources of maladaptive behavior in 'normal' organisms. *Behav. Process.* **2018**, *154*, 4–12. [CrossRef]
32. Studnitz, M.; Bak Jensen, M.; Pedersen, L.J. Why do pigs root and in what will they root? A review on the exploratory behavior of pigs in relation to environmental enrichment. *Appl. Anim. Behav. Sci.* **2007**, *107*, 183–197. [CrossRef]
33. Thodberg, K.; Jensen, K.H.; Herskin, M.S. Nest building and farrowing in sows: Relation to the reaction pattern during stress, farrowing environment and experience. *Appl. Anim. Behav. Sci.* **2002**, *77*, 21–42. [CrossRef]
34. Križman, M.; Dobeic, M. Mobile slaughterhouse: Short overview. *MESO* **2023**, *25*, 237–245. Available online: https://www.meso.hr/wp-content/uploads/2023/05/Mobile-slaughterhouse-Short-overview.pdf (accessed on 8 August 2023). [CrossRef]
35. McCulloch, S.P. A critique of FAWC's five freedoms as a framework for the analysis of animal welfare. *J. Agric. Environ. Ethics* **2013**, *26*, 959–975. [CrossRef]
36. Republic of Slovenia. Decree about animal welfare measure of Rural Development Program of the Republic of Slovenia 2014–2020 in the year 2021. In *Official Gazette of the Republic of Slovenia*; No. 3/21 and 28/21; Republic of Slovenia: Ljubljana, Slovenia, 2021.
37. Li, Y.; Salman, M.; Shen, C.; Yang, H.; Wang, Y.; Jiang, Z.; Edwards, J.; Huang, B. African swine fever in a commercial pig farm: Outbreak investigation and an approach for identifying the source of infection. *Transbound. Emerg. Dis.* **2020**, *67*, 2564–2578. [CrossRef]
38. Mur, L.; Martínez-López, B.; Sánchez-Vizcaíno, J.M. Risk of African swine fever introduction into the European Union through transport-associated routes: Returning trucks and waste from international ships and planes. *BMC Vet. Res.* **2012**, *8*, 149. [CrossRef] [PubMed]
39. Ssematimba, A.; Hagenaars, T.J.; de Wit, J.J.; Ruiterkamp, F.; Fabri, T.H.; Stegeman, J.A.; de Jong, M.C.M. Avian influenza transmission risks: Analysis of biosecurity measures and contact structure in Dutch poultry farming. *Prev. Vet. Med.* **2013**, *109*, 106–115. [CrossRef]
40. Vion CSR Report 2022; Vion Holding N.V.: Boxtel, The Netherlands. Available online: https://www.vionfoodgroup.com/wp-content/uploads/2023/07/CSR-report-Vion.pdf (accessed on 8 August 2023).
41. Council of the European Union. Council Regulation (EC) No 1/2005 of 22 December 2004 on the protection of animals during transport and related operations and amending Directives 64/432/EEC and 93/119/EC and Regulation (EC) No 1255/97. *Off. J. Eur. Union* **2005**, *3*, 1–44.

animals

MDPI

Article

Alternatives to Carbon Dioxide in Two Phases for the Improvement of Broiler Chickens' Welfare during Stunning

Daniel Santiago Rucinque [1], Antonio Velarde [1], Aida Xercavins [1], Aranzazu Varvaró-Porter [1], Troy John Gibson [2], Virginie Michel [3] and Alexandra Contreras-Jodar [1,*]

1 Animal Welfare Program, Institute of Agrifood Research and Technology (IRTA), 17121 Monells, Spain; santiago.rucinque@irta.cat (D.S.R.); antonio.velarde@irta.cat (A.V.); aida.xercavins@irta.cat (A.X.); aranzazu.varvaro@irta.cat (A.V.-P.)

2 Department of Pathobiology and Population Sciences, Royal Veterinary College, University of London, Hatfield AL9 7TA, UK; tgibson@rvc.ac.uk

3 Direction of Strategy and Programmes, French Agency for Food, Environmental and Occupational Health & Safety (ANSES), 14 Rue Pierre et Marie Curie, 94701 Maisons-Alfort, France; virginie.michel@anses.fr

* Correspondence: alexandra.contreras@irta.cat

Simple Summary: Stunning during the slaughter process consists of inducing unconsciousness in animals in order to prevent them from feeling any avoidable pain, distress, or suffering during bleeding and related operations. In an unconscious state, the animal is unable to perceive and respond to any external stimuli, including pain. Currently, the two main stunning methods used commercially in broiler chickens are the electrical waterbath and carbon dioxide in two phases. Although the latter is widely recommended over electrical waterbath stunning, it is still not devoid of risks for animal welfare. For instance, the induction to unconsciousness is not immediate and involves a transitional period during which aversion to the inhalation of carbon dioxide might occur. The present study demonstrates that mixtures of carbon dioxide with nitrogen improve a broiler chicken's welfare during stunning since they result in a more rapid induction of unconsciousness and reduce the aversion experienced, compared to carbon dioxide in two phases, in broiler chickens.

Abstract: This study evaluated the exposure to gas mixtures of carbon dioxide (CO_2) associated with nitrogen (N_2) as alternatives to CO_2 in two phases to improve the welfare of broiler chickens at slaughter. Broilers were exposed to one of three treatments: 40C90C (1st phase: <40% CO_2 for 2 min; 2nd phase: >90% CO_2 and <2% O_2 for 2 min, n = 92), 40C60N (40% CO_2, 60% N_2, and <2% O_2 for 4 min, n = 79), or 20C80N (20% CO_2, 80% N_2, and <2% O_2 for 4 min, n = 72). Brain activity (EEG) was assessed to determine the onset of loss of consciousness (LOC) and death. Behavioural assessment allowed for characterisation of an aversive response to the treatments and confirmed loss of posture (LOP) and motionlessness as behavioural proxies of LOC and brain death in 40C60N and 20N80C. However, the lack of quality of the EEG traces obtained in 40C90C did not allow us to determine the onset of LOC and brain death for this treatment. The onset of LOC in 40C60N was found at 19 s [14–30 s] and in 20C80N at 21 s [16–37 s], whereas a LOP was seen at 53 s [26–156 s] in 40C90C. Birds showed brain death in 40C60N at 64 s [43–108 s] and in 20C80N at 70 s [45–88 s]), while they became motionless in 40C90C at 177 s [89–212 s]. The 40C90C birds not only experienced more events of aversive behaviours related to mucosal irritation, dyspnoea, and breathlessness during induction to unconsciousness but were at risk of remaining conscious when the CO_2 concentration was increased in the 2nd phase (known to cause severe pain). From an animal welfare point of view, 40C60N proved to be the least aversive of the three treatments tested, followed by 20C80N and 40C90C.

Keywords: controlled atmosphere stunning; gas stunning; carbon dioxide; inert gases; nitrogen; unconsciousness; death; aversion; broiler chicken

Citation: Rucinque, D.S.; Velarde, A.; Xercavins, A.; Varvaró-Porter, A.; Gibson, T.J.; Michel, V.; Contreras-Jodar, A. Alternatives to Carbon Dioxide in Two Phases for the Improvement of Broiler Chickens' Welfare during Stunning. *Animals* **2024**, *14*, 486. https://doi.org/10.3390/ani14030486

Academic Editor: Elbert Lambooij

Received: 2 October 2023
Revised: 25 January 2024
Accepted: 26 January 2024
Published: 1 February 2024

1. Introduction

Pre-slaughter stunning is mandatory in the European Union [1] and many other countries worldwide [2]. It consists of inducing unconsciousness in animals in order to prevent them from feeling any avoidable pain, distress, or suffering during bleeding and related operations. In an unconscious state, the animal is unable to perceive and respond to any external stimuli, including pain [3]. To protect animal welfare at slaughter, it is essential to induce fast and effective stunning for enough time so that the animals do not regain consciousness before brain death due to bleeding.

Currently, the two main stunning methods used commercially in broiler chickens are electrical waterbath stunning (WBS) and controlled atmosphere stunning (CAS). WBS entails live bird shackling; pre-stun shocks may occur, and the stunning is not always effective. Therefore, a great concern regarding the bird's welfare exists [4]. For this reason, CAS emerged as an alternative stunning method to WBS [5]. It consists of exposing large numbers of broiler chickens to modified atmosphere environments (e.g., carbon dioxide in two phases) or reducing the atmospheric pressure (LAPS), which induces a gradual loss of consciousness (LOC), and if the duration of the exposure is long enough, it causes death. However, CAS methods are not devoid of risks for animal welfare. For instance, the induction of unconsciousness is not immediate and involves a transitional period, during which negative welfare outcomes may occur [6].

Although WBS is the most common method used in the European Union, the number of slaughterhouses using CAS has dramatically increased during the last few years [7,8]. Most slaughterhouses with CAS use carbon dioxide (CO_2) in two phases, while very few use CO_2 associated with inert gases, and none of them use either inert gases or LAPS at present [8].

Commercial CAS equipment for poultry differs in design, as it is either tunnels, pits, or closed cabinets. In tunnels and pits, birds enter the system in their transport crates or they are uncrated by tilting the container [4] and enter the system on a conveyor belt. In tunnels and pits, the system is pre-filled with gas and birds enter continuously at one end of the system, and, while they are conveyed to the opposite end, they are exposed to different gas concentrations. In closed cabinets, birds enter the system in their transport crates, one batch at a time. Once the birds have been loaded into the system the gas is then added, and it is removed upon completion of the stunning cycle [8].

The physiological principle during the induction of unconsciousness using CO_2 is to induce acidosis and neuronal depression [9]. However, prior to LOC, CO_2 activates chemoreceptors in the mucous membranes of the respiratory tract, which may induce discomfort, pain, and breathlessness, as shown by behaviours indicative of aversion [10]. The degree of aversion depends on the CO_2 concentration. The higher the concentration, the greater the aversion, but the more rapid the LOC [11]. To reduce aversion during the induction of unconsciousness, stunning with CO_2 in two phases is carried out. It consists of exposing broiler chickens to an initial concentration of up to 40% CO_2 until LOC occurs. Thereafter, the CO_2 concentration is increased in the second phase, inducing a deeper state of unconsciousness and then death while birds are unconscious [4].

As an alternative to CO_2 in two phases, the exposure to inert gases (e.g., nitrogen N_2; argon Ar) is expected to reduce aversion. Inert gases are colourless, odourless, tasteless, and non-irritative, and, therefore, it is presumed that they are imperceptible to birds due to the lack of chemoreceptors in their air ways [5,12]. In addition, inert gases displace oxygen (O_2) from atmospheric air, and this ensures that the birds lose consciousness by anoxia [3]. In this sense, the inhalation of inert gases is expected not to cause aversive reactions [13,14]. When birds enter a chamber filled with inert gases, their behaviour does not differ from when they are breathing atmospheric air [13]; they do not withdraw [11] and they barely show behavioural signs of distress [10,15]. Nevertheless, when birds are unconscious, they present severe convulsion expressed as wing flapping [10,13], which may cause self-inflicted injuries (wing fractures) or injuries and pain to the other birds that have not yet lost consciousness [5,16]. However, it is not entirely clear whether the onset of wing flapping is a reflexive reaction occurring after the bird loses consciousness or whether the birds are still conscious and trying to escape from this modified atmosphere [16,17].

The European Food Safety Authority (EFSA) has pointed out that research evaluating stunning methods requires well-controlled studies under laboratory conditions to characterise the animals' responses to the stunning method (i.e., the onset of unconsciousness and death, the magnitude of aversion) [18]. Measuring electrical brain activity by means of electroencephalography (EEG) is the most accurate method to assess the onset and the duration of unconsciousness and time to death [18], and therefore to discern between aversive behaviours that occur before LOC and likely unconscious convulsions. The aim of this study was to assess different gas mixtures of CO_2 and N_2 as potential alternatives to the exposure to CO_2 in two phases for the improvement of animal welfare during the stunning of broiler chickens. To this end, we aimed to compare EEG and behavioural recordings to determine the time to the onset of unconsciousness and death and to characterize the gas aversion response.

2. Materials and Methods

2.1. Experimental Design and Facilities

A total of two hundred and forty-three 39-day-old mixed-sex Ross 308 broiler chickens were transported from a commercial farm to the experimental facilities of the Institute of Research and Technology for Agriculture and Food (IRTA) in Monells (Spain). Birds were weighed (2.42 ± 0.18 kg) and individually identified with numbered leg bands before being allocated randomly to the different treatments. On arrival and after checking their health status and appropriate locomotor behaviour, birds were distributed randomly into seven adjacent lairage pens of 2 m × 1.8 m (35 broiler chickens per pen; stocking density of 23.5 kg/m^2). Each pen was provided with litter material (wood shavings) and feed and water ad libitum throughout the experiment. The pens served as a lairage before slaughter but were not associated with any specific treatment.

The study was carried out at the experimental slaughterhouse of IRTA, located next to the lairage pens. It is equipped with a Dip-lift XL G2 gas stunning system (Butina Aps, Copenhagen, DK) that contains a lift (240 cm × 111 cm × 100 cm) with a perforated floor. The lift descended into the base of a pit at a depth of 290 cm. The pit was pre-filled with gas mixtures through an inlet valve placed at the bottom of the pit. CO_2 and O_2 concentrations were measured through a portable infrared single beam sensor for CO_2 and an electrochemical sensor for O_2 (Dansensor® CheckPoint 3 O_2/CO_2, MOCON Europe A/S, Ringsted, Danmark) using one fixed sounding line placed at a depth of 260 cm and another mobile sounding line to check CO_2 concentrations at different depths.

The experimental study lasted 5 days. On the first day, a subset of broiler chickens was exposed to atmospheric air (AIR, n = 100), serving as the control. These birds were reallocated in similar numbers to one of the three experimental gas treatments. Therefore, from d 2 to d 5, broiler chickens were stunned with one of the following gas treatments: CO_2 in two phases, the 1st phase with <40% CO_2 by volume in air for 2 min followed by the 2nd phase with >90% CO_2 for 2 min (40C90C; n = 92); a gas mixture of 40% CO2 and 60% N_2 with less than 2% residual O_2 for 4 min (40C60N; n = 79); and gas mixture of 20% CO_2 and 80% N_2 with less than 2% residual O_2 for 4 min (20C80N; n = 72). Each day consisted of two sessions: first session from 800 to 1200 h and second session from 1500 to 1900 h, with treatments alternating per session to avoid potential bias (Table 1). Each session consisted of 8 to 11 cycles (dips into the pit). In each cycle, four broiler chickens were placed in the lift and exposed to the treatment. In AIR, the behaviour of the four chickens per cycle was assessed, but in 40C90C, 40C60N, and 20C80N, in one of the four chickens, only the brain activity was assessed, while the other three were used for behavioural assessment (Figure 1). The schedule of the gas stunning treatments applied along the 5 d experimental period to broiler chickens, the number of birds used per cycle, the number of cycles, and birds used for EEG and behavioural assessments per treatment and session are summarised in Table 1. First session dedicated to 40C90C failed at registering the EEG, so an extra session with 40C90C was included at the end of the experimental period (see Table 1).

Table 1. Schedule of the gas stunning treatments applied along the 5 d experimental period to broiler chickens.

Item	Day 1	Day 2		Day 3		Day 4		Day 5
Session	1500 to 1900 h	800 to 1200 h	1500 to 1900 h	800 to 1200 h	1500 to 1900 h	800 to 1200 h	1500 to 1900 h	800 to 1200 h
Treatment	AIR	40C90C	40C60N	40C60N	20C80N	20C80N	40C90C	40C90C
cycles, *n*	25	10	10	11	9	9	8	8
Total birds, *n*	100	30	35	44	36	36	32	30
Birds/cycle	4	3	3–4 *	4	4	4	4	3–4 **
Brain activity assessment, *n*	0	0	5	11	9	9	8	8
Behavioural assessment, *n*	100	30	30	33	27	27	24	22
Cycles, *n*	25	10	10	11	9	9	8	8

* Five cycles with three birds used in each (none for EEG assessment) and five cycles with four birds used in each (one for EEG assessment and three for behavioural assessment). ** Six cycles with four birds used in each (one for EEG assessment and three for behavioural assessment) and two cycles with three birds used in each (one for EEG assessment and two for behavioural assessment).

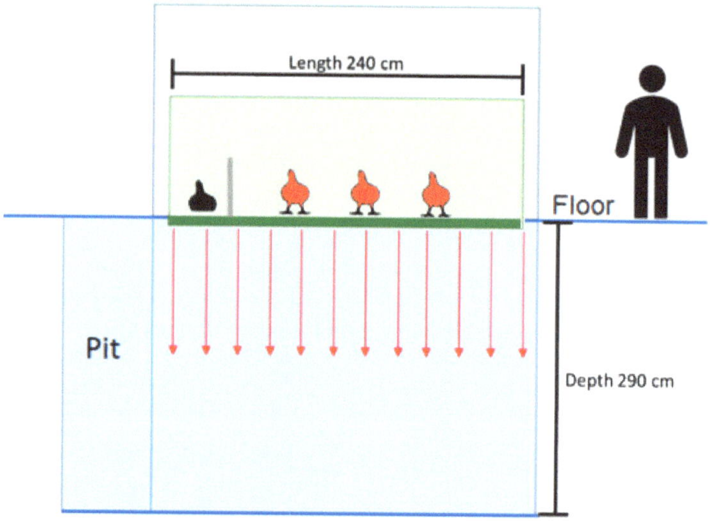

Figure 1. Representation of the distribution of broiler chickens in the pit-type gas stunning system per stunning cycle (dips into the pit). In each cycle, four birds were stunned, one bird had its brain activity recorded via electroencephalography (bird highlighted in black) and three birds were recorded with video cameras to assess their behaviour (birds highlighted in red). The bird whose brain activity was assessed was separated from the rest of the chickens by a transparent methacrylate wall.

The bird used for brain activity assessment was placed on the lift and separated from its three other conspecifics by a transparent methacrylate wall with a floor area of 48 cm × 112 cm (0.53 m^2). The separation was intended to prevent any disturbance to the birds and EEG signal interference from other birds. The three birds monitored for their behavioural activity were placed in the lift with a floor area of 144 cm × 112 cm (1.6 m^2).

The exposure time was considered as being from when the lift started to descend into the pit until the lift arrived at its original position. The duration of exposure to each experimental treatment was determined from pre-trials aimed at assuring death in all birds at the end of the process. Gases used were pure CO_2 and premixed mixtures of CO_2 with N_2 (Freshline gases® for food use, Carburos Metálicos, Barcelona, Spain).

2.2. Gas Concentration Assessment

The pit was gas-filled before the birds entered, and CO_2 and O_2 concentrations were continuously monitored before, during, and after each cycle in all treatments. Gas concentrations were measured with a portable gas analyser (Dansensor® CheckPoint 3 O_2/CO_2, MOCON Europe A/S, DK) every 10 cm vertically from the bottom to top of the pit. This allowed measures of CO_2 and O_2 concentrations at different depths.

In the 1st phase of 40C90C, the CO_2 concentration varied, but was close to 40% and never exceeded that level throughout the cycles. Then, the distance from the top of the pit was registered and the lift descended until reaching that depth (53.1 ± 15.7 cm) to ensure that birds were exposed to the target concentration. During the second phase in 40C90C, and the cycles of 40C60N and 20C80N, the sounding line monitored the gas concentration at 30 cm from the bottom of the pit in order to monitor the gas concentrations at the level of the chickens' heads.

2.3. Brain Activity Assessment

Fifty broiler chickens were randomly selected for electrical brain activity assessment through electroencephalography (EEG). Chickens were distributed into the 40C90C (n = 16), 40C60N (n = 16), and 20C80N (n = 18) treatment groups.

Each bird was prepared prior the exposure to the gas treatment. Firstly, the bird was wrapped with a textile mesh to restrain it wings, body, and legs, leaving the head and neck exposed in order to minimize movement during the EEG recording. Secondly, the chicken's neck was restrained gently (Figure 1). Then, head and neck feathers were shaved with an electric shaver and a gauze pad with topical alcohol was rubbed on the bare skin before subcutaneous administration of local anaesthesia to the head and neck. Local anaesthesia consisted in 0.1 mL of lidocaine 2% subcutaneously injected with an insulin needle and syringe in regions where EEG electrodes were to be placed. Once the skin was desensitized, three 24-gauge stainless steel subdermal needle electrodes (Neuroline Subdermal, Ambu Inc., Glen Burnie, MD, USA) were placed on the head as described in Gibson et al. [19].

The active electrode was inserted ≈6 mm right of midline and ≈3 mm cranial from the end of the comb over the right optic lobe. The reference electrode was inserted over the right rostral aspect of the forebrain, ≈6 mm right of midline and ≈20 mm caudal from the end of comb, and the ground electrode was inserted in the neck in the caudal direction (Figure 2). Electrodes were secured in position with surgical tape (Durapore, 3M, Maplewood, MN, USA). Inter-electrode impedance was established to be between 1.2 and 2.0 kΩ (MkIII Checktrode, UFI, Morro Bay, CA, USA), and electrode leads were further secured with a loose band of surgical auto-fixing tape around the neck (Coeban, 3M).

EEG signals were amplified and filtered with an analogue filter (dual Bio Amp, ADInstruments Ltd., Sydney, Australia) with low- and high-pass filters of 100 and 0.1 Hz, respectively. The analogue signals were digitalized (1 kHz) with a 4/20 PowerLab (ADInstruments Ltd., Sydney, Australia) converter and recorded using a laptop for offline analyses. Pre-treatment EEG signals were collected for 90 s while the bird was on the floor of the lift with the other three birds prior to their descent into the pit to obtain the normal EEG data (i.e., baseline) to compare with post-treatment results. EEG recordings were monitored, saved, and pre-processed using LabChart 8 Pro (v.8.1.21, AD Instruments, Dunedin, New Zealand) [19].

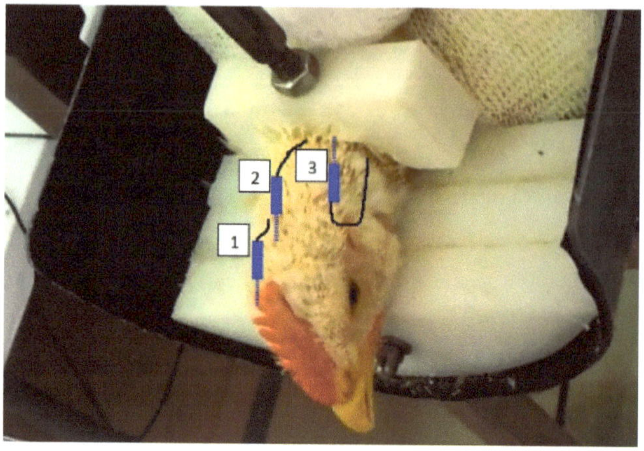

Figure 2. Representation of needle electrodes' positions for brain activity assessment via electroencephalography in feather-shaved broiler chickens with an (1) active electrode, (2) reference electrode, and (3) ground electrode.

Spectral analysis of EEG recordings was used for detecting waveform changes that indicate the onset of unconsciousness. Spectral variables, including total power (Ptot), median frequency (F50), and spectral edge frequency (F95) were computed from the EEG data. Ptot represents the overall area under the power spectrum curve, F50 corresponds to the median frequency of the power spectrum curve, and F95 indicates the frequency at which 95% of the power spectrum curve is located [20]. On the other hand, the brain's electrical activity recorded in the EEG was categorized into different frequency bands: Delta (<4 Hz), Theta (4 to 8 Hz), Alpha (8 to 13 Hz), Beta (13 to 32 Hz), and Gamma (32 to 200 Hz).

In a conscious animal, the Alpha and Beta frequency bands predominate the EEG spectrum. Thus, a decrease in F50 and an increase in Delta frequencies suggest a transition from consciousness to unconsciousness [14,16,21], and, therefore, the change in the relative contributions of F50 and Delta frequencies in the spectrum was used to estimate the loss of consciousness (LOC). The Ptot represents the total power on the spectrum, and a decrease in Ptot is associated with reduced EEG activity. This reduction was observed in the EEG of unconscious broiler chickens by Raj et al. [21] and Sandercock et al. [22], hence, the reduction in Ptot was employed to estimate the onset of death during spectral analysis [14,15]. The isoelectric pattern was observed on the filtered trace as a visual indicator of brain death, as in other studies [14,16,19,23,24].

2.4. Behavioural Assessment

Broiler chickens' behaviour during exposure to the gases was recorded using three video cameras (IP Camera DH-IPC-HDW2231TP-ZS-S2, Zhejiang Dahua Vision Technology Co., Ltd., Hangzhou, China) and a digital voice recorder (Olympus VN-712PC, Olympus imaging Corp., Tokio, Japan). Two video cameras were placed inside the lift on each of the laterals and one in the central part of the lift's ceiling. Video cameras were connected to a digital image recorder (Network video recorder DHI-NVR4108-8P-4KS2/L, Zhejiang Dahua Vision Technology Co., Ltd., Hangzhou, China). Then, the video records and audios were synchronized.

These records helped in assessing the birds' behaviour retrospectively, used by an assessor blinded to the experimental treatments. Behavioural observations were assessed continuously at the individual level; each bird was observed for 4 min (i.e., from the time the lift started to descend into the pit until the end of gas exposure) using BORIS (Behavioural

Observation Research Interactive Software) v.7.13.8 [25] based on the ethogram shown in Table 2.

Loss of posture (LOP) was considered a behavioural indicator of the onset of unconsciousness [10,26,31,32]; therefore, behaviours appearing before LOP were considered voluntary behaviours related to aversion (e.g., pain, distress, breathlessness) as the birds were still conscious during gas exposure, while behaviours appearing after LOP were considered related to convulsions or any other involuntary movements. For this reason, behaviours were separated and annotated into two different groups: those occurring before and after LOP. Motionless was considered the behavioural indicator of brain death [24,27,28,30].

Table 2. Ethogram used to assess the behaviour of broiler chickens subjected to different experimental gas stunning treatments.

Behaviour	Description	Adapted from
Loss of posture	Cessation of standing, with the head resting against either the floor or wall of the gas stunning system.	[10,26]
Motionlessness	Limp carcass, with the bird being completely still including the cessation of visible breathing movements.	[27]
Sitting	Legs underneath the body and wings relaxed against body.	[28]
Standing	On their feet with the body fully or partly lifted off of the ground.	[24]
Walking	Moving forward at a regular pace.	[29]
Ataxia	Uncoordinated walking with exaggerated lateral movement or fluttering when standing to maintain posture.	[10]
Deep inhalation	Wide open-mouth breathing with neck extension.	[10]
Gasping	Opening and closing mouth without neck extension and with reduced frequency compared to physiological breathing.	[10]
Head shaking	Rapid side-to-side movement of the head, which occurs whilst the bird is standing, walking, or sitting.	[10]
Jumping	Any vertical movement from a plantar stance, resulting in both feet leaving contact with the floor.	[10]
Leg paddling	Leg movements in the air or towards the ground depending on the body position of the bird. It can also be determined by an alternating upwards and downwards movement of the body if bird is lying sternal.	[28]
Wing flapping	Bouts of fast, short flapping, rapid movement of the wings.	[10,30]
High-pitch vocalisations	Single or repeated short and loud shrieking (screaming).	[4]

2.5. Data Pre-Processing and Statistical Analyses

EEG recordings were pre-processed using LabChart 8 Pro (v.8.1.21, AD Instruments, Dunedin, NZ). First, EEG recordings were digitally filtered to remove noise interference (band pass: 1 to 30 Hz). Then, epochs of 1 s, from baseline and during the 4 min of gas exposure, were selected for spectral analysis. Data were set at 1K Fast Fourier transformed, Hamming windowed, 50% window overlap, and zero frequency removed. For each bird, the following spectral data variables were calculated from each epoch: total power (Ptot, μV2), spectral median frequency (F50, Hz), and contribution of Delta frequency (1–4 Hz) to Ptot (%). The median value from the baseline was calculated for all variables at the individual level. Data generated were exported and analysed using Microsoft Excel 2016 (Microsoft Corporation, Redmond, WA, USA). Then, F50 values under 4 Hz on the baseline were removed in order to discard low-frequency artefacts caused by bird movements.

Onset of LOC was calculated as the mean time at which F50 decreased below 50% [24,27,28] and the Delta frequency increased above 65% in comparison to the baseline value in four consecutive epochs. Brain death was determined via spectral analysis to be when Ptot decreased by 90% in comparison to the baseline values in four consecutive epochs [21,27] and visually when the trace was isoelectric, representing an almost flat line with very low Ptot (<2.5 μV). This is a pattern on an EEG related to a permanent state of unconsciousness or brain death [19,24,27,28,30,33,34]. Spectral and visual analyses have been well-established in previous studies [27,28,33,35–37], allowing for a more accurate estimation of the time to death and its relationship with behaviour.

At the end, no EEG trace from 40C90C could be used due to the low quality of the records. Time to LOC and brain death fulfilled the normality and homoscedasticity conditions in the remaining gas treatments (i.e., 40C60N and 20C80N). Therefore, Student's *t*-test was used for the comparison of the two means.

Behavioural data pre-processing, statistical analyses, and plots were performed using R software v.4.3.2. [38]. For all the statistical analyses, significance was declared at $p < 0.05$.

The analysis of behavioural measurements comprised the time to onset of LOP and motionlessness, the proportion of broilers that performed the rest of the behaviours listed in the ethogram, as well as the number of events per bird, and total duration of each, both before and after LOP, per treatment. In order to avoid potential pseudo-replication, all data except for the proportion of broilers were analysed using mixed models containing the fixed effect of the gas treatment (40C90C, 40C60N, and 20C80N) and the cycle as the random effect, using the nlme package [39]. The proportion of broilers that performed a certain behaviour between treatments was compared by means of Pearson's Chi-squared test.

3. Results

3.1. Gas Concentration Assessment

Broiler chickens subjected to 40C90C were exposed to a CO_2 concentration below 40% by volume in atmospheric air during the first phase, in all cycles (38.1 ± 0.1%). During the second phase, CO_2 was kept higher than 90% (92.2 ± 0.6%) and residual O_2 was lower than 2% by volume (1.0 ± 0.1%). On the other hand, broiler chickens subjected to gas mixtures of CO_2 with inert gases were exposed to CO_2 concentrations at 36.3 ± 1.1% in 40C60N and 18.0 ± 0.3% in 20C80N, while the O_2 mean value was below 2% by volume during the 4 min of exposure (40C60N: 1.6 ± 0.3%; 20C80N: 1.9 ± 0.3%). However, the anoxic atmosphere (<2% of O_2) was steadier over time in 40C60N compared to 20C80N.

3.2. Brain Activity Assessment

Brain activity before and during the gas stunning procedure was recorded via EEG in 50 broiler chickens, generating one trace per bird (40C90C: $n = 16$; 40C60N: $n = 16$ and 20C80N: $n = 18$). Twenty-seven out of these 50 EEG records were discarded: the first 10 records from 40C90C due to the low quality of the electrodes used, one of them due to interference from eyelid movement preventing the selection of several 1 s epochs on the baseline; two due to disconnection of the EEG equipment during the exposure to the treatment; and 24 due to recording issues resulting in low-quality records (further explained in discussion section). Hence, EEG analysis was performed on the 23 remaining records (40C90C: $n = 0$; 40C60N: $n = 14$; 20C80N: $n = 9$).

The time to the onset of LOC and brain death for the 40C60N and 20C80N treatments is summarized in Table 3. The time to the onset of LOC did not differ significantly between 40C60N and 20C80N. However, the time to brain death was similar between the two treatments when the isoelectric pattern was visually identified ($p > 0.05$) but statistically different when spectral analysis was performed ($p < 0.05$).

Table 3. Time to onset of loss of consciousness and death, determined by electroencephalography of broiler chickens submitted to different experimental gas stunning treatments.

Analysis	Outcome	Treatments			
		40C90C	40C60N	20C80N	*p*-Value
Spectral	Loss of consciousness, s *	NA	25.7 ± 7.0	20.7 ± 6.6	0.144
Spectral	Death, s **	NA	65.8 ± 14.1 [b]	122 ± 53.2 [a]	0.048
Visual	Death, s ***	NA	69.8 ± 11.9	66.3 ± 8.1	0.456

40C90C: CO_2 in two phases entailed a first phase with <40% CO_2 by volume in air for 2 min and a second phase with >90% CO_2 for 2 min (n = 0); 40C60N: a gas mixture of 40% CO_2 and 60% nitrogen (N_2) with less than 2% O_2 for 4 min (n = 14); and 20C80N: a gas mixture of 20% CO_2 and 80% N_2 with less than 2% O_2 for 4 min (n = 9). *, based on spectral analysis (reduction of 50% of F50 or increase of 65% of Delta contribution). **, based on spectral analysis (reduction of 90% Ptot). ***, not based on spectral analysis but visual (isoelectric pattern). NA means not available. Different superscripts in the same row indicate a significant statistical difference between treatments ($p < 0.05$).

3.3. Behavioural Assessment

3.3.1. Behavioural Assessment of Loss of Posture and Motionlessness

The time to the onset of LOP and motionlessness with respect to the three experimental treatments is summarized in Table 4. Broiler chickens exposed to 40C60N and 20C80N took a similar amount of time to lose posture (21.0 ± 4.5 s, $p = 0.357$) but, significantly, 2.8-fold less time compared to 40C90C (59.2 ± 21.9 s, $p < 0.001$). Likewise, 40C60N and 20C80N took a similar amount of time to remain motionless (68.2 ± 10.3 s, $p = 0.282$) but, significantly, 2.5-fold less time compared to 40C90C (168.8 ± 27.2, $p < 0.001$). It is noteworthy that the range of time to LOP and motionlessness was broader in 40C90C than in 40C60N and 20C80N, while the 40C60N and 20C80N broiler chickens showed less variability. In addition, two broiler chickens exposed to 40C90C lost posture at 144 and 156 s (after 2 min); therefore, they were still conscious when the lift descended to a CO_2 concentration higher than 40% during the second phase. Furthermore, the latest onset of motionlessness was, in all cases, before the end of the exposure (240 s), indicating that all birds were dead before the end of the process.

Table 4. Time to onset of loss of posture and motionlessness of broiler chickens, expressed as mean (min–max), when submitted to different experimental gas stunning treatments.

Behaviour	Treatment			SE	*p*-Value
	40C90C	40C60N	20C80N		
Loss of posture	59.2 (26.0–156.5] [a]	19.8 (14.0–30.8] [b]	22.3 (15.8–37.0] [b]	2.7	<0.001
Motionless	168.8 (89.0–212.7] [a]	66.1 (43.0–108.0] [b]	70.4 (45.2–88.5] [b]	3.5	<0.001

40C90C: CO_2 in two phases entailed a first phase with <40% CO_2 by volume in air for 2 min and a second phase with >90% CO_2 for 2 min (n = 76, stunned in groups of three except for two times that birds were stunned in groups of two); 40C60N: a gas mixture of 40% CO_2 and 60% N_2 with less than 2% O_2 for 4 min (n = 63, stunned in groups of three); and 20C80N: a gas mixture of 20% CO_2 and 80% N_2 with less than 2% O_2 for 4 min (n = 54, stunned in groups of three). Different superscripts in the same row indicate a significant statistical difference between treatments ($p < 0.05$).

3.3.2. Behavioural Assessment before Loss of Posture

A general overview of individual broiler chickens exposed to AIR and gas treatments is shown in Figures 3 and 4, respectively.

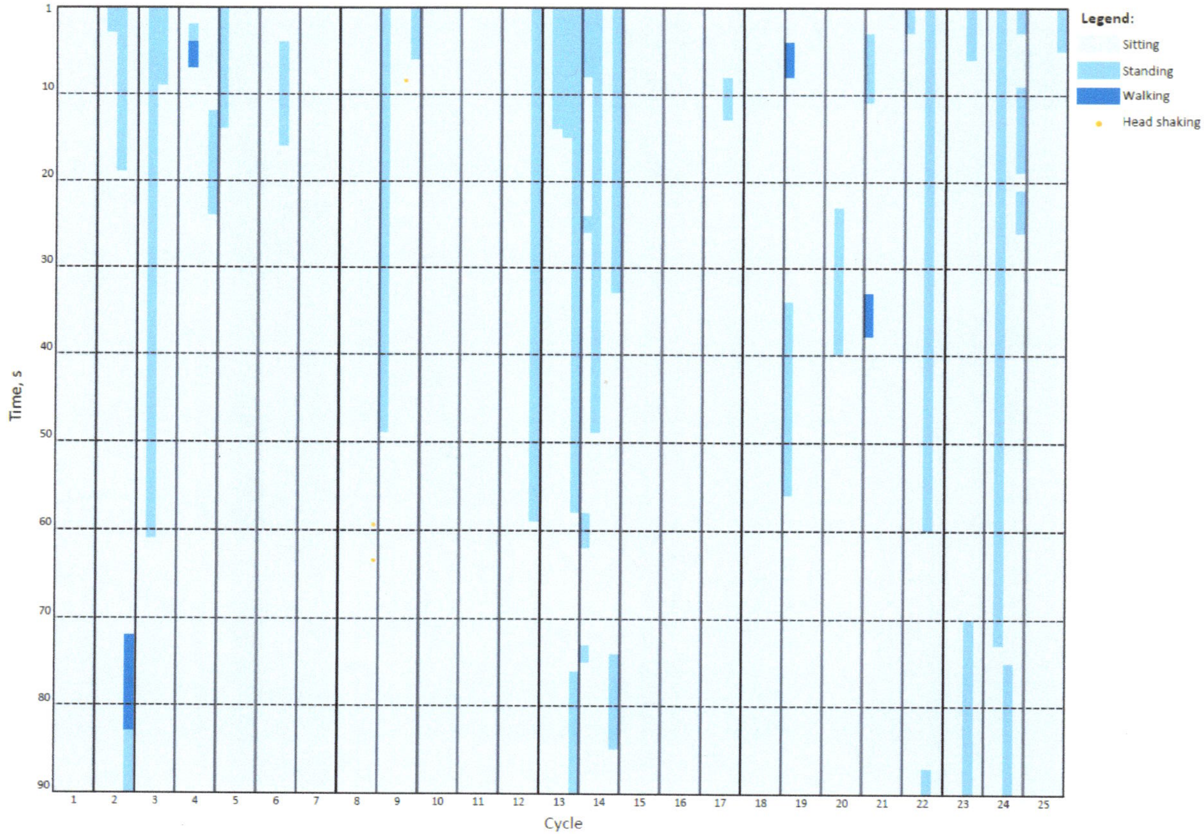

Figure 3. Behavioural plot of broiler chickens (n = 100) exposed to atmospheric air in a pit-type gas stunning system. The graphical plot shows the behaviour of the birds in 1 s bins. Segments of 10s appear as horizontal dashed lines, whereas cycles (dips into the pit) are displayed as vertical lines. Four birds were used per cycle and each bird's behaviour is represented by coloured vertical segments based on the colour coding shown in the legend.

(A) **(B)** **(C)**

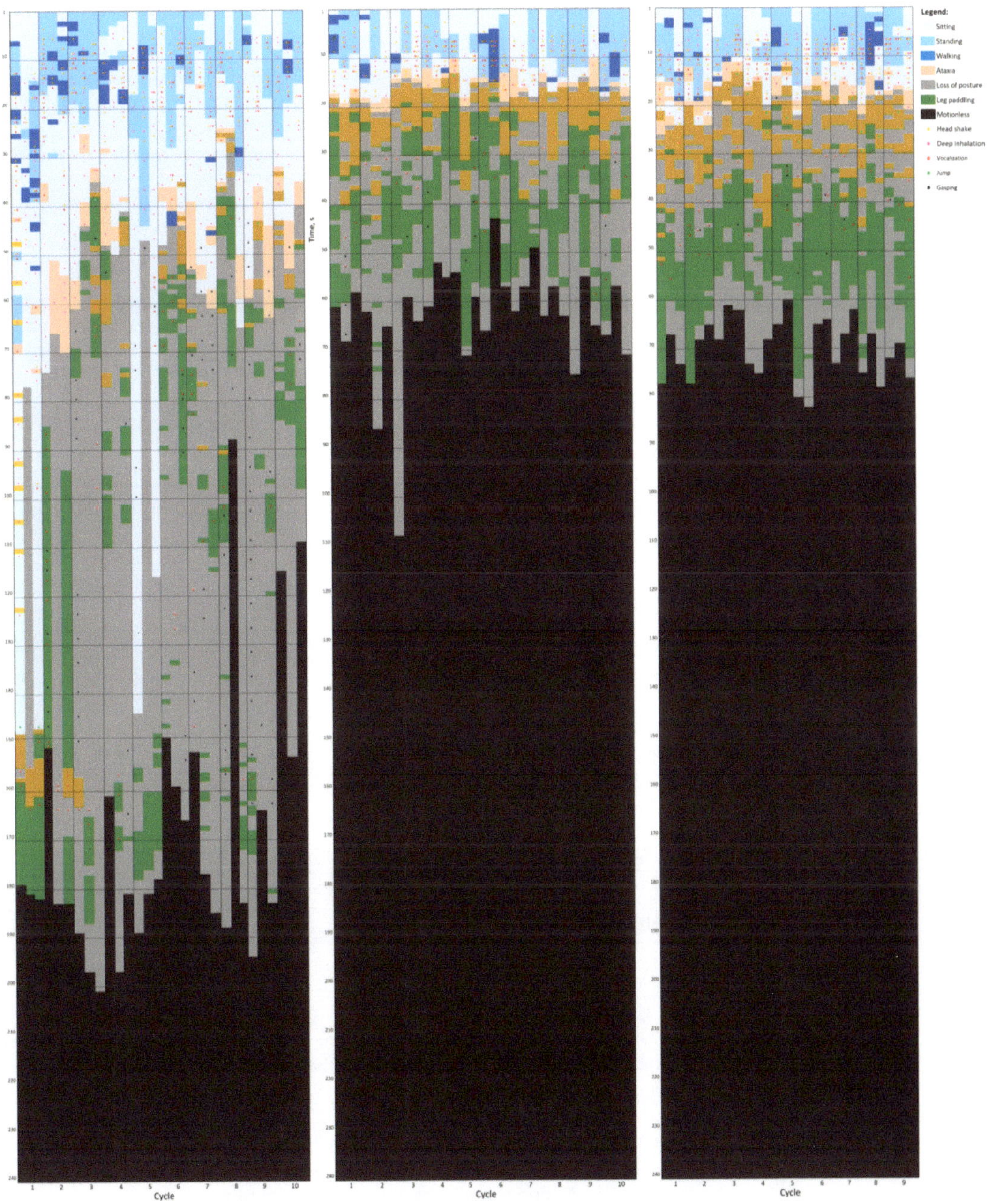

Figure 4. Behavioural plot of a sample of broiler chickens exposed to different gas stunning treatments: **(A)** CO_2 in two phases, the 1st phase with <40% CO_2 by volume in air for 2 min and 2nd phase with >90% CO_2 for 2 min (n = 30); **(B)** gas mixture of 40% CO_2 and 60% nitrogen (N_2) with less than 2% O_2 for 4 min (n = 30); and **(C)** gas mixture of 20% CO_2 and 80% N_2 with less than 2% O_2 for 4 min in a

pit-type gas stunning system (n = 27). The graphical plot shows the behaviour of the birds in 1 s bins. Segments of 10s appear as horizontal dashed lines whereas cycles (dips into the pit) are displayed as vertical lines. Three birds were used per cycle and each bird's behaviour is represented by coloured vertical segments based on the colour coding shown in the legend.

The proportion of birds performing the behaviours, the number of events per bird, and the total duration before LOP, according to the experimental treatment used, are shown in Table 5.

Table 5. Proportion of broiler chickens that showed different behaviours and the number of events per individual, expressed as mean (min–max), when inhaling the experimental gas stunning treatments. Birds were tested before losing posture.

		Treatment				
Item	**Behaviour**	**40C90C**	**40C60N**	**20C80N**	**SE**	***p*-Value**
Proportion, (%)	Head shaking	76/76 (100%)	63/63 (100%)	54/54 (100%)	-	1.000
	Deep inhalation	76/76 (100%)	63/63 (100%)	54/54 (100%)	-	1.000
	HPV	42/76 (55.3%)	43/63 (68.3%)	53/54 (98.1%)	-	0.101
	Gasping	20/76 (26.3%)	0/63 (0.0%)	0/54 (0.0%)	-	-
	Sitting	75/76 (98.7%)	58/63 (92.1%)	50/54 (92.6%)	-	0.951
	Standing	71/76 (93.4%)	51/63 (81.0%)	48/54 (88.9%)	-	0.848
	Walking	45/76 (59.2%)	25/63 (39.7%)	28/54 (51.9%)	-	0.413
	Ataxia	63/76 (82.9%)	55/63 (87.3%)	48/54 (88.8%)	-	0.297
	Wing flapping	20/76 (26.3%) ^{c}	35/63 (55.5%) ^{b}	43/54 (79.6%) ^{a}	-	0.001
Events/bird, n	Head shaking	5.4 (1–12)	4.8 (1–9)	5.1 (1–9)	0.3	0.069
	Deep inhalation	9.4 (2–18) ^{a}	3.9 (1–9) ^{b}	4.8 (0–9) ^{b}	0.4	<0.001
	HPV	2.9 (0–12) ^{a}	2.0 (0–6) ^{b}	3.3 (0–11) ^{a}	0.4	0.001
	Gasping	0.7 (0–10)	0 (0–0)	0 (0–0)	0.2	-
	Sitting	2.1 (1–6) ^{a}	1.3 (0–3) ^{b}	1.6 (0–4) ^{ab}	0.2	<0.001
	Standing	2.0 (0–7) ^{a}	1.1 (0–3) ^{b}	1.3 (0–4) ^{b}	0.2	0.004
	Walking	1.2 (0–7) ^{a}	0.6 (0–3) ^{b}	0.7 (0–3) ^{b}	0.2	<0.001
	Wing flapping	0.4 (0–4) ^{c}	0.6 (0–2) ^{b}	1.0 (0–3) ^{a}	0.1	<0.001
Total duration, s	Sitting	35.9 (3.4–135.1) ^{a}	7.4 (0–19.5) ^{b}	6.9 (0–22.8) ^{b}	2.7	<0.001
	Standing	13.4 (0.0–37.9) ^{a}	7.5 (0.0–22.2) ^{b}	8.5 (0–18.2) ^{b}	1.1	<0.001
	Walking	2.6 (0.0–11.8) ^{a}	1.5 (0.0–10.5) ^{b}	1.5 (0.0–9.3) ^{b}	0.5	0.024
	Ataxia	4.6 (0.0–19.7) ^{a}	2.7 (0.0–7.1) ^{b}	3.6 (0.0–8.2) ^{ab}	0.5	<0.001
	Wing flapping	0.7 (0.0–6.8) ^{c}	1.1 (0.0–6.2) ^{b}	2.3 (0.0–6.5) ^{a}	0.3	0.021

40C90C: CO_2 in two phases, the first phase with <40% CO_2 by volume in air for 2 min and the second phase with >90% CO_2 for 2 min (n = 76, stunned in threes except for two times that birds were stunned in groups of two); 40C60N: gas mixture of 40% CO_2 and 60% N_2 with less than 2% O_2 for 4 min (n = 63, stunned in threes); and 20C80N: gas mixture of 20% CO_2 and 80% N_2 with less than 2% O_2 for 4 min (n = 54, stunned in threes). HPV means high-pitch vocalisation. Different superscripts in the same line indicate significant statistical differences between treatments ($p < 0.05$).

Sitting, standing, walking, and head shaking were behaviours that broiler chickens performed when exposed to AIR and to all experimental gas treatments before LOP. However, in AIR, all birds sat (100%) and only some stood (30%) at a certain point in time and very few walked (4%), while the proportion of birds that stood and walked increased significantly in all three gas treatments (see Table 5) ($p < 0.05$). In AIR, 2 out of 100 broilers showed head shaking once or twice and the proportion of birds performing head shaking in AIR differed significantly from the experimental gas treatments (2% vs. 100%; $p < 0.001$). No broiler chickens exposed to AIR exhibited ataxia, deep inhalation, gasping, jumping, wing flapping, or high-pitch vocalisations (HPVs).

When considering only the three gas treatments (i.e., leaving AIR aside), all birds showed head shaking and deep inhalation, while gasping was only performed by some birds in 40C90C and never in 40C60N and 20C80N. When comparing the number of events per behaviour and bird, there was a tendency towards fewer head shakes in 40C60N and 20C80N compared to 40C90C ($p < 0.10$). In addition, deep inhalation events were dramatically reduced in 40C60N and 20C80N compared to 40C90C ($p < 0.001$) but were still present in 40C60N, with a tendency towards fewer events than in 20C80N ($p = 0.088$). The lowest prevalence of vocalizations was found in 40C90C, but the number of HPVs was reduced in 40C60N compared to 40C90C and 20C80N ($p < 0.001$), and was similar between 40C90C and 20C80N ($p = 0.254$).

Except for wing flapping, the other behaviours assessed occurred in a similar proportion of birds in all gas treatments ($p > 0.05$). However, the 40C60N and 20C80N birds spent significantly less time sitting, standing, and walking ($p < 0.05$) and had fewer events per birds compared to 40C90C ($p < 0.01$). The proportion of broilers showing wing flapping was higher and the duration of wing flapping was longer in 20C80N compared to 40C60N and 40C90C ($p < 0.01$), and in 40C60N compared to the 40C90C treatment ($p < 0.01$). However, it was the broilers exposed to 40C60N that showed the lowest number of events compared to 40C90C and 20C80N. Ataxia in 40C60N was significantly shorter compared to 40C90C ($p < 0.001$), and in 40C60N it tended to be shorter than in 20C80N ($p = 0.075$), while 20C80N also tended to have shorter ataxia than 40C90C ($p = 0.062$).

The order in which each of the behaviours appeared for the first time before LOP is shown in Figure 5. As can be observed, there is a pattern, in which the first behaviours displayed in response to gas treatments are head shaking or deep inhalation. Next, the birds begin to walk, vocalize, become ataxic, and eventually flap their wings. Gasping was only observed in 40C90C, and it was the last behaviour observed before LOP.

3.3.3. Behavioural Assessment after Loss of Posture

The proportion of birds performing different behaviours, the number of events per bird, and their total duration after LOP, according to the experimental treatment used, are shown in Table 6.

Between LOP and motionlessness, the behaviours observed were gasping, jumping, leg flapping, wing flapping, and HPV. Leg paddling was the most commonly observed behaviour, occurring in almost all birds regardless of gas treatment. Although the proportion of birds showing leg paddling was similar between treatments, 40C90C showed significantly fewer events per bird compared to 20C80N ($p = 0.041$), but a similar number to 40C60N ($p = 0.203$) and, moreover, with a shorter total duration compared to 40C60N and 20C80N ($p < 0.01$).

The lowest proportion of birds performing gasping was found in 40C60N and 20C80N, along with a lower number of gasps per animal, compared to 40C90C ($p < 0.001$). The proportion of birds showing wing flapping was similar between treatments ($p > 0.05$). However, broilers in 40C90C performed fewer wing flapping events compared to 40C60N ($p < 0.001$), and both 40C90C and 40C60N performed fewer events than 20C80N ($p < 0.001$). In addition, 40C90C was the treatment during which broiler chickens performed wing flapping for the least amount of time ($p < 0.001$), while 20C80N had the longest total duration of wing flapping ($p < 0.001$). The proportion of birds that vocalized was similar between treatments and so was the number of vocalizations per bird ($p > 0.05$). Jumping was the least-observed behaviour in all treatments, although the proportion of birds jumping was lower in both 40C90C and 40C60N compared to 20C80N ($p < 0.01$), and 40C90C tended to show fewer events than 40C60N ($p = 0.099$) and significantly fewer than 40C60N ($p < 0.001$).

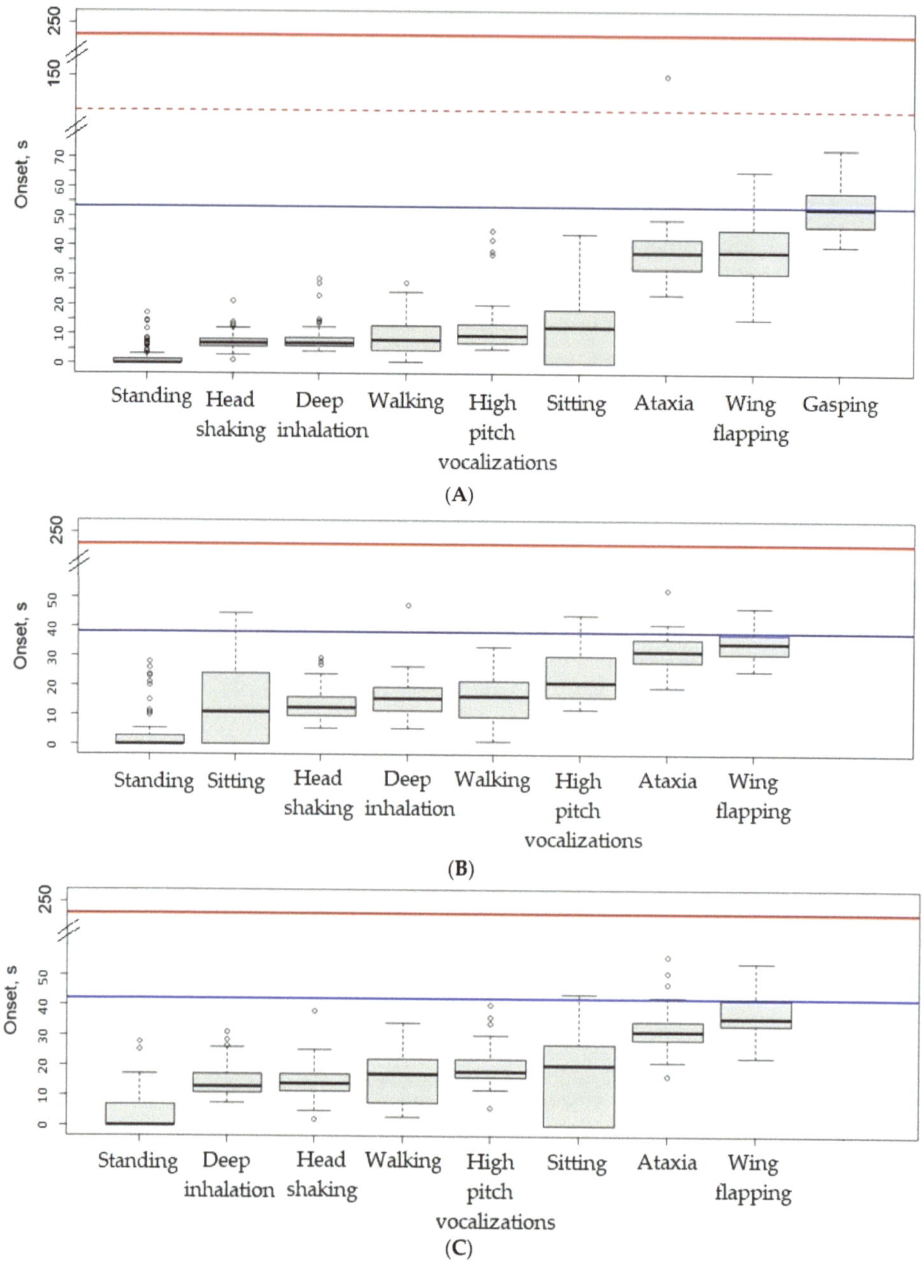

Figure 5. Boxplot of time to the onset of the different behaviours in broiler chickens before loss of posture, in sequence, after the exposure to different gas mixtures. Blue line represents median time to loss of posture, dotted red line means the second phase of 40C90C treatment, and red line indicates end of the period of exposure. (**A**) 40C90C: CO_2 in two phases, the first phase with <40% CO_2 by volume in air for 2 min and the second phase with >90% CO_2 for 2 min (n = 76, stunned in groups of three except for two times that birds were stunned in groups of two). (**B**) 40C60N: gas mixture of 40% CO_2 and 60% N_2 with less than 2% O_2 for 4 min (n = 63, stunned in threes). (**C**) 20C80N: gas mixture of 20% CO_2 and 80% N_2 with less than 2% O_2 for 4 min (n = 54, stunned in threes).

Table 6. Proportion, number of events/bird, and total duration of behaviours observed in broiler chickens exposed to different experimental gas stunning treatments, expressed as mean [min–max], tested after losing posture.

Item	Behaviour	Treatment			SE	*p*-Value
		40C90C	40C60N	20C80N		
Proportion, *n* (%)	Gasping	43/76 (56.6%) ^{a}	14/63 (22.2%) ^{b}	14/54 (25.9%) ^{b}	-	0.009
	Jumping	60/76 (7.9%) ^{b}	7/63 (10.5%) ^{b}	14/54 (25.0%) ^{a}	-	0.003
	Leg paddling	75/76 (98.7%)	63/63 (100%)	54/54 (100%)	-	0.998
	Wing flapping	53/76 (69.7%)	61/63 (96.8%)	53/54 (98.1%)	-	0.320
	HPV	65/76 (85.5%)	53/63 (84.1%)	53/54 (98.1%)	-	0.818
Events/bird, *n*	Gasping	2.5 (0–16) ^{a}	0.4 (1–3) ^{b}	0.3 (0–3) ^{b}	0.4	<0.001
	Jumping	0.1 (0–2) ^{b}	0.2 (0–3) ^{ab}	0.4 (0–3) ^{a}	0.1	<0.001
	Leg paddling	3.9 (0–10) ^{b}	4.3 (2–8) ^{ab}	4.6 (0–9) ^{a}	0.3	0.041
	Wing flapping	1.5 (0–0) ^{c}	2.4 (0–7) ^{b}	3.1 (0–7) ^{a}	0.3	<0.001
	HPV	1.8 (0–10)	1.5 (0–7)	1.8 (0–4)	0.3	0.297
Total duration, s	Leg paddling	14.3 (1.5–36.5) ^{b}	18.2 (4.0–39.3) ^{a}	19.0 (2.0–36.9) ^{a}	1.3	<0.001
	Wing flapping	0.7 (0–6.8) ^{c}	1.1 (0–6.2) ^{b}	2.3 (0–6.5) ^{a}	0.3	<0.001

40C90C: CO_2 in two phases, the first phase with <40% CO_2 by volume in air for 2 min and the second phase with >90% CO_2 for 2 min (*n* = 76, stunned in threes except for two times that birds were stunned in groups of two); 40C60N: a gas mixture of 40% CO_2 and 60% N_2 with less than 2% O_2 for 4 min (*n* = 63, stunned in threes); and 20C80N: a gas mixture of 20% CO_2 and 80% N_2 with less than 2% O_2 for 4 min (*n* = 54, stunned in threes). HPV means high pitch vocalisation. Different superscripts in the same line indicate significant statistical differences between treatments ($p < 0.05$).

3.3.4. Relationships between EEG and Behaviour

The time it took for F50 reduction fell within the range of the LOP. Hence, it was considered that the LOP was the beginning of the LOC. In the 40C60N treatment, the F50 values remained below 50% of baseline during the entire gas exposure period. In contrast, the F50 values of the 20C80N treatment exhibited an increase over the 50% baseline after LOP and remained elevated until the end of the gas exposure period.

During the gas exposure, there was a 65% increase in the proportion of Delta frequencies compared to the baseline. In the EEGs of birds exposed to 20C80N, the proportion of Delta frequencies initially increased but then decreased after 75 s of gas exposure. In the 20C80N treatment, there was an increase in the proportion of Beta frequencies after LOP. The decrease in Delta and increase in Beta frequencies observed in the 20C80N treatment are not biologically relevant, as they corresponded to the times of the isoelectric pattern. The proportion of Gamma frequencies, which are associated with background noise, remained below 2%. The 65% increase in the proportion of Delta frequencies coincided with the time of statistical dispersion until LOP in both the 40C60N and 20C80N treatments.

As expected, the baseline power spectrum in 40C60N and 20C80N demonstrated a greater dominance of frequencies above 4 Hz, suggesting a state of consciousness. In contrast, the isoelectric power spectrum in the 40C60N treatment exhibited no power, suggesting permanent unconsciousness or brain death. In the case of the 20C80N treatment, the power spectrum during the isoelectric pattern had minimal proportion of 25 Hz. These results explain the increase in both F50 and Beta frequencies observed during the isoelectric pattern of the 20C80N treatment. However, during the isoelectric pattern, the residual power from the EEG is biologically insignificant. Hence, the power spectrum confirmed the non-biological relevance of this increase to the 20C80N treatment.

4. Discussion

4.1. Gas Mixture Concentration and Stability Assessment

The concentrations of both CO_2 and O_2 stayed stable during the cycles, and within the limits required by the European Union regulation for the protection of animals at the time of killing, across each gas treatment [1].

4.2. Brain Activity Assessment

The purpose of stunning is to induce a temporary or permanent disruption of brain function, rendering the animal unconscious and insensitive to pain. In contrast to electrical and mechanical stunning methods, gas stunning does not immediately result in a loss of consciousness. Instead, there is a delay between initiation and the onset of unconsciousness, known as the induction phase.

EEG is considered the most accurate approach for assessing the transition from consciousness to unconsciousness [22,23,28,30,32,40]. However, its accuracy depends on the quality of the signal. Factors influencing signal quality include the type of electrode, the quality of the recording equipment, the sample rate used by the equipment, signal amplification, noise filtering, the size of the animal's brain, and the placement of the electrodes (i.e., subcutaneous or intracranial). Hence, there is no established threshold to determine LOC in poultry subjected to CAS. During anaesthesia monitoring, it is feasible to establish a threshold for the unconscious state using an index, such as the Bispectral Index (BIS) [20]. The BIS is computed from spectral variables (Ptot, F50, F95, and frequencies) within an algorithm, generating a numerical value ranging from 0 to 100, where 0 signifies death and 100 indicates consciousness. The threshold for unconsciousness is defined as a BIS below 60 [20,41]. However, the practical application of the BIS is not feasible for small farmed animals in the context of slaughter due to the absence of an amplifier of the signal, noise filtering, the small size of their brain, and contamination of signal due to muscular activity [20,41].

Additionally, the movements of conscious animals can be restricted but not eliminated. In fact, even blinking can lead to the exclusion of records. Another factor influencing the quality of the outcomes is the type of analysis conducted, which relies on the evaluator's expertise, particularly in subjective analyses like the visual examination of filtered records. The reasons for losing all data from the EEG in the 40C90C treatment are still unknown, but perhaps the quality of the electrodes used on the first day of the experiment could have been the cause. In accordance with authors' experience, a change in a batch of commercial electrodes can have a negative impact on the signal quality, due to the quality of the metals used in the needle, the welding, or the wire. The use of costumed electrodes can be an alternative as an attempt to increase the signal quality. The use of costumed electrodes had been used for EEG in animals with the smallest of brains, like fish, with satisfactory results [42].

The time elapsed until the onset of LOC in 40C60N and 20C80N was determined using EEG spectral variables. Specifically, the onset of LOC was identified as when the F50 decreased below 50% and the Delta frequencies increased to above 65% of the baseline. It is important to highlight that the chosen threshold values in this study differ from those used in previous research studies. For example, in broiler chickens exposed to LAPS, LOC was determined when the F50 dropped below 7 Hz, compared to baseline values of 20 Hz [24], or when the F50 was reduced to below 75% of the baseline [27]. The different thresholds used in our study compared to LAPS are attributed to the type of electrodes used. In our study, subcutaneous electrodes were used, while other authors [24] used implanted electrodes placed through the skull, resulting in outputs on a different scale. Although subcutaneous electrodes provide lower-quality EEG recordings compared to implanted electrodes, they are less invasive for the animals as they do not require surgery or recovery time prior to the experiment. In the present study, despite the use of subcutaneous electrodes, the contribution of Gamma frequencies to the Ptot remained at an average of 0.1%, indicating irrelevant interference from background noise [27]. This confirms the quality of the EEG recordings and the reliability of the obtained results in determining the time elapsed until the onset of LOC and death.

General anaesthesia serves as an ideal model for understanding the EEG waveform alterations observed during the transition from a conscious to an unconscious state [23]. During general anaesthesia stages, there is a predominance of the Delta and Theta frequency bands [22]. F50 is particularly sensitive to changes in lower frequencies, while F95 is more responsive to shifts towards higher frequencies [20]. Reductions in the Ptot and F50 are well-established indicators correlated with clinical signs of a loss of consciousness and anaesthesia in animals [43,44].

In the study conducted by Sandercock et al. [22], a reduction in F50 was observed during inhalational anaesthesia, using a face mask with a sevoflurane vaporizer at a concentration of 8%, in hens and turkeys. So far, there are no studies that have reported a specific threshold for Delta frequency proportions. Previous reports have described the analysis of the Delta frequency on EEG as a visual change in the EEG trace, without an objective threshold [19,33]. Therefore, a decrease below 50% of the F50 and an increase above 65% of the Delta frequency proportion can be considered potential indicators of LOC during EEG recording in broiler chickens exposed to gas stunning.

The time elapsed until the onset of LOC found in 40C60N and 20C80N was similar to those reported in previous studies [16,17]. McKeegan et al. [16] reported that the onset of LOC in broiler chickens exposed to a gas mixture of 40% CO_2 and 60% N_2 was at 23 ± 4 s. In our study, it occurred, interestingly, at a similar timing, at 25.7 ± 7 s. This slight difference may be due to the use of a pit pre-filled with gas instead of the closed cabinet (not pre-filled but flushed) used in McKeegan [16] and, therefore, the reduction in available O_2 was reported to be not instantaneous; thus, there was a slight delay in reaching the desired modified atmosphere. Coenen et al. [17] reported a LOC in broiler chickens exposed to a gas mixture of 30% CO_2 and 70% N_2 in a tunnel system at 34 ± 12 s (determined by visual analysis of EEG waveforms when the trace was isoelectric).

The visual interpretation of EEGs can be subjective, and various studies have employed different waveform patterns to determine the LOC, such as suppressed, isoelectric, high-amplitude low-frequency (HALF), and transitional states [16,19,45]. For this reason, we only use visual analysis as a proxy for death, relying on the consolidated indicator of the isoelectric pattern [19,24,27,28,30,33,34]. In addition to visual analysis, spectral analysis provides a quantitative approach for assessing the LOC by generating numerical results. The utilization of numerical variables derived from EEGs has gained attention in similar experimental studies in recent years [23,27,35,40]. This complementary approach enhances the accuracy and reliability of assessing LOC.

Permanent unconsciousness or brain death was assessed using EEG through visual analysis and spectral analysis. The average time for reaching irreversible loss of brain function and brain stem death, and observing the isoelectric pattern visually, was estimated to be 69.8 ± 11.9 s in the 40C60N treatment and 66.3 ± 8.1 s in the 20C80N treatment. The high variability observed in 20C80N for death via the spectral analysis may be attributed to signal degradation. Signal degradation refers to slight changes in frequencies when there is a low Ptot. These changes can influence the frequency contribution and the Ptot itself. Hence, determining the point of death may take longer. However, since the Ptot is low, these changes do not represent any biological significance and rather indicate signal degradation. In such cases, visualizing the isoelectric pattern could provide a more accurate assessment. Both visual and spectral analysis methods demonstrated the irreversible cessation of brain activity and brain stem function [19,22,23,27]. In similar studies on broiler chickens exposed to gas stunning in an experimental chamber using 40% CO_2 and 60% N_2 via flushing, the time to death estimated by a visual analysis of the EEG was 67.8 ± 4.6 s [16].

Raj et al. [46] reported a slightly different onset time of isoelectric EEG at 58 ± 2.3 s in broiler chickens exposed to a gas mixture (30% CO_2 + 60% Argon + 10% air) in a pre-filled box, which may be attributed to the time it takes for the lift to descend into the pit (23 s) and reach the target gas concentration. In contrast, McKeegan et al. [16] visually observed the onset of death at 80.7 s through the isoelectric pattern on the EEG and the absence of motion in broiler chickens exposed to 40% CO_2 and 60% N_2 in a gas-flushed closed cabinet. If the exposure time to the gas is insufficient to induce brain death and the birds can breathe atmospheric air, they may quickly regain consciousness [12]. Our results suggested that a four-minute exposure to the 40C60N and 20C80N gas mixtures induced permanent unconsciousness in all birds, eliminating the possibility of them regaining consciousness.

The integration of the visual and spectral analysis of EEG proved to be a reliable method for accurately estimating brain death in broiler chickens exposed to gas stunning.

The present study intended to record EEG traces from 50 broiler chickens. However, 28 out of 50 (56%) EEG records were unsuitable for analysis due to the loss of electrodes during gas exposure or non-readable EEG activity. This decrease in the sample size is common when EEG is performed. Previous studies have reported a loss of readable EEGs in 9–71% of animals [33,42,47].

4.3. Behavioural Assessment

4.3.1. Behavioural Assessment of Loss of Posture and Motionlessness

Broiler chickens differed in the elapsed time to LOP and motionlessness according to the experimental gas treatment they were subjected to. In particular, exposure to 40C60N or 20C80N not only drastically reduced the time until the birds lost posture and became motionless, but there was also much less inter-individual variability in the time to LOP compared to 40C90C. The high inter-individual variability in the time to LOP in 40C90C represents a serious welfare risk if the birds still remain conscious when the first phase has finished, and certain birds are therefore exposed to more than 40% CO_2 during the second phase while conscious, as was the case for 2 out of 76 broiler chickens in the present study. It is known that the inhalation of concentrations of above 40% CO_2 in conscious chickens generates a very painful mucosal stimulus [15,48].

4.3.2. Behavioural Assessment before Loss of Posture

Since LOP is considered the behavioural indicator of the onset of LOC [10,26,31,49], all the behaviours observed before the LOP occur in birds that are conscious and therefore may be potential indicators of pain, distress, or dyspnoea caused by the inhalation of the gas or gas mixture. In order to discern whether the descent into the pit per se caused aversive behaviours in broiler chickens and caused confusion in the results, first, the behaviour of the birds in the gas stunning equipment was assessed while they were breathing atmospheric air (AIR), and then again when subjected to experimental treatments. In AIR, most chickens remained sitting, while some were standing, few walked, and very few exhibited head shaking once or twice while descending into the pit. Although these behaviours were also observed in the other three experimental treatments, the proportion of birds performing sitting and walking was higher in the gas mixtures compared to AIR, which may indicate a behavioural change pattern that might be related to fear. The cause of head shaking observed in only two birds in the AIR group remains unclear. Unlike the head shaking observed in the gas mixtures, the head shaking in AIR was not associated with a sound of sneezing. It is possible that these birds are sensitive to new stimuli [11,50], like the descent of the lift, or that they are attempting to self-activate after a period of rest [48]. The almost minimal proportion of birds exhibiting head shaking in the AIR treatment, along with the absence of other considered aversive behaviours (i.e., deep inhalation, gasping, wing flapping, HPVs, ataxia) suggests that descending into the pit did not induce aversion in broiler chickens, unlike the behaviour that was observed during the three experimental treatments.

Head shaking and deep inhalation was performed at least once by all birds from all gas treatments, and head shaking was associated with a sneezing sound, but gasping was only observed in some birds exposed to 40C90C. Head shaking, deep inhalation, and gasping are associated with mucosal irritation, dyspnoea, and breathlessness ("air hunger") and, thus, a reduction in welfare during gas stunning [27]. Head shaking and deep inhalation were the first aversive behaviours displayed during the stunning process. While head shaking is associated with an unpleasant stimulus caused by the activation of chemoreceptors sensitive to CO_2 in respiratory tract, deep inhalation and gasping are associated with hyperventilation during CO_2 stunning; however, these responses have also been observed with the inhalation of inert gases alone, such as argon [49], and are related to respiratory distress [26]. So far, no scientific study on gas stunning in broilers has included vocalisations as a behaviour to be assessed in the ethogram [15,16,30,51]. However, the vocalisations recorded in this study are particularly high-pitched and suggestive of fear and/or pain, if heard before LOP.

Comparing the three experimental gas treatments, both 40C60N and 20C80N showed fewer transitions between locomotor behaviours (sitting, standing, and walking) and these were of a shorter duration than 40C90C, because the time to LOP in these treatments was 2.8-fold shorter (40C60N and 20C80N: 21.0 $\pm$ 4.5 s vs. 40C90C: 59.2 $\pm$ 21.9 s), so the likelihood of repeating these locomotor behaviours is lower. Despite this shorter time to LOP, both 20C80N and 40C60N had a similar proportion of birds displaying head shaking, deep inhalation, and HPVs. In addition, 20C80N had similar number of head shakes and HPVs but fewer deep inhalations and no gasping compared to 40C90C. In contrast, 40C60N showed a tendency to have less head shaking, less HPVs, and no gasping compared to 40C90C. Therefore, both 20C80N and 40C60N seems to cause less aversion than 40C90C.

The proportion of chickens that flapped their wings before LOP varied depending on the experimental treatment. Nitrogen-containing gas mixtures caused a higher proportion of birds to flap their wings before losing posture than 40C90C. The higher the nitrogen concentration in the gas mixture, the higher the proportion of birds showing wing flapping. This could be due to the fact that anoxic environments lead to an increase in wing flapping, although CO_2 has an anaesthetic effect on birds and, therefore, when a higher CO_2 concentration and anoxic environment are combined (as in 40C60N compared to 20C80N) it can result in a calmer induction of unconsciousness. The occurrence of wing flapping in the gas stunning equipment is a welfare concern, since it may cause injuries and pain to the other birds that have not yet lost consciousness [5,16].

In addition, in 40C60N or 20C80N there was no risk that birds inhaled CO_2 concentrations above 40%, as can happen in birds that are still conscious when the 2nd phase of 40C90C starts. Therefore, in 40C60N or 20C80N the experience of severe pain in the mucosa is mitigated.

Taking all this into consideration, it seems that the fastest induction to unconsciousness was achieved with 40C60N, which was also the least aversive gas mixture. Moreover, the risk of inhaling a CO_2 concentration above 40% (known to cause severe pain in conscious birds due to the activation of chemoreceptors in mucous membranes), as can occur in 40C90C, is mitigated. On the other hand, 20C80N also offers a rapid induction of unconsciousness (of a similar time to 40C60N); it appears to be less aversive than 40C90C but slightly more so than 40C60N, and there is also no risk of inhaling a CO_2 concentration above 40%. Therefore, from an animal welfare point of view, 40C60N appears to be the least aversive of the three treatments tested as our experimental conditions.

4.3.3. Behavioural Assessment after Loss of Posture

Since LOP was the behavioural indicator of the onset of unconsciousness, behaviours after LOP do not represent a welfare concern (e.g., pain, distress, breathlessness) as the bird is presumed to be unconscious. The behaviours observed may be related to convulsions or other involuntary movements rather than aversive behaviours [27,33].

The convulsions in broiler chickens exposed to the three experimental treatments were expressed as wing flapping, leg paddling, and jumping due to uncontrolled muscle jerks. Leg paddling and jumping were behaviours observed only after LOP, but leg paddling was observed in almost all birds at this stage. Therefore, leg paddling seems to be the most reliable indicator of convulsions and unconsciousness.

Gasping and HPVs were performed both before and after LOP in gas treatments, indicating that consciousness is not required for their performance. In relation to HPVs, birds do not need to contract the vocal cords in their larynx to produce sound, unlike terrestrial mammals. Additionally, the eight air sacs present in birds are expansions of the respiratory tree. During convulsions, the muscular jerks can lead to the passing of air from the air sac to the lung and then through the syrinx, causing vocalizations in unconscious birds. HPVs after LOP are presumed to be a consequence of air movement through the syrinx caused by convulsions (muscle jerks). One might think that HPVs could also suggest pain or aversion after LOP. However, during the experiment, even once the dead chickens were removed from the pit and piled up, we heard squeaks due to the compression exerted by the weight of one carcass on top of another.

The number of HPVs heard after LOP was higher in the 20C80N treatment, which is consistent with the higher number of wing flapping events observed in this treatment. Similarly, as they occurred before LOP, anoxic environments also lead to an increase in wing flapping and leg paddling after LOP, but the higher the CO_2 concentration, the lower the convulsions expressed as wing flapping due to the anaesthetic effect of CO_2 on birds. The occurrence of wing flapping in the gas stunning equipment may be a welfare concern since it may cause injuries and pain to other birds that have not yet lost consciousness [5,16]. This finding is consistent with a study by Gent et al. [10], which reported a longer duration of wing flapping after the LOP in broiler chickens exposed to N_2 (19.7 s, on average) compared to CO_2 (7.1 s, on average). Further research could investigate the effect of different gas mixtures on the meat quality of broiler chickens subjected to stunning.

4.4. *Relationship between Brain Activity and Behavioural Assessment*

The relationship between LOC and death events in the EEG activity and behavioural observations in unrestrained animals is crucial for interpreting behavioural indicators. Nonetheless, there was no statistical correlation between LOC and LOP in this study because brain activity and behaviour were not assessed in the same animal ($r = -0.091$; $p = 0.779$). Benson et al. [31] surgically implanted wireless EEG electrodes into broiler chickens and, once recovered, the chickens were unrestricted and exposed to isoflurane anaesthesia to correlate their LOC determined by brain activity with their LOP. However, this correlation was not statistically significant ($r = 0.150$; $p > 0.05$). The authors concluded that the assessment of LOP has a certain inaccuracy as it depends on the subjectivity of the observer, who defines when the bird ceases to maintain a sitting position or neck tension. In addition, the determination of LOP was sometimes hindered in cases where birds were huddled against the walls of the chamber, artificially providing them with support. However, they concluded that LOP can be utilized as a proxy for the onset of LOC.

In the present study, the range of time to LOC and brain death (i.e., an isoelectric pattern during EEG) was similar to the range of time to LOP and motionlessness in 40C60N and 20C80N, respectively. Therefore, the results suggest that LOP and motionlessness can be used as behavioural indicators to estimate LOC and brain death, respectively, when EEG recording is not a possibility (e.g., in commercial slaughterhouses or depopulation conditions). The comparison between LOP and LOC was carried out only in the 40C60N and 20C80N treatments. We considered 40C60N the less aversive treatment, due to the smaller proportion of birds performing vocalizations before LOP and the faster LOC.

The use of video cameras to monitor the behaviour of birds during stunning procedures under commercial conditions is of the utmost importance, as well as operators being trained to detect indicators of loss of consciousness, death, aversion, and convulsion. However, in current CAS designs, even if video cameras are present, it is not always possible to monitor the behaviour of the animals, as the birds are often stunned directly in their transport containers, where the observer's visibility and the mobility of the birds is greatly reduced. In current designs where birds are removed from their transport containers before being introduced into the gas stunning systems, the monitoring of the behaviour of the birds is also hindered as they are usually stunned in tunnel-type systems where the birds are conveyed from one end of the system to the other while exposed to gas concentrations, so the images observed via video camera of a specific individual are of a very limited amount of time.

5. Conclusions

The exposure of broiler chickens to 40C90C, 40C60N, or 20C80N does not induce immediate unconsciousness. Regardless of the gas mixture tested, all broiler chickens experienced aversion during the induction of a loss of consciousness. The exposure to 40C60N and 20C80N not only decreased dramatically the time to the induction of LOC and death, but also did so with less variability in the elapsed time between individuals compared to 40C90C. The 40C90C birds not only experienced more aversion during the induction of LOC but were also at risk of remaining conscious when the CO_2 concentration was increased in the 2nd phase. From an animal welfare point of view, 40C60N was the least aversive of the three treatments tested, followed by 20C80N and 40C90C. Further research is required to explore alternative gases or gas mixtures that can minimize or eliminate aversive responses during the induction of unconsciousness to improve broiler chicken welfare during stunning.

Author Contributions: Conceptualization, A.V. and V.M.; methodology, D.S.R., A.V., A.X., A.V.-P., T.J.G., V.M. and A.C.-J.; data acquisition, A.V., A.V.-P., A.X., T.J.G. and A.C.-J.; formal analysis, D.S.R., T.J.G. and A.C.-J.; statistical analysis, D.S.R. and A.C.-J.; writing—original draft preparation, D.S.R. and A.C.-J.; writing—review and editing, A.V., T.J.G., V.M. and A.C.-J. All authors have read and agreed to the published version of the manuscript.

Funding: This research was funded by the EURCAW-Poultry-SFA. European Commission Grant number: SANTE/EURC/2021-2022/SI2.871763.

Institutional Review Board Statement: The animal study protocol was approved by the Institutional or Ethics Committee of IRTA (protocol code 11653/2022) on the 14 December 2022.

Informed Consent Statement: Not applicable.

Data Availability Statement: Data sharing upon request.

Acknowledgments: The authors appreciate the professional technical assistance and the essential food-grade gas provision of the Carburos Metálicos-Air Products Group, as well as Marc Bagaria for his tireless help during the experimental period.

Conflicts of Interest: The funders had no role in the design of the study; in the collection, analyses, or interpretation of data; in the writing of the manuscript; or in the decision to publish the results. The authors declare no conflicts of interest.

References

1. European Union Council Regulation (EC). *No 1099/2009 of 24 September 2009 on the Protection of Animals at the Time of Killing*; Council of the European Union: Brussels, Belgium, 2009.
2. United Kingdom the Welfare of Animals (Slaughter or Killing) Regulations. 1995. Available online: https://www.legislation.gov.uk/uksi/1995/731/made (accessed on 7 March 2023).
3. EFSA (European Food Safety Authority). Opinion of the Scientific Panel on Animal Health and Welfare (AHAW) on a Request from the Commission Related to Welfare Aspects of the Main Systems of Stunning and Killing the Main Commercial Species of Animals. *EFSA J.* **2004**, *45*, 1–29. [CrossRef]
4. EFSA AWAH Panel (ESFA Panel on Animal Health and Animal Welfare); Nielsen, S.S.; Alvarez, J.; Bicout, D.J.; Calistri, P.; Depner, K.; Drewe, J.A.; Garin-Bastuji, B.; Gonzales Rojas, J.L.; Gortázar Schmidt, C.; et al. Slaughter of Animals: Poultry. *EFSA J.* **2019**, *17*, e05849. [CrossRef]
5. Berg, C.; Raj, M. A Review of Different Stunning Methods for Poultry—Animal Welfare Aspects (Stunning Methods for Poultry). *Animals* **2015**, *5*, 1207–1219. [CrossRef]
6. Verhoeven, M.T.W.; Gerritzen, M.A.; Kluivers-Poodt, M.; Hellebrekers, L.J.; Kemp, B. Validation of Behavioural Indicators Used to Assess Unconsciousness in Sheep. *Res. Vet. Sci.* **2015**, *101*, 144–153. [CrossRef]
7. European Commission. Report from the Commission to the European Parliament and the Council on the Various Stunning Methods for Poultry. Available online: https://food.ec.europa.eu/system/files/2016-10/aw_practice_slaughter_com_2013_915_report_en.pdf (accessed on 31 January 2023).
8. EURCAW-Poultry-SFA Main Welfare Aspects of Stunning Chicken by Exposure to Controlled Atmosphere. Available online: https://zenodo.org/record/7382889#.Y8-a-8nMI2x (accessed on 20 February 2023).
9. Martoft, L.; Lomholt, L.; Kolthoff, C.; Rodriguez, B.E.; Jensen, E.W.; Jørgensen, P.F.; Pedersen, H.D.; Forslid, A. Effects of CO_2 Anaesthesia on Central Nervous System Activity in Swine. *Lab. Anim.* **2002**, *36*, 115–126. [CrossRef]

10. Gent, T.C.; Gebhardt-Henrich, S.; Schild, S.-L.A.; Rahman, A.A.; Toscano, M.J. Evaluation of Poultry Stunning with Low Atmospheric Pressure, Carbon Dioxide or Nitrogen Using a Single Aversion Testing Paradigm. *Animals* **2020**, *10*, 1308. [CrossRef]
11. McKeegan, D.E.F.; McIntyre, J.; Demmers, T.G.M.; Wathes, C.M.; Jones, R.B. Behavioural Responses of Broiler Chickens during Acute Exposure to Gaseous Stimulation. *Appl. Anim. Behav. Sci.* **2006**, *99*, 271–286. [CrossRef]
12. Velarde, A.; Raj, M. *Animal Welfare at Slaughter*, 1st ed.; Velarde, A., Raj, M., Eds.; 5m Publishing: Sheffield, UK, 2016.
13. Webster, A.B.; Fletcher, D.L. Assessment of the Aversion of Hens to Different Gas Atmospheres Using an Approach-Avoidance Test. *Appl. Anim. Behav. Sci.* **2004**, *88*, 275–287. [CrossRef]
14. Raj, A.B.M. Aversive Reactions of Turkeys to Argon, Carbon Dioxide and a Mixture of Carbon Dioxide and Argon. *Vet. Rec.* **1996**, *138*, 592–593. [CrossRef] [PubMed]
15. Wooley, S.C.; Gentle, M.J. Physiological and Behavioural Responses of the Domestic Hen to Hypoxia. *Res. Vet. Sci.* **1988**, *45*, 377–382. [CrossRef]
16. McKeegan, D.; McIntyre, J.; Demmers, T.; Lowe, J.; Wathes, C.; van den Broek, P.; Coenen, A.; Gentle, M. Physiological and Behavioural Responses of Broilers to Controlled Atmosphere Stunning: Implications for Welfare. *Anim. Welf.* **2007**, *16*, 409–426. [CrossRef]
17. Coenen, A.M.L.; Lankhaar, J.; Lowe, J.C.; McKeegan, D.E.F. Remote Monitoring of Electroencephalogram, Electrocardiogram, and Behavior during Controlled Atmosphere Stunning in Broilers: Implications for Welfare. *Poult. Sci.* **2009**, *88*, 10–19. [CrossRef] [PubMed]
18. EFSA (European Food Safety Authority). Guidance on the Assessment Criteria for Applications for New or Modified Stunning Methods Regarding Animal Protection at the Time of Killing. *EFSA J.* **2018**, *16*, e05343. [CrossRef]
19. Gibson, T.J.; Rebelo, C.B.; Gowers, T.A.; Chancellor, N.M. Electroencephalographic Assessment of Concussive Non-Penetrative Captive Bolt Stunning of Turkeys. *Br. Poult. Sci.* **2018**, *59*, 90–95. [CrossRef] [PubMed]
20. Tonner, P.H.; Bein, B. Classic Electroencephalographic Parameters: Median Frequency, Spectral Edge Frequency Etc. *Best. Pract. Res. Clin. Anaesthesiol.* **2006**, *20*, 147–159. [CrossRef]
21. Raj, A.B.M.; O'Callaghan, M.; Knowles, T.G. The Effects of Amount and Frequency of Alternating Current Used in Water Bath Stunning and of Slaughter Methods on Electroencephalograms in Broilers. *Anim. Welf.* **2006**, *15*, 7–18. [CrossRef]
22. Sandercock, D.A.; Auckburally, A.; Flaherty, D.; Sandilands, V.; McKeegan, D.E.F. Avian Reflex and Electroencephalogram Responses in Different States of Consciousness. *Physiol. Behav.* **2014**, *133*, 252–259. [CrossRef]
23. Hernandez, E.; James, F.; Torrey, S.; Widowski, T.; Schwean-Lardner, K.; Monteith, G.; Turner, P.V. Evaluation of Brain Death in Laying Hens During On-Farm Killing by Cervical Dislocation Methods or Pentobarbital Sodium Injection. *Front. Vet. Sci.* **2019**, *6*, 297. [CrossRef] [PubMed]
24. Martin, J.E.; Christensen, K.; Vizzier-Thaxton, Y.; McKeegan, D.E.F. Effects of Light on Responses to Low Atmospheric Pressure Stunning in Broilers. *Br. Poult. Sci.* **2016**, *57*, 585–600. [CrossRef] [PubMed]
25. Friard, O.; Gamba, M. BORIS: A Free, Versatile Open-Source Event-Logging Software for Video/Audio Coding and Live Observations. *Methods Ecol. Evol.* **2016**, *7*, 1325–1330. [CrossRef]
26. Gerritzen, M.A.; Lambooij, E.; Hillebrand, S.J.W.; Lankhaar, J.A.C.; Pieterse, C. Behavioral Responses of Broilers to Different Gaseous Atmospheres. *Poult. Sci.* **2000**, *79*, 928–933. [CrossRef]
27. EFSA AHAW Panel (EFSA Panel on Animal Health and Welfare); More, S.; Bicout, D.; Bøtner, A.; Butterworth, A.; Calistri, P.; Depner, K.; Edwards, S.; Garin-Bastuji, B.; Good, M.; et al. Low Atmospheric Pressure System for Stunning Broiler Chickens. *EFSA J.* **2017**, *15*, e05056. [CrossRef]
28. Martin, J.E.; Christensen, K.; Vizzier-Thaxton, Y.; Mitchell, M.A.; McKeegan, D.E.F. Behavioural, Brain and Cardiac Responses to Hypobaric Hypoxia in Broiler Chickens. *Physiol. Behav.* **2016**, *163*, 25–36. [CrossRef]
29. Lou, M.E.; Kleinhenz, M.D.; Schroeder, R.; Lechtenberg, K.; Montgomery, S.; Coetzee, J.F.; Viscardi, A.V. Evaluating the Utility of a CO_2 Surgical Laser for Piglet Tail Docking to Reduce Behavioral and Physiological Indicators of Pain and to Improve Wound Healing: A Pilot Study. *Appl. Anim. Behav. Sci.* **2022**, *254*, 105720. [CrossRef]
30. Gerritzen, M.A.; Reimert, H.G.M.; Hindle, V.A.; Verhoeven, M.T.W.; Veerkamp, W.B. Multistage Carbon Dioxide Gas Stunning of Broilers. *Poult. Sci.* **2013**, *92*, 41–50. [CrossRef]
31. Benson, E.R.; Alphin, R.L.; Rankin, M.K.; Caputo, M.P.; Kinney, C.A.; Johnson, A.L. Evaluation of EEG Based Determination of Unconsciousness vs. Loss of Posture in Broilers. *Res. Vet. Sci.* **2012**, *93*, 960–964. [CrossRef] [PubMed]
32. Raj, A.B.M.; Wotton, S.B.; Gregory, N.G. Changes in the Somatosensory Evoked Potentials and Spontaneous Electroencephalogram of Hens during Stunning with Carbon Dioxide and Argon Mixture. *Br. Vet. J.* **1992**, *148*, 147–156. [CrossRef] [PubMed]
33. Gerritzen, M.A.; Lambooij, B.; Reimert, H.; Stegeman, A.; Spruijt, B. On-Farm Euthanasia of Broiler Chickens: Effects of Different Gas Mixtures on Behavior and Brain Activity. *Poult. Sci.* **2004**, *83*, 1294–1301. [CrossRef]
34. Johnson, C.B.; Gibson, T.J.; Stafford, K.J.; Mellor, D.J. Pain Perception at Slaughter. *Anim. Welf.* **2012**, *21*, 113–122. [CrossRef]
35. Martin, J.E.; Sandilands, V.; Sparrey, J.; Baker, L.; Dixon, L.M.; McKeegan, D.E.F. Welfare Assessment of Novel On-Farm Killing Methods for Poultry. *PLoS ONE* **2019**, *14*, e0212872. [CrossRef]
36. Raj, A.B.M.; O'Callaghan, M. Effects of Electrical Water Bath Stunning Current Frequencies on the Spontaneous Electroencephalogram and Somatosensory Evoked Potentials in Hens. *Br. Poult. Sci.* **2004**, *45*, 230–236. [CrossRef]
37. Raj, A.; O'Callaghan, M.; Hughes, S. The Effects of Amount and Frequency of Pulsed Direct Current Used in Water Bath Stunning and of Slaughter Methods on Spontaneous Electroencephalograms in Broilers. *Anim. Welf.* **2006**, *15*, 19–24. [CrossRef]

38. R Core Team. *R: A Language and Environment for Statistical Computing*; R Foundation for Statistical Computing: Vienna, Austria. Available online: https://www.r-project.org/ (accessed on 27 December 2023).
39. Pinheiro, J.; Bates, D.; DebRoy, S.; Sarkar, D. R Core Team Nlme: Linear and Nonlinear Mixed Effects Models. R Package Version 3.1-164. Available online: https://cran.r-project.org/web/packages/nlme/index.html (accessed on 2 January 2024).
40. Hernandez, E.; James, F.; Torrey, S.; Widowski, T.; Schwean-Lardner, K.; Monteith, G.; Turner, P.V. Electroencephalographic, Physiologic and Behavioural Responses during Cervical Dislocation Euthanasia in Turkeys. *BMC Vet. Res.* **2019**, *15*, 132. [CrossRef] [PubMed]
41. Martin-Jurado, O.; Vogt, R.; Kutter, A.P.N.; Bettschart-Wolfensberger, R.; Hatt, J.-M. Effect of Inhalation of Isoflurane at End-Tidal Concentrations Greater than, Equal to, and Less than the Minimum Anesthetic Concentration on Bispectral Index in Chickens. *Am. J. Vet. Res.* **2008**, *69*, 1254–1261. [CrossRef]
42. Lambooij, E.; Kloosterboer, R.J.; Gerritzen, M.A.; van de Vis, J.W. Assessment of Electrical Stunning in Fresh Water of African Catfish (*Clarias gariepinus*) and Chilling in Ice Water for Loss of Consciousness and Sensibility. *Aquaculture* **2006**, *254*, 388–395. [CrossRef]
43. Martín-Cancho, M.F.; Lima, J.R.; Luis, L.; Crisóstomo, V.; López, M.A.; Ezquerra, L.J.; Carrasco-Jiménez, M.S.; Usón-Gargallo, J. Bispectral Index, Spectral Edge Frequency 95% and Median Frequency Recorded at Varying Desflurane Concentrations in Pigs. *Res. Vet. Sci.* **2006**, *81*, 373–381. [CrossRef] [PubMed]
44. Martín-Cancho, M.F.; Lima, J.R.; Luis, L.; Crisóstomo, V.; Ezquerra, L.J.; Carrasco, M.S.; Usós-Gargallo, J. Bispectral Index, Spectral Edge Frequency 95%, and Median Frequency Recorded for Various Concentrations of Isoflurane and Sevoflurane in Pigs. *Am. J. Vet. Res.* **2003**, *64*, 866–873. [CrossRef]
45. McKeegan, D.E.F.; Reimert, H.G.M.; Hindle, V.A.; Boulcott, P.; Sparrey, J.M.; Wathes, C.M.; Demmers, T.G.M.; Gerritzen, M.A. Physiological and Behavioral Responses of Poultry Exposed to Gas-Filled High Expansion Foam. *Poult. Sci.* **2013**, *92*, 1145–1154. [CrossRef]
46. Raj, A.B.M.; Wotton, S.B.; McKinstry, J.L.; Hillebrand, S.J.W.; Pieterse, C. Changes in the Somatosensory Evoked Potentials and Spontaneous Electroencephalogram of Broiler Chickens during Exposure to Gas Mixtures. *Br. Poult. Sci.* **1998**, *39*, 686–695. [CrossRef] [PubMed]
47. Velarde, A.; Ruiz-de-la-Torre, J.L.; Roselló, C.; Fàbrega, E.; Diestre, A.; Manteca, X. Assessment of Return to Consciousness After Electrical Stunning in Lambs. *Anim. Welf.* **2002**, *11*, 333–341. [CrossRef]
48. Webster, A.B.; Fletcher, D.L. Reactions of Laying Hens and Broilers to Different Gases Used for Stunning Poultry. *Poult. Sci.* **2001**, *80*, 1371–1377. [CrossRef]
49. Lambooij, E.; Gerritzen, M.A.; Engel, B.; Hillebrand, S.J.W.; Lankhaar, J.; Pieterse, C. Behavioural Responses during Exposure of Broiler Chickens to Different Gas Mixtures. *Appl. Anim. Behav. Sci.* **1999**, *62*, 255–265. [CrossRef]
50. Hughes, B.O. Headshaking in Fowls: The Effect of Environmental Stimuli. *Appl. Anim. Ethol.* **1983**, *11*, 45–53. [CrossRef]
51. Gerritzen, M.; Lambooij, B.; Reimert, H.; Stegeman, A.; Spruijt, B. A Note on Behaviour of Poultry Exposed to Increasing Carbon Dioxide Concentrations. *Appl. Anim. Behav. Sci.* **2007**, *108*, 179–185. [CrossRef]

animals

MDPI

Article

The Effects of Electrolytic Multivitamins and Neomycin on Antioxidant Capacity and Intestinal Damage in Transported Lambs

Cui Xia [1,2,†], Chunhui Duan [1,†], Conghui Chen [1], Xinyu Yang [2], Yingjie Zhang [1], Yueqin Liu [1,*,‡] and Yuzhong Ma [2,*,‡]

1 College of Animal Science and Technology, Agricultural University of Hebei, Baoding 071001, China; xiacui950114@163.com (C.X.); duanchh211@126.com (C.D.); cchpluto@163.com (C.C.); zhangyingjie66@126.com (Y.Z.)

2 College of Veterinary Medicine, Agricultural University of Hebei, Baoding 071001, China; b18803308783@163.com

* Correspondence: liuyueqin66@126.com (Y.L.); dkma@hebau.edu.cn (Y.M.)

† These authors have contributed equally to this work and should be regarded as co-first authors.

‡ These authors have contributed equally to this work.

Simple Summary: Transportation stress can lead to a decrease in immune function and induce various oxidative stresses, which affect health and productive performance. In this experiment, lambs were fed diets containing electrolytic multivitamins and neomycin. The activities of the antioxidant enzymes SOD, GSH-Px, and CAT, and the levels of MDA and T-AOC in sera, were examined. At the same time, the activities of antioxidant enzymes SOD and GSH-Px, and the levels of MDA and T-AOC in the jejunum and colon, as well as the mRNA expressions of SOD, CAT, Nrf2, HO-1, Keap1, IL-1, IL-2, IL-12, Bax, Bcl-2, and Caspase3 in the jejunum and colon, were measured. In addition, the contents of IgA, IgG, IgM, and sIgA in the jejunum and colon were examined. It was found that road transport can decrease the antioxidant capacity and contents of immunoglobulin and increase the expression levels of inflammatory factors and apoptosis in the jejunum and colon of lambs. Electrolytic multivitamins had a better effect on improving antioxidant activity, inhibiting the expression of inflammatory factors in lambs, and potentially reducing the expression levels of apoptotic factors and oxidative damage to the jejunum and colon.

Citation: Xia, C.; Duan, C.; Chen, C.; Yang, X.; Zhang, Y.; Liu, Y.; Ma, Y. The Effects of Electrolytic Multivitamins and Neomycin on Antioxidant Capacity and Intestinal Damage in Transported Lambs. *Animals* **2024**, *14*, 824. https://doi.org/10.3390/ani14060824

Academic Editor: Elbert Lambooij

Received: 13 February 2024
Revised: 1 March 2024
Accepted: 2 March 2024
Published: 7 March 2024

Abstract: Transport stress can cause damage to animals. In this experiment, 60 four-month-old lambs were randomly divided into three groups: CG (basal diet), EG (basal diet + 375 mg/d/lamb electrolytic multivitamin), and NG (basal diet + 200 mg/d/lamb neomycin). The results were as follows: during road transport, in all groups, the levels of SOD, T-AOC, and GSP-Px, and mRNA expressions of CAT, SOD, Nrf2, HO-1, and Bcl-2 in the jejunum and colon decreased ($p < 0.01$). However, mRNA expressions of Keap1, IL-1β, IL-2, IL-12, Bax, and Caspase3 in the jejunum and colon and the level of MDA increased ($p < 0.01$). The concentrations of IgA, IgG, and sIgA in the jejunum and colon also decreased ($p < 0.01$). In the EG and NG, the levels of SOD ($p < 0.05$) and T-AOC ($p < 0.01$) increased, and the level of MDA decreased ($p < 0.01$). However, in the jejunum, the levels of SOD and T-AOC, the concentrations of IgA and IgG, and mRNA expression of Bcl-2 increased ($p < 0.05$). mRNA expressions of IL-1, IL-2, and Caspase 3 ($p < 0.05$), and mRNA expression of IL-12 ($p < 0.01$) decreased. In the colon, SOD activity and the concentration of sIgA increased ($p < 0.01$). The level of MDA and mRNA expressions of IL-2 and Caspase 3 also decreased ($p < 0.05$). In the jejunum and colon, mRNA expression of SOD ($p < 0.05$) and mRNA expression of Nrf2 increased ($p < 0.01$). mRNA expression of Keap1 ($p < 0.05$) and Bax ($p < 0.01$) decreased. In summary, road transport can cause a decrease in antioxidant activity and immunity of lambs and an increase in oxidative damage. Electrolytic multivitamins and neomycin can improve immune function and potentially reduce oxidative damage to the jejunum and colon.

Keywords: transport stress; lamb; antioxidant; intestinal injury; electrolytic multivitamin; neomycin

1. Introduction

Transportation is an essential link in husbandry production; it is also one of the most common stressors in animals. Research shows that Japanese black cattle experienced transport stress after they were transported for 6 h [1]. Various transport stressors can lead to various infectious diseases [2] and even death [3] in animals, which brings economic losses and restricts the development of husbandry. Under stress, the body will produce more free radicals, which can lead to oxidative damage in animals. The intestine is an important site for digestion, absorption, and nutrient metabolism; the stability of intestinal function is crucial for the body. However, the intestine can easily become a target organ for stress. Large amounts of reactive oxygen species (ROS) accumulate in intestinal epithelial cells during stress, which disrupts the antioxidant system and leads to oxidative stress. Wei et al. [4] showed that intestinal oxidative stress occurred, the antioxidant system was destroyed, intestinal barrier function was damaged, and the incidence rate increased after pigs experienced transport stress. In addition, oxidative stress can regulate the expression of apoptosis genes, mediate apoptosis, and damage the intestinal barrier of animals [5,6]. It increases the expression of inflammatory factors, inhibits immune function [7], and affects normal metabolism.

At present, only a limited number of papers investigating the effects of transport stress on antioxidant capacity and intestinal damage in lambs are available. Research has shown that a certain amount of vitamins could effectively alleviate transport stress [8,9]. Transport stress causes infectious disease and increases the incidence rate. Antibiotics are the first choice for the treatment of bacterial diseases [10]. A lack of research is observed in the literature regarding the effects and mechanisms of electrolytic multivitamins and neomycin on antioxidant capacity and intestinal damage in transported lambs. Regarding antioxidant capacity and intestinal damage, research is mainly focused on cattle [11] and pigs [4]. Our previous research found that electrolytic multivitamins and neomycin could potentially affect the intestinal barrier function of lambs [12]. The purpose of this study is to evaluate the effects of electrolytic multivitamins and neomycin on antioxidant capacity and intestinal damage in transported lambs and explore the possible mechanisms. It provides new ideas for improving antioxidant capacity and reducing intestinal damage in transported lambs.

2. Materials and Methods

2.1. Experimental Design and Feeding Management

A total of 60 lambs (average live weight, 18.99 ± 0.54 kg, 4 months old) were randomly divided into three equal groups. The control group (CG) was fed a basal diet, and the composition of the basal diet is shown in Table 1. This experiment was based on previous reports [13,14]; preliminary experiments were conducted in the early stage. According to the preliminary experimental results, the best time for adding electrolytic multivitamins was from 2 days before transportation to 7 days after transportation, and the best time for adding neomycin was from 0 to 7 days after transportation. The electrolytic multivitamin group (EG) was fed a basal diet with 375 mg/d of electrolytic multivitamin per lamb (Anhui Lingmu Biotechnology Co., Ltd., Fuyang, China); the main components and contents of electrolytic multivitamins are shown in Table 2. The neomycin group (NG) was fed a basal diet of 200 mg/d of neomycin per lamb (Henan Weilong Veterinary Medicine Co., Ltd., Kaifeng, China). The experimental lambs were transported for 8 h from 09:00 to 17:00 on 28 July. The day of transportation was recorded as the first day of the experiment (day 0). Five lambs in each group were randomly selected on days −2, 0, 7, and 14, then 5 mL of jugular venous blood was collected for separating sera at 3500 r/min for 5 min; it was placed in sterile tubes and stored at −20 °C. After blood collection, lambs were anesthetized with a dose of 0.2 mg/kg of xylazine (Hebei Mingeng Biotechnology Co., Ltd.,

Shijiazhuang, China) on day −2, 0, 7, and 14, and exsanguinated. The middle of the jejunal and colonic tissues were cut and collected, then rinsed the contents with physiological saline. The tissues were placed in sterile tubes and frozen in liquid nitrogen. The approval number for this experiment was 2021023.

Table 1. Ingredients and nutrient composition of the basal diet (dry matter basis).

Ingredients	Content %	Nutrient Composition	Content
Cracked corn	55.00	Neutral Detergent Fibers (%)	33.33
Soybean meal	20.00	Crude Protein (%)	18.06
Peanut seeding	12.50	Acidic Detergent Fibers (%)	14.49
Peanut meal	9.00	Metabolizable Energy (MJ/kg) [2]	12.50
Premix [1]	2.50	Ca (%)	0.75
NaCl	0.50	*p* (%)	0.31
Baking soda	0.50		

[1] Per kg premix contained the following: vitamin A 15,356 IU, vitamin D 4300 IU, vitamin E 50 mg, Fe 88.70 mg, Zn 70.90 mg, Mn 51.80 mg, Cu 13.75 mg, Se 0.23 mg, I 1.50 mg, and Co 0.49 mg. [2] Metabolizable energy was a calculated value, while the others were measured values.

Table 2. The main components and contents of electrolytic multivitamins.

Components	Contents
Vitamin C	≥2000 IU/kg
Vitamin B2	≥750 mg/kg
Vitamin A	30,000–5,000,000 IU/kg
Vitamin D3	75,000–20,000,000 IU/kg
Vitamin E	≥500 IU/kg
Vitamin B1	≥500 mg/kg
Water	≤10%
Folate	30 mg/kg
Taurine	20,000 mg/kg
Zn	1000 mg/kg
Mn	1000 mg/kg
Fe	1000 mg/kg
Cu	600 mg/kg

The lambs were bought from Zhangjiakou Lanhai Livestock Breeding Co., Ltd. (4.0°37′ N 115°03′ E and ~1300–1600 m above sea level, Zhangjiakou, China). The vehicle was a truck with a density of 1.7 m^2/lamb. The speed of the vehicle was 30–75 km/h. On the day of transportation, the outdoor temperature was 18–29 °C, and the relative humidity was 64%.

The pens and breeding tools were disinfected thoroughly. Each group of lambs was raised separately from 1 July to 30 July. Before feeding, the electrolytic multivitamins and neomycin were mixed thoroughly with a small amount of basal diet to ensure drugs were consumed completely by each lamb so all lambs could eat and drink freely. Lambs were fasted for 10 h before transportation, and fresh water was freely provided until transportation. Only water was given within 12 h after transportation, and a basal diet was provided after 12 h. The basal diet was included by the sheep farm; lambs were fed twice a day at 7:00 and 17:00 to ensure that they all had leftover feed in their feed tanks every day.

2.2. Determination of Antioxidant Capacity in Sera and Intestines

The indicators of the antioxidant capacity, such as superoxide dismutase (SOD), glutathione peroxidase (GSH-Px), catalase (CAT), malondialdehyde (MDA), and total antioxidant capacity (T-AOC) were measured according to the instructions of the detection kit (Nanjing Jiancheng Bioengineering Institute, Nanjing, China).

2.3. Determination of Intestinal Immunoglobulins

The concentrations of immunoglobulin A (IgA), immunoglobulin G (IgG), secretory immunoglobulin A (sIgA), and immunoglobulin M (IgM) in the jejunum and colon were measured by the enzyme-linked immunosorbent assay (ELISA) detection kit (Nanjing Jiancheng Bioengineering Institute, Nanjing, China).

2.4. mRNA Expressions Analysis by Quantitative Real-Time PCR

Total RNA was isolated from the jejunum and colon (50 mg) with Trizol reagent (TransGen Biotech, Beijing, China). The concentration of total RNA in the samples was evaluated using a spectrophotometer (NanoDrop-2000, Thermo Fisher Scientific, Waltham, MA, USA) at 260 nm and 280 nm, respectively. Ratios of absorption (260:280 nm) between 1.8 and 2.0 for all samples were accepted as "pure" for RNA. Reverse transcription of RNA into cDNA by the Reverse Transcription Kit was performed according to the instructions (Transgen Biotech, Beijing, China). Primer sequences for IL-1β, IL-2, IL-12, CAT, SOD, Nrf2, Keap1, HO-1, Bax, Bcl 2, and Caspase-3 in the jejunum and colon were designed using the GenBank database from the National Center for Biotechnology Information and software of primer 5 design. RT-qPCR was performed with SYBR Green Premix Es Taq (Takara, Beijing, China) using the Step One Plus real-time PCR system (Applied Biosystems (Thermo-Fisher Scientific, Shanghai, China)) on 96-well plates with 20 μL total reaction volume, which included 10 μL of SYBR Green Premix, 1 μL of cDNA, 0.4 μL of forward primers, 0.4 μL of reverse primers, and 8.2 μL of double-distilled water. Each reaction was run in duplicate. The PCR cycling protocol included one cycle of pre-incubation at 95 °C for 2 min, 40 cycles of denaturation at 95 °C for 5 s, and annealing at 60 °C for 30 s; each cycle increased by 0.5 °C at 65–95 °C. β-actin was used as a reference, and the relative expression of the target gene to β-actin was determined using the $2^{-\Delta\Delta Ct}$ method, as described by Livak [15]. Primers for RT-qPCR were synthesized by the Beijing Genomics Institution (BGI Genomics Co., Ltd., Beijing, China; Table 3).

Table 3. Gene primer information.

Primer		Sequence (5′→3′)	Length (bp)	Annealing Temperature (°C)	Accession Number
IL-1β	F	ACAGATGAAGAGCTGCACCC	161	58	443,539
	R	AGACATGTTCGTAGGCACGG			
IL-2	F	GTTGCAAACGGTGCACCTAC	122	58	443,401
	R	GAGAGCTTGAGGTTCTCGGG			
IL-12	F	GCTTCGCAGCCTCCTCC	136	59	443,472
	R	CCTCAGCAGGTTTTGGGAGT			
CAT	F	TTGCGGGCCATCTGAAAG	101	53	100,307,035
	R	AAGAGCCTGGATGCGGGAG			
SOD	F	GGTAAACACGGTTTCCAT	261	53	692,639
	R	CAAGTCATCAGGGTCAGC			
Nrf2	F	AAGGTCCTCCCCATCCATGA	193	58	443,276
	R	GCTCAACCCAGACTGGAGAC			
Keap1	F	CGTGGAGACAGAAACGTGGA	159	59	101,113,845
	R	CAGGTGTCTGTGTCTGGGTC			
HO-1	F	CGATGGGTCCTCACACTCAG	75	59	101,120,910
	R	GCACACTCGCATTCACATGG			
Bax	F	TGCCAGCAAACTGGTGCTCAA	243	60	443,059
	R	GCACTCCAGCCACAAAGATGGT			
Bcl-2	F	CGCTGAAGCGAAGCTGTAGA	92	60	101,119,602
	R	CGTTGAGCCTGAAAGCTGTTT			
Caspase3	F	AATGCAAGAAGCAGGGCACCCA	155	60	443,031
	R	GGGTTACAGCGATGCAGAAGGTTCA			
β-actin	F	CAGTCGGTTGGATCGAGCAT	151	59	443,340
	R	AGAAGGAGGGTGGCTTTTGG			

2.5. Statistical Analysis

All data were analyzed using single-factor ANOVA in SPSS 21.0 statistical software [16] to determine differences in indicators when different treatments were administered at the same time versus the same treatment at different times. Duncan's method was used for multiple comparisons. The General Linear Model (GLM) was used to analyze the effects of treatments and time interaction. Effects $p \leq 0.05$ were considered different. Values were expressed as "Mean ± Standard Error" (M ± SEM).

3. Results

3.1. Effects of Different Treatments on the Antioxidant Capacity of Transported Lambs

3.1.1. Effects of Different Treatments on the Antioxidant Capacity in Sera of Transported Lambs

As shown in Table 4, after lambs were transported, the levels of SOD and T-AOC in the CG and NG decreased ($p < 0.01$), while the level of MDA increased ($p < 0.01$). On day 0, compared with the CG, the SOD activity in the EG ($p < 0.05$), as well as the activity of T-AOC in the EG, increased ($p < 0.01$). On day 14, the level of MDA in the EG was lower than that in the CG ($p < 0.05$), and the level of T-AOC in the EG was higher than those in the CG and NG ($p < 0.01$). There was no difference in the levels of MDA in the CG and NG on day 7, and SOD and T-AOC in the NG on day 14 compared with those on day −2d ($p > 0.05$).

3.1.2. Effects of Different Treatments on the Antioxidant Capacity of Jejunum and Colon in Transported Lambs

As shown in Table 5, by transportation, the levels of GSH-Px and T-AOC in the jejunum and colon of the EG and NG, as well as the SOD activity in the jejunum and colon of each group, decreased ($p < 0.01$). The level of MDA in the jejunum of each group, as well as those in the colon of the CG and NG, increased ($p < 0.01$). On day 0, the level of MDA in the jejunum of the EG was lower than that of the CG ($p < 0.01$). In the colon, the SOD activity of the EG was higher than those of the CG and NG ($p < 0.01$), and the MDA of the EG was lower than those of the CG and NG ($p < 0.01$). On day 7, in the jejunum, the SOD activity of the EG was higher than that of the CG ($p < 0.05$), and the level of MDA was lower than that of the CG ($p < 0.01$). The SOD activity in the colon of the EG and NG was higher than that of the CG ($p < 0.01$). On day 14, the level of T-AOC in the jejunum of the EG and NG was higher than that of the CG ($p < 0.05$), and the SOD activity in the colon of the EG and NG was higher than that of the CG ($p < 0.01$). The levels of MDA in the jejunum of the EG, T-AOC in the jejunum of the NG, and MDA in the colon of the CG and NG on day 7 had no significant changes compared with those on day −2 ($p > 0.05$). In the jejunum, there was no difference in GSH-Px activity of the CG and NG, SOD activity of the EG, and MDA in the NG on day 14 compared with those on day −2 ($p > 0.05$). In the colon, the GSH-Px activity of the NG and SOD of the EG and NG on day 14 had no significant changes compared with those on day −2 ($p > 0.05$).

3.1.3. Effects of Different Treatments on the mRNA Expression of Antioxidant Factors in the Jejunum and Colon of Transported Lambs

As shown in Table 6, by transportation, the mRNA expression of CAT in the jejunum and colon of each group, SOD in the jejunum of each group, and SOD in the colon of the CG and NG decreased ($p < 0.01$). On day 7, the mRNA expression of SOD in the jejunum of the EG was higher than that of the CG ($p < 0.01$). On day 14, the mRNA expressions of SOD in the jejunum and colon of the EG were higher than those of the CG ($p < 0.05$). On day 7, there was no difference in the mRNA expression of SOD in the jejunum of the EG compared with that on day −2 ($p > 0.05$). On day 14, the mRNA expressions of CAT in the jejunum and colon of the EG, as well as the mRNA expression of SOD in the jejunum of the NG, had no significant changes compared with those on day −2 ($p > 0.05$).

Table 4. Effect of different treatments on the antioxidant capacity in sera of transported lambs.

Items	Groups	Time					*p*-Values		
		−2d	0d	7d	14d	Total	Time	Treatment	Interaction
SOD (U/mL)	CG	204.20 ± 10.77^{a}	142.58 ± 6.97^{Bb}	159.40 ± 10.27^{b}	169.65 ± 9.46^{b}	168.96 ± 6.75	<0.010	0.017	0.501
	EG	196.81 ± 13.63	175.45 ± 8.53^{A}	186.04 ± 10.37	198.41 ± 10.22	189.18 ± 5.41			
	NG	197.00 ± 10.70^{a}	153.35 ± 7.54^{ABb}	161.68 ± 10.89^{b}	181.49 ± 9.62^{ab}	173.38 ± 5.95			
	Total	199.34 ± 6.36	157.13 ± 5.51	169.04 ± 6.48	183.18 ± 6.10				
MDA (nmol/mL)	CG	6.64 ± 0.38^{b}	10.20 ± 1.08^{a}	8.79 ± 0.70^{ab}	8.13 ± 0.41^{Aab}	8.44 ± 0.44	<0.010	0.019	0.695
	EG	6.79 ± 0.53	8.14 ± 0.69	7.35 ± 0.73	6.48 ± 0.35^{B}	7.19 ± 0.31			
	NG	6.84 ± 0.63^{b}	9.53 ± 0.53^{a}	8.15 ± 0.58^{ab}	7.44 ± 0.29^{ABb}	7.99 ± 0.33			
	Total	6.76 ± 0.28	9.29 ± 0.49	8.10 ± 0.39	7.35 ± 0.26				
GSH-Px (U/mL)	CG	131.84 ± 8.30	108.50 ± 5.78	117.82 ± 6.08	123.99 ± 6.04	120.54 ± 3.62	0.069	0.165	0.925
	EG	130.60 ± 7.18	124.05 ± 8.96	129.97 ± 6.76	133.40 ± 6.87	129.51 ± 3.53			
	NG	126.86 ± 7.37	115.95 ± 5.46	122.81 ± 3.86	127.79 ± 6.37	123.36 ± 2.91			
	Total	129.77 ± 4.12	116.17 ± 4.07	123.53 ± 3.33	128.39 ± 3.59				
CAT (U/mL)	CG	0.77 ± 0.03	0.58 ± 0.11	0.70 ± 0.24	0.74 ± 0.08	0.70 ± 0.06	0.159	0.266	0.918
	EG	0.75 ± 0.02	0.72 ± 0.18	0.95 ± 0.33	1.08 ± 0.11	0.87 ± 0.12			
	NG	0.70 ± 0.05	0.48 ± 0.16	0.65 ± 0.12	0.96 ± 0.25	0.70 ± 0.07			
	Total	0.74 ± 0.02	0.59 ± 0.09	0.77 ± 0.14	0.92 ± 0.11				
T-AOC (U/mL)	CG	0.29 ± 0.02^{a}	0.17 ± 0.01^{Bc}	0.21 ± 0.01^{bc}	0.23 ± 0.02^{Bb}	0.23 ± 0.01	<0.010	<0.010	0.347
	EG	0.28 ± 0.02	0.23 ± 0.01^{A}	0.25 ± 0.03	0.30 ± 0.04^{A}	0.27 ± 0.01			
	NG	0.28 ± 0.02^{a}	0.21 ± 0.01^{Ab}	0.22 ± 0.01^{b}	0.25 ± 0.03^{Bab}	0.24 ± 0.01			
	Total	0.28 ± 0.01	0.21 ± 0.00	0.23 ± 0.01	0.26 ± 0.01				

Note: [abc] Different lowercase letters superscripted in the same row indicate significant differences between different treatments ($p < 0.05$). [AB] The superscript values in the same column with different uppercase letters indicate significant differences at different times ($p < 0.05$). The same letter or no letter indicates no significant difference ($p > 0.05$). There is no meaning between uppercase and lowercase letters.

Table 5. Effect of different treatments on the jejunal and colonic antioxidant capacity of transported lambs.

Items		Groups	Time					p-Values		
			−2d	0d	7d	14d	Total	Time	Treatment	Interaction
Jejunum	GSH-Px (U/g)	CG	4272.54 ± 424.49 [a]	2677.12 ± 212.66 [b]	3124.78 ± 316.62 [b]	3304.09 ± 284.91 [ab]	3344.63 ± 222.09	<0.010	<0.440	0.968
		EG	4322.18 ± 429.24	2794.13 ± 284.62	3526.44 ± 315.13	3955.46 ± 482.23	3649.55 ± 237.75			
		NG	4499.46 ± 359.46 [a]	2932.41 ± 212.41 [b]	3285.34 ± 310.11 [b]	3583.22 ± 390.33 [ab]	3575.11 ± 223.47			
		Total	4364.72 ± 205.73	2801.22 ± 125.06	3312.19 ± 167.48	3614.26 ± 218.49				
	SOD (U/g)	CG	1202.02 ± 132.21 [a]	630.72 ± 90.16 [b]	752.42 ± 41.34 [Bb]	811.85 ± 69.81 [b]	849.25 ± 74.98	<0.010	0.017	0.883
		EG	1276.40 ± 105.72 [a]	840.96 ± 68.96 [b]	979.44 ± 59.99 [Ab]	1025.74 ± 88.25 [ab]	1030.39 ± 59.00			
		NG	1340.50 ± 105.97 [a]	719.87 ± 61.30 [b]	827.62 ± 46.85 [ABb]	915.56 ± 67.26 [b]	950.89 ± 77.61			
		Total	1272.64 ± 61.02	730.52 ± 48.11	853.16 ± 41.71	917.72 ± 48.84				
	MDA (nmol/g)	CG	46.76 ± 2.83 [c]	81.85 ± 3.68 [Aa]	70.22 ± 3.26 [Ab]	61.74 ± 4.20 [b]	65.14 ± 4.14	<0.010	<0.010	0.247
		EG	45.25 ± 1.01 [b]	61.49 ± 2.34 [Ba]	54.52 ± 3.58 [Bab]	46.70 ± 3.80 [b]	57.21 ± 3.46			
		NG	43.80 ± 1.83 [c]	71.10 ± 4.87 [ABa]	61.50 ± 2.92 [ABab]	52.44 ± 4.61 [bc]	51.99 ± 2.32			
		Total	45.27 ± 1.10	71.48 ± 3.49	62.08 ± 2.80	53.62 ± 3.04				
	T-AOC (μmol/mL)	CG	0.20 ± 0.01 [a]	0.14 ± 0.01 [b]	0.15 ± 0.01 [b]	0.16 ± 0.01 [Bb]	0.16 ± 0.01	<0.010	0.019	0.939
		EG	0.21 ± 0.01	0.16 ± 0.02	0.17 ± 0.03	0.19 ± 0.01 [A]	0.18 ± 0.01			
		NG	0.21 ± 0.01 [a]	0.17 ± 0.01 [b]	0.17 ± 0.01 [ab]	0.20 ± 0.01 [Aab]	0.19 ± 0.01			
		Total	0.21 ± 0.00	0.15 ± 0.01	0.16 ± 0.01	0.18 ± 0.01				

Table 5. *Cont.*

Items		Groups	Time					*p*-Values		
			−2d	0d	7d	14d	Total	Time	Treatment	Interaction
Colon	GSH-Px (U/g)	CG	4132.23 ± 297.75 ^{a}	2827.76 ± 193.04 ^{b}	3084.56 ± 193.57 ^{b}	3289.40 ± 216.07 ^{b}	3333.49 ± 176.96	<0.010	0.075	0.933
		EG	4176.81 ± 377.81	3375.81 ± 204.78	3552.28 ± 312.19	3963.06 ± 320.70	3766.99 ± 163.45			
		NG	4165.16 ± 351.27 ^{a}	3008.08 ± 211.41 ^{b}	3195.69 ± 146.68 ^{b}	3470.47 ± 200.25 ^{ab}	3459 ± 85 ± 167.25			
		Total	4158.07 ± 172.08	3070.55 ± 129.72	3277.51 ± 134.21	3574.31 ± 161.03				
	SOD (U/g)	CG	1165.63 ± 83.32 ^{a}	623.04 ± 30.36 ^{Bb}	694.00 ± 69.47 ^{Bb}	774.66 ± 43.42 ^{Bb}	814.33 ± 68.29	<0.010	<0.010	0.157
		EG	1335.44 ± 64.47^{a}	1012.40 ± 62.18 ^{Ab}	1021.90 ± 59.88 ^{Ab}	1282.88 ± 90.73 ^{Aa}	1014.72 ± 57.87			
		NG	1194.94 ± 89.52 ^{a}	774.05 ± 42.83 ^{Bc}	946.69 ± 78.70 ^{Abc}	1143.19 ± 43.21 ^{Aab}	1163.16 ± 53.59			
		Total	1232.01 ± 47.74	803.16 ± 61.34	887.53 ± 60.63	1066.91 ± 82.12				
	MDA (nmol/g)	CG	39.04 ± 3.93 ^{b}	61.31 ± 3.71 ^{Aa}	49.20 ± 4.28 ^{ab}	44.33 ± 3.94 ^{b}	48.47 ± 3.01	<0.010	0.042	0.494
		EG	38.52 ± 4.43	45.31 ± 3.34 ^{B}	41.73 ± 3.69	38.67 ± 4.06	41.06 ± 1.86			
		NG	38.28 ± 4.50 ^{b}	60.08 ± 4.06 ^{Aa}	45.12 ± 4.21 ^{b}	41.03 ± 3.48 ^{b}	46.13 ± 3.08			
		Total	38.61 ± 2.15	55.57 ± 3.17	45.35 ± 2.30	41.34 ± 2.09				
	T-AOC (μmol/mL)	CG	0.20 ± 0.01 ^{a}	0.15 ± 0.01 ^{b}	0.14 ± 0.01 ^{b}	0.14 ± 0.01 ^{b}	0.16 ± 0.01	<0.010	0.537	0.995
		EG	0.20 ± 0.02	0.16 ± 0.02	0.16 ± 0.02	0.15 ± 0.01	0.17 ± 0.01			
		NG	0.20 ± 0.01 ^{a}	0.15 ± 0.01 ^{b}	0.15 ± 0.01 ^{b}	0.14 ± 0.01 ^{b}	0.16 ± 0.01			
		Total	0.20 ± 0.01	0.15 ± 0.01	0.15 ± 0.01	0.14 ± 0.01				

Note: ^{abc} Different lowercase letters superscripted in the same row indicate significant differences between different treatments ($p < 0.05$). ^{AB} The superscript values in the same column with different uppercase letters indicate significant differences at different times ($p < 0.05$. The same letter or no letter indicates no significant difference ($p > 0.05$). There is no meaning between uppercase and lowercase letters.

Table 6. Effects of different treatments on the mRNA expressions of antioxidant factors in the jejunum and colon of transported lambs.

Items		Groups	Time					p-Value		
			−2d	0d	7d	14d	Total	Time	Treatment	Interaction
Jejunum	CAT	CG	1.00 ± 0.00^{a}	0.65 ± 0.05^{b}	0.70 ± 0.05^{b}	0.75 ± 0.07^{b}	0.78 ± 0.04	<0.010	0.091	0.915
		EG	1.00 ± 0.00^{a}	0.75 ± 0.06^{b}	0.80 ± 0.07^{b}	0.90 ± 0.07^{ab}	0.86 ± 0.03			
		NG	1.00 ± 0.00^{a}	0.69 ± 0.08^{b}	0.74 ± 0.07^{b}	0.79 ± 0.06^{b}	0.80 ± 0.04			
		Total	1.00 ± 0.00	0.69 ± 0.04	0.75 ± 0.04	0.81 ± 0.04				
	SOD	CG	1.00 ± 0.00^{a}	0.54 ± 0.08^{b}	0.56 ± 0.07^{Bb}	0.70 ± 0.05^{Bb}	0.70 ± 0.05	<0.01	<0.01	0.123
		EG	1.00 ± 0.00^{a}	0.72 ± 0.09^{b}	0.89 ± 0.05^{Aab}	1.10 ± 0.09^{Aa}	0.93 ± 0.04			
		NG	1.00 ± 0.00^{a}	0.58 ± 0.09^{b}	0.69 ± 0.08^{ABb}	0.91 ± 0.09^{ABa}	0.79 ± 0.05			
		Total	1.00 ± 0.00	0.61 ± 0.05	0.71 ± 0.05	0.90 ± 0.06				
Colon	CAT	CG	1.00 ± 0.00^{a}	0.67 ± 0.07^{b}	0.70 ± 0.06^{b}	0.74 ± 0.07^{b}	0.78 ± 0.04	<0.010	0.076	0.893
		EG	1.00 ± 0.00^{a}	0.76 ± 0.06^{b}	0.81 ± 0.05^{b}	0.89 ± 0.05^{ab}	0.86 ± 0.03			
		NG	1.00 ± 0.00^{a}	0.71 ± 0.08^{b}	0.75 ± 0.03^{b}	0.79 ± 0.06^{b}	0.81 ± 0.03			
		Total	1.00 ± 0.00	0.71 ± 0.04	0.75 ± 0.03	0.81 ± 0.04				
	SOD	CG	1.00 ± 0.00^{a}	0.66 ± 0.08^{b}	0.74 ± 0.07^{b}	0.76 ± 0.04^{Bb}	0.79 ± 0.04	<0.01	0.011	0.736
		EG	1.00 ± 0.00	0.82 ± 0.10	0.88 ± 0.08	0.96 ± 0.05^{A}	0.92 ± 0.03			
		NG	1.00 ± 0.00^{a}	0.73 ± 0.05^{b}	0.79 ± 0.06^{b}	0.86 ± 0.05^{ABb}	0.85 ± 0.03			
		Total	1.00 ± 0.00	0.74 ± 0.04	0.80 ± 0.04	0.86 ± 0.03				

Note: ^{ab} Different lowercase letters superscripted in the same row indicate significant differences between different treatments ($p < 0.05$). ^{AB} The superscript values in the same column with different uppercase letters indicate significant differences at different times ($p < 0.05$). The same letter or no letter indicates no significant difference ($p > 0.05$). There is no meaning between uppercase and lowercase letters.

3.1.4. Effects of Different Treatments on mRNA Expression of Key Factors in the Nrf2 Pathway in the Jejunum and Colon of Transported Lambs

As shown in Table 7, by transportation, the mRNA expressions of Nrf2 and HO-1 in the jejunum and colon of each group decreased ($p < 0.01$). The mRNA expressions of Keap1 in the jejunum of the CG and NG, as well as the mRNA expression of Keap1 in the colon of each group, increased ($p < 0.01$). On day 0, the mRNA expression of Nrf2 in the jejunum of the EG was higher than those of the CG and NG ($p < 0.01$), while Keap1 in the jejunum of the EG was lower than that of the CG ($p < 0.05$). On day 7, the mRNA expression of Keap1 in the colon of the EG was lower than that of the CG ($p < 0.05$). On day 14, the mRNA expressions of Nrf2 in the jejunum and colon of the EG were higher than that of the CG ($p < 0.01$).

On day 7, the mRNA expressions of Keap1 in the jejunum of the CG and NG, as well as the mRNA expression of Keap1 in the colon of the EG, had no significant changes compared with those on day −2 ($p > 0.05$). On day 14, there was no difference in the mRNA expressions of Nrf2 in the jejunum and colon of the EG and NG compared with those on day −2 ($p > 0.05$). The mRNA expressions of HO-1 in the jejunum and colon of the EG had no significant change compared with that on day −2 ($p > 0.05$). There was no significant difference in the mRNA expressions of Keap1 in the colon of the CG and NG compared with that on day −2 ($p > 0.05$).

3.2. Effects of Different Treatments on the Intestinal Immune Function of Transported Lambs

3.2.1. Effects of Different Treatments on Intestinal Immunoglobulin of Transported Lambs

As shown in Table 8, by transportation, the concentrations of IgA, IgG, and sIgA in the jejunum and colon of the CG and NG, as well as the concentration of sIgA in the jejunum of the EG decreased ($p < 0.01$). The concentration of sIgA in the jejunum of the EG on day 0, as well as the concentration of sIgA in the colon of the EG, was higher than those in the CG and NG on days 0, 7, and 14 ($p < 0.01$). The concentrations of IgG in the jejunum of the EG on day 7 and IgA in the jejunum of the EG on day 14 were higher than those of the CG ($p < 0.05$). The concentrations of sIgA in the jejunum of the EG and NG were higher than that of the CG on days 7 and 14 ($p < 0.01$).

On day 7, there was no significant difference in the concentrations of IgA and IgG in the jejunum and colon of the NG and sIgA in the EG compared with those on day −2 ($p > 0.05$). On day 14, the concentration of IgG in the jejunum and colon of the CG had no significant change compared with those on day −2 ($p > 0.05$). There was no significant difference in the concentration of sIgA in the jejunum of the NG and IgA in the colon of the CG compared with those on day −2 ($p > 0.05$).

Table 7. Effects of different treatments on mRNA expressions of key factors in the Nrf2 pathway in the jejunum and colon of transported lambs.

Items		Groups	Time					*p*-Values		
			−2d	0d	7d	14d	Total	Time	Treatment	Interaction
Jejunum	Nrf2	CG	1.00 ± 0.00^{a}	0.51 ± 0.03^{Bc}	0.61 ± 0.08^{c}	0.82 ± 0.06^{Bb}	0.73 ± 0.04	<0.010	<0.010	0.144
		EG	1.00 ± 0.00^{a}	0.72 ± 0.04^{Ab}	0.81 ± 0.09^{b}	1.10 ± 0.05^{Aa}	0.91 ± 0.04			
		NG	1.00 ± 0.00^{a}	0.55 ± 0.03^{Bc}	0.71 ± 0.06^{b}	0.95 ± 0.06^{ABa}	0.80 ± 0.04			
		Total	1.00 ± 0.00	0.59 ± 0.03	0.71 ± 0.04	0.96 ± 0.04				
	HO-1	CG	1.00 ± 0.00^{a}	0.57 ± 0.04^{c}	0.67 ± 0.04^{c}	0.80 ± 0.06^{b}	0.76 ± 0.04	<0.010	0.073	0.849
		EG	1.00 ± 0.00^{a}	0.64 ± 0.04^{b}	0.75 ± 0.06^{b}	0.92 ± 0.05^{a}	0.83 ± 0.04			
		NG	1.00 ± 0.00^{a}	0.61 ± 0.03^{c}	0.69 ± 0.03^{c}	0.82 ± 0.07^{b}	0.78 ± 0.04			
		Total	1.00 ± 0.00	0.61 ± 0.02	0.71 ± 0.03	0.85 ± 0.04				
	Keap1	CG	1.00 ± 0.00^{b}	1.36 ± 0.06^{Aa}	1.20 ± 0.10^{ab}	1.07 ± 0.07^{b}	1.16 ± 0.04	<0.01	0.040	0.753
		EG	1.00 ± 0.00	1.13 ± 0.05^{B}	1.09 ± 0.07	0.96 ± 0.06	1.05 ± 0.03			
		NG	1.00 ± 0.00^{b}	1.24 ± 0.05^{ABa}	1.15 ± 0.07^{ab}	1.03 ± 0.08^{b}	1.11 ± 0.03			
		Total	1.00 ± 0.00	1.25 ± 0.04	1.15 ± 0.05	1.02 ± 0.04				
Colon	Nrf2	CG	1.00 ± 0.00^{a}	0.63 ± 0.04^{b}	0.65 ± 0.04^{b}	0.70 ± 0.04^{Bb}	0.75 ± 0.04	<0.010	<0.010	0.179
		EG	1.00 ± 0.00^{a}	0.76 ± 0.05^{b}	0.81 ± 0.08^{b}	0.99 ± 0.07^{Aa}	0.89 ± 0.03			
		NG	1.00 ± 0.00^{a}	0.65 ± 0.05^{c}	0.72 ± 0.05^{bc}	0.85 ± 0.08^{ABab}	0.80 ± 0.04			
		Total	1.00 ± 0.00	0.68 ± 0.03	0.73 ± 0.03	0.85 ± 0.04				
	HO-1	CG	1.00 ± 0.00^{a}	0.60 ± 0.06^{c}	0.70 ± 0.06^{bc}	0.83 ± 0.05^{b}	0.78 ± 0.04	<0.010	0.255	0.977
		EG	1.00 ± 0.00^{a}	0.68 ± 0.07^{c}	0.80 ± 0.05^{bc}	0.90 ± 0.05^{ab}	0.84 ± 0.03			
		NG	1.00 ± 0.00^{a}	0.64 ± 0.05^{b}	0.74 ± 0.07^{b}	0.89 ± 0.06^{b}	0.82 ± 0.04			
		Total	1.00 ± 0.00	0.64 ± 0.03	0.74 ± 0.03	0.87 ± 0.03				
	Keap1	CG	1.00 ± 0.00^{c}	1.59 ± 0.13^{a}	1.31 ± 0.04^{Ab}	1.21 ± 0.09^{bc}	1.28 ± 0.06	<0.010	0.011	0.741
		EG	1.00 ± 0.00^{b}	1.37 ± 0.12^{a}	1.09 ± 0.05^{Bb}	1.03 ± 0.05^{b}	1.12 ± 0.04			
		NG	1.00 ± 0.00^{c}	1.44 ± 0.09^{a}	1.21 ± 0.06^{ABb}	1.08 ± 0.07^{bc}	1.18 ± 0.05			
		Total	1.00 ± 0.00	1.47 ± 0.07	1.20 ± 0.04	1.11 ± 0.04				

Note: ^{abc} Different lowercase letters superscripted in the same row indicate significant differences between different treatments ($p < 0.05$). ^{AB} The superscript values in the same column with different uppercase letters indicate significant differences at different times ($p < 0.05$). The same letter or no letter indicates no significant difference ($p > 0.05$). There is no meaning between uppercase and lowercase letters.

Table 8. Effects of different treatments on intestinal immunoglobulin in transpcrted lambs.

Items		Groups	Time					p-Values		
			−2d	0d	7d	14d	Total	Time	Treatment	Interaction
Jejunum	IgA (mg/g)	CG	2.43 ± 0.22 [a]	1.37 ± 0.17 [b]	1.49 ± 0.11 [b]	1.71 ± 0.13 [Bb]	1750.19 ± 142.58	<0.010	0.011	0.567
		EG	2.40 ± 0.24	1.91 ± 0.20	2.15 ± 0.25	2.41 ± 0.16 [A]	2217.40 ± 110.16			
		NG	2.35 ± 0.25 [a]	1.34 ± 0.23 [b]	1.78 ± 0.20 [ab]	2.00 ± 0.13 [ABab]	1866.17 ± 140.95			
		Total	2393.11 ± 119.24	1539.97 ± 136.78	1805.70 ± 136.18	2039.57 ± 122.67				
	IgG (mg/g)	CG	703.33 ± 52.35 [a]	489.48 ± 27.65 [c]	538.86 ± 33.17 [Bbc]	631.33 ± 36.03 [ab]	590.75 ± 29.80	<0.010	0.035	0.613
		EG	697.38 ± 55.10	554.22 ± 35.75	705.96 ± 34.92 [A]	736.46 ± 44.44	673.50 ± 97.47			
		NG	709.14 ± 45.48 [a]	521.16 ± 38.85 [b]	636.52 ± 33.27 [ABab]	681.68 ± 56.48[a]	637.12 ± 99.60			
		Total	703.28 ± 25.63	521.62 ± 19.58	627.11 ± 29.54	683.16 ± 27.83				
	IgM (mg/g)	CG	18.70 ± 3.5	10.45 ± 2.52	12.40 ± 2.40	14.76 ± 3.31	14.08 ± 1.57	0.129	0.075	0.903
		EG	18.48 ± 3.26	12.61 ± 1.95	14.98 ± 2.28	17.25 ± 3.81	19.07 ± 1.42			
		NG	18.53 ± 3.41	16.62 ± 2.67	18.93 ± 3.07	22.18 ± 2.93	15.83 ± 1.42			
		Total	18.57 ± 1.70	13.23 ± 1.50	15.44 ± 4.84	18.07 ± 2.01				
	sIgA (μg/g)	CG	342.43 ± 21.75 [a]	167.20 ± 8.63 [Bc]	167.88 ± 14.77 [Bc]	290.37 ± 14.18 [Bb]	214.97 ± 24.05	<0.010	<0.010	0.558
		EG	358.40 ± 28.81 [a]	258.26 ± 16.80 [Ab]	330.23 ± 26.01 [Aab]	378.44 ± 29.20 [Aa]	331.33 ± 17.56			
		NG	350.60 ± 23.22[a]	197.13 ± 18.47 [Bc]	267.51 ± 15.26 [Ab]	300.48 ± 20.62 [Aab]	278.93 ± 18.77			
		Total	350.47 ± 12.60	207.53 ± 15.42	255.21 ± 25.55	323.10 ± 17.80				
	IgA (mg/g)	CG	2.43 ± 0.19 [a]	1.52 ± 0.13 [b]	1.67 ± 0.17 [b]	1.94 ± 0.18 [ab]	1.89 ± 0.13	<0.010	0.116	0.853
		EG	2.35 ± 0.21	1.88 ± 0.21	2.15 ± 0.22	2.42 ± 0.26	2.20 ± 0.11			
		NG	2.31 ± 0.25 [a]	1.50 ± 0.10 [b]	1.76 ± 0.10 [ab]	2.10 ± 0.17 [a]	1.92 ± 0.12			
		Total	2.36 ± 0.11	1.63 ± 0.10	1.86 ± 0.11	2.15 ± 0.13				

Table 8. *Cont.*

Items		Groups	Time					*p*-Values		
			−2d	0d	7d	14d	Total	Time	Treatment	Interaction
Colon	IgG (mg/g)	CG	747.80 ± 88.62 ^{a}	510.64 ± 40.44 ^{b}	552.42 ± 32.93 ^{b}	665.01 ± 39.02 ^{ab}	618.97 ± 36.64	<0.010	0.551	0.996
		EG	743.44 ± 71.48	560.51 ± 44.91	606.16 ± 46.22	732.72 ± 53.97	660.71 ± 33.51			
		NG	747.18 ± 70.10 ^{a}	527.66 ± 38.62 ^{b}	581.19 ± 38.43 ^{ab}	706.58 ± 45.80 ^{a}	640.65 ± 34.36			
		Total	746.14 ± 38.60	532.94 ± 21.96	579.92 ± 21.25	701.44 ± 25.33				
	IgM(mg/g)	CG	17.69 ± 2.87	12.29 ± 3.17	14.81 ± 2.46	16.42 ± 1.97	15.30 ± 1.29	0.366	0.174	0.972
		EG	17.70 ± 3.22	13.81 ± 2.83	16.02 ± 2.33	18.36 ± 2.59	19.07 ± 1.34			
		NG	17.84 ± 2.54	17.83 ± 2.58	19.54 ± 3.03	21.08 ± 3.72	16.47 ± 1.29			
		Total	17.74 ± 1.45	14.64 ± 1.66	16.79 ± 1.49	18.62 ± 1.58				
	sIgA (μg/g)	CG	355.95 ± 21.74 ^{a}	188.01 ± 18.03 ^{Bc}	233.59 ± 18.733 ^{Bbc}	293.41 ± 17.34 ^{Bb}	267.74 ± 20.70	<0.010	<0.010	0.138
		EG	388.19 ± 33.88	320.89 ± 21.49 ^{A}	395.63 ± 26.92 ^{A}	454.04 ± 29.97 ^{A}	389.69 ± 18.55			
		NG	365.47 ± 27.14 ^{a}	233.09 ± 16.86 ^{Bc}	302.30 ± 19.00 ^{Bbc}	337.17 ± 24.88 ^{Bb}	309.51 ± 17.70			
		Total	369.87 ± 14.81	247.33 ± 21.67	310.51 ± 25.90	361.54 ± 26.77				

Note: ^{abc} Different lowercase letters superscripted in the same row indicate significant differences between different treatments ($p < 0.05$). ^{AB} The superscript values in the same column with different uppercase letters indicate significant differences at different times ($p < 0.05$). The same letter or no letter indicates no significant difference ($p > 0.05$). There is no meaning between uppercase and lowercase letters.

3.2.2. Effects of Different Treatments on mRNA Expression of Inflammatory Factors in Jejunum and Colon of Transported Lambs

As shown in Table 9, by transportation, the mRNA expressions of IL-1β and IL-2 in the jejunum of each group, as well as the mRNA expressions of IL-12 in the jejunum and IL-1β, IL-2, and IL-12 in the colon of the CG and NG increased ($p < 0.01$). The mRNA expressions of IL-1β in the jejunum of the EG on day 14 and IL-2 in the colon of the EG on day 0 were lower than those of the CG and NG ($p < 0.05$). On day 14, the mRNA expressions of IL-2 in the jejunum of the EG and NG ($p < 0.05$), as well as the mRNA expression of IL-12 in the jejunum of the EG and NG was lower than those of the CG ($p < 0.01$).

Table 9. Effects of different treatments on mRNA expression of inflammatory factors in the jejunum and colon of transported lambs.

Items		Groups	Time					*p*-Values		
			−2d	0d	7d	14d	Total	Time	Treatment	Interaction
Jejunum	IL-1β	CG	1.00 ± 0.00 ^{b}	1.30 ± 0.08 ^{a}	1.26 ± 0.09 ^{a}	1.19 ± 0.03 ^{Aa}	1.27 ± 0.07			
		EG	1.00 ± 0.00 ^{b}	1.22 ± 0.09 ^{a}	1.13 ± 0.04 ^{ab}	0.98 ± 0.06 ^{Bb}	1.19 ± 0.07	<0.010	0.030	0.678
		NG	1.00 ± 0.00 ^{c}	1.27 ± 0.07 ^{a}	1.22 ± 0.08 ^{ab}	1.12 ± 0.03 ^{Abc}	1.20 ± 0.06			
		Total	1.00 ± 0.00	1.46 ± 0.09	1.28 ± 0.07	1.15 ± 0.06				
	IL-2	CG	1.00 ± 0.00 ^{b}	1.58 ± 0.18 ^{a}	1.41 ± 0.04 ^{a}	1.36 ± 0.04 ^{Aa}	1.34 ± 0.06			
		EG	1.00 ± 0.00 ^{bc}	1.37 ± 0.08 ^{a}	1.26 ± 0.09 ^{ab}	1.07 ± 0.08 ^{Bbc}	1.18 ± 0.05	<0.010	0.013	0.612
		NG	1.00 ± 0.00 ^{b}	1.43 ± 0.06 ^{a}	1.36 ± 0.07 ^{a}	1.15 ± 0.04 ^{Bb}	1.24 ± 0.04			
		Total	1.00 ± 0.00	1.46 ± 0.07	1.34 ± 0.04	1.19 ± 0.04				
	IL-12	CG	1.00 ± 0.00 ^{c}	1.57 ± 0.07 ^{a}	1.49 ± 0.08 ^{a}	1.25 ± 0.07 ^{Ab}	1.33 ± 0.05			
		EG	1.00 ± 0.00 ^{b}	1.38 ± 0.11 ^{a}	1.25 ± 0.05 ^{a}	0.87 ± 0.07 ^{Bb}	1.13 ± 0.05	<0.010	<0.010	0.286
		NG	1.00 ± 0.00 ^{b}	1.41 ± 0.10 ^{a}	1.37 ± 0.10 ^{a}	1.01 ± 0.09 ^{Bb}	1.20 ± 0.06			
		Total	1.00 ± 0.00	1.45 ± 0.05	1.37 ± 0.05	1.05 ± 0.06				
Colon	IL-1β	CG	1.00 ± 0.00 ^{b}	1.47 ± 0.16 ^{a}	1.21 ± 0.12 ^{ab}	1.08 ± 0.11 ^{ab}	1.19 ± 0.04			
		EG	1.00 ± 0.00	1.44 ± 0.16	1.39 ± 0.13	1.26 ± 0.15	1.08 ± 0.03	<0.010	0.531	0.957
		NG	1.00 ± 0.00 ^{b}	1.46 ± 0.15 ^{a}	1.23 ± 0.11 ^{ab}	1.12 ± 0.05 ^{ab}	1.15 ± 0.03			
		Total	1.00 ± 0.00	1.27 ± 0.04	1.20 ± 0.04	1.10 ± 0.03				
	IL-2	CG	1.00 ± 0.00 ^{c}	1.50 ± 0.05 ^{Aa}	1.40 ± 0.07 ^{ab}	1.19 ± 0.12 ^{bc}	1.27 ± 0.05			
		EG	1.00 ± 0.00	1.23 ± 0.03 ^{B}	1.05 ± 0.16	1.06 ± 0.05	1.09 ± 0.04	<0.010	0.023	0.449
		NG	1.00 ± 0.00 ^{b}	1.46 ± 0.08 ^{Aa}	1.16 ± 0.21 ^{ab}	1.03 ± 0.03 ^{b}	1.16 ± 0.07			
		Total	1.00 ± 0.00	1.40 ± 0.04	1.20 ± 0.09	1.09 ± 0.05				
	IL-12	CG	1.00 ± 0.00 ^{b}	1.44 ± 0.11 ^{a}	1.28 ± 0.12 ^{b}	1.14 ± 0.09 ^{ab}	1.22 ± 0.06			
		EG	1.00 ± 0.00	1.25 ± 0.11	1.11 ± 0.12	1.03 ± 0.07	1.10 ± 0.05	<0.010	0.169	0.960
		NG	1.00 ± 0.00 ^{b}	1.34 ± 0.10 ^{a}	1.19 ± 0.11 ^{ab}	1.07 ± 0.10 ^{ab}	1.15 ± 0.05			
		Total	1.00 ± 0.00	1.35 ± 0.06	1.19 ± 0.07	1.08 ± 0.05				

Note: ^{abc} Different lowercase letters superscripted in the same row indicate significant differences between different treatments ($p < 0.05$). ^{AB} The superscript values in the same column with different uppercase letters indicate significant differences at different times ($p < 0.05$). The same letter or no letter indicates no significant difference ($p > 0.05$). There is no meaning between uppercase and lowercase letters.

On day 7, the mRNA expressions of IL-1β and IL-2 in the jejunum of the EG, IL-1β, and IL-12 in the colon of the CG and NG, and IL-2 in the colon of the NG had no significant changes compared with those on day −2 ($p > 0.05$). On day 14, there was no significant difference in the mRNA expressions of IL-1β, IL-2, and IL-12 in the jejunum of the NG, IL-12 in the jejunum of the EG, and IL-2 in the colon of the CG compared with those on day −2 ($p > 0.05$).

3.3. *Effects of Different Treatments on Apoptosis Factors in the Jejunum and Colon of Transported Lambs*

As shown in Table 10, by transportation, the mRNA expressions of Bax in the jejunum and colon and Caspase3 in the jejunum of each group increased ($p < 0.01$), while the mRNA expression of Bcl-2 decreased ($p < 0.01$). On day 0, the mRNA expressions of Bax in the jejunum and colon of the EG were lower than that of the CG ($p < 0.01$), and Caspase3 in the jejunum of the EG was lower than those of the CG and NG ($p < 0.01$). On day 7, the mRNA expressions of Bax in the jejunum and colon of the EG and Caspase3 in the colon of the EG were lower than those of the CG ($p < 0.01$). On day 14, the mRNA expression of Bcl-2 in the jejunum of the EG was higher than that of the CG ($p < 0.05$).

Table 10. The effect of different treatments on apoptosis factors in the jejunum and colon of transported lambs.

Items		Groups	Time					*p*-Values		
			−2d	0d	7d	14d	Total	Time	Treatment	Interaction
Jejunum	Bax	CG	1.00 ± 0.00 ^{c}	1.56 ± 0.06 ^{Aa}	1.36 ± 0.06 ^{Ab}	1.25 ± 0.04 ^{b}	1.29 ± 0.05	<0.010	<0.010	0.275
		EG	1.00 ± 0.00 ^{b}	1.25 ± 0.05 ^{Ba}	1.10 ± 0.05 ^{Bab}	1.02 ± 0.08 ^{b}	1.09 ± 0.03			
		NG	1.00 ± 0.00 ^{c}	1.41 ± 0.10 ^{ABa}	1.26 ± 0.05 ^{ABab}	1.11 ± 0.11 ^{bc}	1.20 ± 0.05			
		Total	1.00 ± 0.00	1.41 ± 0.05	1.24 ± 0.04	1.13 ± 0.05				
	Bcl-2	CG	1.00 ± 0.00 ^{a}	0.66 ± 0.03 ^{c}	0.72 ± 0.04 ^{c}	0.89 ± 0.04 ^{Bb}	0.82 ± 0.03	<0.010	0.012	0.360
		EG	1.00 ± 0.00 ^{a}	0.72 ± 0.04 ^{b}	0.80 ± 0.05 ^{b}	1.08 ± 0.06 ^{Aa}	0.90 ± 0.04			
		NG	1.00 ± 0.00 ^{a}	0.70 ± 0.04 ^{b}	0.77 ± 0.05 ^{b}	0.98 ± 0.04 ^{ABa}	0.86 ± 0.03			
		Total	1.00 ± 0.00	0.69 ± 0.02	0.76 ± 0.03	0.98 ± 0.03				
	Caspase3	CG	1.00 ± 0.00 ^{c}	1.57 ± 0.07 ^{Aa}	1.32 ± 0.14 ^{ab}	1.20 ± 0.13 ^{bc}	1.27 ± 0.06	<0.010	0.010	0.646
		EG	1.00 ± 0.00 ^{c}	1.26 ± 0.03 ^{Ba}	1.14 ± 0.07 ^{b}	1.00 ± 0.04 ^{c}	1.10 ± 0.03			
		NG	1.00 ± 0.00 ^{c}	1.46 ± 0.05 ^{Aa}	1.23 ± 0.11 ^{b}	1.14 ± 0.09 ^{bc}	1.21 ± 0.05			
		Total	1.00 ± 0.00	1.43 ± 0.04	1.23 ± 0.06	1.11 ± 0.05				
Colon	Bax	CG	1.00 ± 0.00 ^{c}	1.65 ± 0.07 ^{Aa}	1.35 ± 0.06 ^{Ab}	1.16 ± 0.15 ^{bc}	1.29 ± 0.06	<0.010	<0.010	0.231
		EG	1.00 ± 0.00 ^{bc}	1.32 ± 0.06 ^{Ba}	1.08 ± 0.07 ^{Bb}	0.88 ± 0.03 ^{c}	1.07 ± 0.04			
		NG	1.00 ± 0.00 ^{c}	1.50 ± 0.07 ^{ABa}	1.19 ± 0.05 ^{ABb}	0.98 ± 0.05 ^{c}	1.17 ± 0.05			
		Total	1.00 ± 0.00	1.49 ± 0.05	1.21 ± 0.04	1.00 ± 0.04				
	Bcl-2	CG	1.00 ± 0.00 ^{a}	0.76 ± 0.05 ^{b}	0.84 ± 0.08 ^{ab}	0.92 ± 0.06 ^{ab}	0.88 ± 0.03	<0.010	0.250	0.981
		EG	1.00 ± 0.00 ^{a}	0.82 ± 0.05 ^{b}	0.93 ± 0.05 ^{ab}	1.02 ± 0.07 ^{a}	0.94 ± 0.03			
		NG	1.00 ± 0.00 ^{a}	0.80 ± 0.03 ^{b}	0.89 ± 0.06 ^{ab}	0.98 ± 0.07 ^{a}	0.92 ± 0.03			
		Total	1.00 ± 0.00	0.80 ± 0.02	0.89 ± 0.03	0.98 ± 0.04				
	Caspase3	CG	1.00 ± 0.00 ^{b}	1.33 ± 0.13 ^{a}	1.26 ± 0.06 ^{Aa}	1.15 ± 0.08 ^{ab}	1.19 ± 0.05	<0.010	0.026	0.730
		EG	1.00 ± 0.00	1.15 ± 0.07	1.05 ± 0.06 ^{B}	1.00 ± 0.06	1.05 ± 0.03			
		NG	1.00 ± 0.00 ^{b}	1.31 ± 0.07 ^{a}	1.12 ± 0.07 ^{ABab}	1.09 ± 0.09 ^{b}	1.13 ± 0.04			
		Total	1.00 ± 0.00	1.26 ± 0.06	1.14 ± 0.04	1.08 ± 0.04				

Note: ^{abc} Different lowercase letters superscripted in the same row indicate significant differences between different treatments ($p < 0.05$). ^{AB} The superscript values in the same column with different uppercase letters indicate significant differences at different times ($p < 0.05$). The same letter or no letter indicates no significant difference ($p > 0.05$). There is no meaning between uppercase and lowercase letters.

On day 7, the mRNA expressions of Bax in the jejunum and colon of the EG, Bcl-2 in the colon of each group, and Caspase3 in the colon of the NG had no significant changes compared with those on day −2 ($p > 0.05$). On day 14, in the jejunum, there was no difference in the mRNA expressions of Bax in the NG, Bcl-2 in the EG and NG, and Caspase3 in the NG compared with those on day −2 ($p > 0.05$). In the colon, the mRNA expressions of Bax of the CG and NG and Caspase3 of the CG had no significant changes compared with those on day −2 ($p > 0.05$).

4. Discussion

Transportation is an important link in the development of the sheep industry; during transportation, the metabolic level increases to resist the damage caused by stress, which leads to excessive production of ROS in cells and oxidative damage to the body. The degree of oxidative damage can be determined by measuring the activity of the antioxidant enzymes and the level of MDA [17]. Earlier research showed that the activities of GSH Px, T-SOD, CAT, and T-AOC reduced significantly, and the MDA levels increased significantly after animals experienced heat stress [18,19]. The concentration of MDA reflects the degree of oxidative damage, which is one of the important indicators that reflect oxidative stress [20]. The decrease in T-AOC level indicates an imbalance in the production and clearance of free radicals in the body. The SOD, CAT, and GSH-Px are the first line of defense against antioxidant stress. In this experiment, by road transport, the levels of GSH-Px, CAT, SOD, and T-AOC of lambs decreased significantly, while the level of MDA increased significantly. The results showed that the antioxidant capacity decreased significantly, and oxidative damage increased after lambs were transported; road transport may cause oxidative stress in lambs. The levels of the antioxidant enzymes and MDA in sera and intestines, and the expressions of SOD and CAT in intestines showed consistent trends; road transport may inhibit the activity of the antioxidant enzyme in lambs by downregulating their relative expressions, which accelerated lipid peroxidation reactions in cells and tissues. Road transport led to large amounts of free radicals produced; SOD, CAT, and GSH-Px were consumed greatly to eliminate free radicals, which accelerated lipid peroxidation in the body, so the levels of GSH-Px, CAT, SOD, and T-AOC decreased, and MDA increased, potentially damaging intestinal structure and function [21]. Other research shows that moderate vitamins can inhibit the damage of oxidative stress and maintain the redox balance of the body [22,23]. In this experiment, using electrolytic multivitamins and neomycin treatments, the levels of SOD and T-AOC increased after lambs were transported, while the level of MDA decreased. The results showed that electrolytic multivitamins and neomycin could improve the antioxidant capacity of transported lambs. In addition, the levels of the antioxidant enzymes and MDA of the EG could recover to the levels before transportation. As expected, there was an increase in the levels of SOD and T-AOC in sera, SOD and T-AOC in the jejunum, and GSH-Px, SOD, and T-AOC in the colon in the CG on day 14 compared to those during transportation, but they did not recover to the levels before transportation. The result showed that road transport still affected the antioxidant capacity of lambs on day 14. Electrolytic multivitamins and neomycin could potentially improve the antioxidant capacity of lambs, and the effect of electrolytic multivitamins was better; the possible reason was that the vitamins and electrolyte ions in electrolytic multivitamins had a synergistic effect, which potentially enhanced the activity of the antioxidant enzymes and improved the antioxidant capacity.

The Nrf2/Keap1/HO-1 pathway plays a crucial role in maintaining the redox balance. It is important to resist oxidative stress. Under normal conditions, Nrf2 and Keap1 inhibit each other mutually; when the body experiences oxidative stress, Nrf2 dissociates from Keap1, which interacts with antioxidant response elements, activates downstream related antioxidant factor HO-1, and promotes the antioxidant effects of the antioxidant enzyme [24–26]. In this experiment, by road transport, the mRNA expressions of Nrf2 in the jejunum and colon decreased significantly, while the mRNA expression of Keap1 increased. It showed that the activation of the Nrf2 pathway in the jejunum and colon was inhibited after lambs were transported, which led to oxidative stress in the jejunum and colon of lambs. Other research found that vitamins could induce activation of the Nrf2 antioxidant pathway, which also prevented oxidative stress and inflammation [27,28]. In this experiment, by treatment of electrolytic multivitamins, the mRNA expressions of Nrf2 in the jejunum and colon increased, while Keap1 in the jejunum and colon decreased; it showed that electrolytic multivitamins were beneficial for activating the Nrf2 signaling pathway and inhibiting the oxidative stress potentially. In addition, it found that the expressions of Nrf2, HO-1, and Keap1 in the EG could recover to the levels before transportation. However, on day 14, the mRNA expressions of Nrf2 and HO-1 in the jejunum and colon of the CG did not recover to the levels before transportation, which indicated that the oxidative stress caused by road transport in the jejunum and colon of lambs still existed on day 14; electrolytic multivitamins had a better effect on potentially reducing antioxidant system damage in transported lambs, which could protect the normal physiological function of the intestines.

Immunoglobulin participates in humoral immunity, and it is often used to evaluate the strength of immune function, which is one of the important indicators that reflects the level of disease resistance. sIgA is the main immunoglobulin in the intestines. It interacts with intestinal microorganisms and enhances the immune and barrier functions of intestinal mucosa to maintain homeostasis of the body [29]. Research showed that the levels of IgA, IgG, and IgM in serum reduced significantly after cows experienced heat stress [30,31]. Weaning stress led to a decrease in the content of sIgA in the intestines of piglets, which reduced the ability to resist pathogenic microorganisms [32]. The results indicated that stress affected the immunity of animals. In this experiment, by road transport, the concentrations of IgA, IgG, and sIgA in the jejunum and colon of lambs decreased significantly. It indicated that the intestinal immune function of lambs was inhibited by road transport, and the mucosal barrier function of the intestines may be disrupted. The concentration of IgG decreased significantly after transportation, which may be related to the HPA axis, which was activated by road transport. A significant decrease was found in the concentrations of IgA and sIgA. It may be due to road transport causing damage to the intestinal mucosal barrier, which affects the metabolism and absorption of nutrients in the intestine, thereby inhibiting the secretion of IgA and sIgA. Other research showed that vitamins could participate in humoral immunity and promote immunoglobulin synthesis [33,34]; adding electrolytes to feed could effectively alleviate heat stress in cows [35]. In this experiment, the concentrations of IgA and IgG in the jejunum increased significantly after treatment with electrolytic multivitamins. This study also found that the concentration of IgA in the jejunum and sIgA in the jejunum and colon of the CG on day 14 did not recover to the levels before transportation. It showed that electrolytic multivitamins had a better effect on potentially enhancing the mucosal immune function of the jejunum and colon and promoting humoral immunity of lambs. It may be caused by electrolytes and vitamins in electrolytic multivitamins. sIgA is the first line of defense against external pathogens, which can eliminate pathogens through non-specific immunity [36]. The concentration of sIgA in the jejunum and colon increased significantly after adding electrolytic multivitamins and neomycin; the intestinal mucosal immunity of lambs improved potentially, it could prevent pathogenic bacteria from adhering to the intestinal tract to cause intestinal infection and reduce the incidence rate of lambs.

Under stress, the intestine is prone to becoming a target organ. Endotoxins, symbiotic bacteria, and pathogenic bacteria can enter the bloodstream through tight epithelial connections. It promotes the secretion of inflammatory factors and causes an imbalance between the pro-inflammatory and anti-inflammatory systems [37]. The abnormal expression of inflammatory factors is related to the immune function of the body. Large amounts of IL-1β could damage tissues and reduce the immunity of animals. IL-2 is produced by antigen-stimulated lymphocytes, which reflects the immune function of animals indirectly. IL-12 is a determinant of Th1 cellular immune response, which promotes the production of IFN-γ to exert antiviral effects. Wang et al. [38] found that the levels of IL-2 and IL-12 in rats increased significantly after they experienced heat stress for 7 days at 32 °C, heat stress-induced inflammatory response in rats. In this experiment, the mRNA expressions of IL-1β, IL-2, and IL-12 in the jejunum and colon of lambs increased significantly after transportation. It indicated that road transport caused an imbalance in the inflammatory systems of the lambs. The possible reason was that lambs were influenced by stress factors; the body increased the secretion of IL-1β, IL-2, and IL-12 to participate in immune responses and mediate inflammatory responses by upregulating their expressions, it could potentially enhance immune function and alleviate damage caused by road transport [39]. Other research showed that vitamins could regulate the inflammatory response by controlling the release of anti-inflammatory and pro-inflammatory factors, thereby reducing the level of inflammatory factors [40]. In this experiment, by treatments of electrolytic multivitamin and neomycin, the mRNA expressions of IL-2 and IL-12 in the jejunum decreased significantly; electrolytic multivitamin also inhibited the mRNA expressions of IL-1 β in the jejunum and IL-2 in the colon. With the extension of time, the mRNA expressions of IL-1β, IL-2, and IL-12 in the jejunum and colon of the EG recovered firstly to the levels before transportation. However, the mRNA expressions of IL-1β, IL-2, and IL-12 in the jejunum of the CG on day 14 did not recover to the levels before transportation. It indicated that electrolytic multivitamins and neomycin could effectively downregulate the expressions of IL-2 and IL-12 in the jejunum and colon, electrolytic multivitamins had a better effect on alleviating inflammatory response potentially, there still was serious inflammatory reaction in the jejunum of lambs possibly on day 14 if no treatments to lambs, it may be related to electrolytic multivitamin could potentially activate the Nrf2/Keap1/HO-1 signaling pathway.

Apoptosis is a necessary physiological phenomenon for maintaining tissue development and function in the body. However, cell apoptosis happens excessively and could cause serious damage to tissues and organs [41]. Histiocytes will experience cell disintegration and fragmentation when histiocyte occurs apoptosis; the collapsed cells exist in tissues and organs with the form of apoptotic bodies, which can induce Caspase-3 activation, trigger cascade reaction, and initiate the process of apoptosis [42]. Bcl-2 can exert anti-apoptotic effects through various pathways, such as inhibiting cytochrome C and apoptosis-inducing factors [43]. Activated Bax can antagonize Bcl-2 and prevent Bcl-2 from exerting anti-apoptotic effects [44]. Wei et al. [45] found that the protein expressions of Bax and Caspase 3 increased, while the expression of Bcl-2 decreased after mice experienced heat stress. In this experiment, by road transport, the mRNA expressions of Bax in the jejunum and colon, as well as the mRNA expression of Caspase3 in the jejunum of lambs increased significantly, while the mRNA expression of Bcl-2 decreased significantly. The results showed that road transport promoted the abnormal apoptosis process of the intestine, which aggravated the overall degree of damage to the intestine. Other research showed that vitamins could effectively promote the expression of Bax and inhibit the expressions of Caspase3 and Bcl-2 in cells, which slowed apoptosis [46,47]. In this experiment, electrolytic multivitamins inhibited the mRNA expressions of Bax and Caspase3 in the jejunum, and colon promoted the mRNA expression of Bcl-2 in the jejunum. In addition, the mRNA expressions of Bax, Bcl-2, and Caspase3 in the jejunum and colon of the EG recovered firstly to the levels before transportation. On day 14, the mRNA expressions of Bax and Bcl-2 in the jejunum of the CG did not recover to the levels before transportation. The results showed that electrolytic multivitamins had a better effect on enhancing the anti-apoptotic ability of cells. It could enhance intestinal resistance to damage and help maintain normal intestinal function.

5. Conclusions

Road transport led to a decrease in the antioxidant capacity and immunoglobulin levels of lambs, while the levels of inflammatory factors and apoptosis in the jejunum and colon of lambs increased. It also inhibited the activation of the Nrf2 pathway in the jejunum and colon and even exacerbated damage to the jejunum and colon. On day 14, the levels of the antioxidant capacity, immunity, and apoptosis of lambs did not recover to the levels before transportation. Electrolytic multivitamins had a better effect on potentially improving the antioxidant level of lambs and inhibiting the release of inflammatory factors. It also potentially reduced the expression of apoptotic factors and alleviated damage in the jejunum and colon.

Author Contributions: Conceptualization: C.X. and C.D.; formal analysis: C.C., X.Y. and Y.Z.; investigation: C.X., C.D., C.C., X.Y. and Y.Z.; writing—original draft: C.X. and C.D.; writing—review and editing: C.X.; project administration: Y.Z., Y.L. and Y.M.; funding acquisition: Y.Z., Y.L. and Y.M. All authors have read and agreed to the published version of the manuscript.

Funding: This research was supported by the National Sheep Industry Technology System for Fluff (CARS39-23) and the 2022 Basic Scientific Research Project of the Agricultural University of Hebei (KY2022029).

Institutional Review Board Statement: The experimental animal protocol of this study was approved by the Ethics Committee of Laboratory Animals of Hebei Agricultural University; the Ethics Committee's name was the Ethics Committee of Laboratory Animals of Hebei Agricultural University. The approval code was 2021023, and the approval date was 26 March 2021. The relevant methods used in this test were implemented in accordance with relevant guidelines and regulations.

Informed Consent Statement: Written informed consent has been obtained from the patient(s) to publish this paper.

Data Availability Statement: The data that support the findings of this study are available from the corresponding author upon reasonable request at any time.

Acknowledgments: We thank the Animal Science and Technology Laboratory of Hebei Agricultural University for providing technical support.

Conflicts of Interest: The authors declare no conflicts of interest.

References

1. Ishizaki, H.; Kariya, Y. Road transportation stress promptly increases bovine peripheral blood absolute NK cell counts and cortisol levels. *J. Vet. Med. Sci.* **2010**, *72*, 747–753. [CrossRef] [PubMed]
2. Stull, C.L.; Spier, S.J.; Aldridge, B.M.; Blanchard, M.; Stott, J.L. Immunological response to long-term transport stress in mature horses and effects of adaptogenic dietary supplementation as an immunomodulator. *Equine Vet. J.* **2004**, *36*, 583–589. [CrossRef] [PubMed]
3. Swallow, J.; Anderson, D.; Buckwell, A.C.; Harris, T.; Hawkins, P.; Kirkwood, J. Guidance on the transport of laboratory animals. *Lab. Anim.* **2005**, *39*, 1–39. [CrossRef]
4. Wei, H.K.; Xue, H.X.; Zhou, Z.X.; Peng, J. A carvacrol-thymol blend decreased intestinal oxidative stress and influenced selected microbes without changing the messenger RNA levels of tight junction proteins in jejunal mucosa of weaning piglets. *Animal* **2017**, *11*, 193–201. [CrossRef] [PubMed]
5. Niture, S.K.; Khatri, R.; Jaiswal, A.K. Regulation of Nrf2—An update. *Free. Radic. Biol. Med.* **2014**, *66*, 36–44. [CrossRef] [PubMed]
6. Zou, Y.; Wang, J.; Peng, J.; Wei, H. Oregano Essential Oil Induces SOD1 and GSH Expression through Nrf2 Activation and Alleviates Hydrogen Peroxide-Induced Oxidative Damage in IPEC-J2 Cells. *Oxidative Med. Cell. Longev.* **2016**, *2016*, 5987183. [CrossRef] [PubMed]
7. Li, F.; Shah, A.M.; Wang, Z.; Peng, Q.; Hu, R. Effects of Land Transport Stress on Variations in Ruminal Microbe Diversity and Immune Functions in Different Breeds of Cattle. *Animals* **2019**, *9*, 599. [CrossRef] [PubMed]
8. Gupta, D.; Ashutosh, M.; Kashyap, G.; Punetha, M.; Patel, B.; Para, I. Physiological response to transportation stress at different flocking densities in hot humid and winter seasons in goats and assessment of pretreatment of Vitamin C, Electrolyte and Jaggery. *Indian J. Anim. Res.* **2019**, *53*, 299–306. [CrossRef]
9. Panaretou, B.; Jones, G.W. Yeast models for amyloid disease. *Essays Biochem.* **2014**, *56*, 85–97.
10. Zhao, Y.; Zhao, R.; Yang, W.; Du, F.; Zhang, H. Establishment of liver and kidney injury model in SD rats induced by neomycin sulphate soluble powder. *J. Henan Agric. Univ.* **2019**, *42*, 100–103. (In Chinese)
11. Chirase, N.K.; Greene, L.W.; Purdy, C.W.; Loan, R.W.; Auvermann, B.W.; Parker, D.B.; Walborg, E.F.; Stevenson, D.E.; Xu, Y.; Klaunig, J.E. Effect of transport stress on respiratory disease, serum antioxidant status, and serum concentrations of lipid peroxidation biomarkers in beef cattle. *Am. J. Vet. Res.* **2004**, *65*, 860–864. [CrossRef] [PubMed]
12. Xia, C.; Duan, C.; Chen, C.; Yang, X.; Zhang, Y.; Liu, Y.; Ma, Y. Effects of electrolyte multivitamins and neomycin on immunity and intestinal barrier function in transported lambs. *Animals* **2024**, *14*, 177. [CrossRef]
13. Wang, B.L.; He, F.H. Management of cattle during transportation. *Rev. Monogr.* **2019**, *49*, 43–44. (In Chinese)
14. Junxiao, M. Therapeutic Effect of Neomycin Sulphate Soluble Powder in Piglets with Whitescour, and Its Safety Evaluation and Residues. Master's Thesis, Nanjing Agricutural University, Nanjing, China, 2023. (In Chinese).
15. Livak, K.J.; Schmittgen, T.D. Analysis of relative gene expression data using real-time quantitative PCR and the 2(-Delta Delta C(T)) Method. *Methods* **2001**, *25*, 402–408. [CrossRef] [PubMed]
16. Wang, K.; Zhang, H.; Han, Q.; Lan, J.; Chen, G.; Cao, G.; Yang, C. Effects of astragalus and ginseng polysaccharides on growth performance, immune function and intestinal barrier in weaned piglets challenged with lipopolysaccharide. *J. Anim. Physiol. Anim. Nutr.* **2020**, *104*, 1096–1105. [CrossRef] [PubMed]
17. Zhou, Y.; Hu, X.; Zhong, S.; Yu, W.; Wang, J.; Zhu, W.; Yang, T.; Zhao, G.; Jiang, Y.; Li, Y. Effects of Continuous LPS Induction on Oxidative Stress and Liver Injury in Weaned Piglets. *Vet. Sci.* **2022**, *10*, 22. [CrossRef] [PubMed]
18. Zhang, C.; Chen, K.; Zhao, X.; Geng, Z. Protective effects of resveratrol against high ambient temperature-induced spleen dysplasia in broilers through modulating splenic redox status and apoptosis. *J. Sci. Food Agric.* **2018**, *98*, 5409–5417. [CrossRef]
19. Shi, L.; Xu, Y.; Mao, C.; Wang, Z.; Guo, S.; Jin, X.; Yan, S.; Shi, B. Effects of heat stress on antioxidant status and immune function and expression of related genes in lambs. *Int. J. Biometeorol.* **2020**, *64*, 1–12. [CrossRef]
20. Cecerska-Heryć, E.; Surowska, O.; Heryć, R.; Serwin, N.; Napiontek-Balińska, S.; Dołęgowska, B. Are antioxidant enzymes essential markers in the diagnosis and monitoring of cancer patients—A review. *Clin. Biochem.* **2021**, *93*, 1–8. [CrossRef]
21. Song, Z.; Tong, G.; Xiao, K.; Jiao, L.F.; Ke, Y.; Hu, C. L-cysteine protects intestinal integrity, attenuates intestinal inflammation and oxidant stress, and modulates NF-kappaB and Nrf2 pathways in weaned piglets after LPS challenge. *Innate Immun.* **2016**, *22*, 152–161. [CrossRef]
22. Blaner, W.S.; Shmarakov, I.O.; Traber, M.G. Vitamin A and Vitamin E: Will the Real Antioxidant Please Stand Up? *Annu. Rev. Nutr.* **2021**, *41*, 105–131. [CrossRef] [PubMed]
23. Zhang, R.; Zheng, Y.; Hu, F.; Meng, X.; Lv, B.; Lao, K. Effect of (m)VD-hemopressin against Aβ1-42-induced oxidative stress and apoptosis in mouse hippocampal neurons. *Peptides* **2020**, *124*, 170185. [CrossRef]
24. Yi, G.; Ding, J.; Zhao, F.; Liu, X. Effect of soybean peptides against hydrogen peroxide induced oxidative stress in HepG2 cells via Nrf2 signaling. *Food Funct.* **2020**, *11*, 2725–2737. [CrossRef] [PubMed]
25. Mi, S.S.; Hye, Y.J.; Hwan, K.S. Role of the Nrf2-ARE pathway in liver diseases. *Oxidative Med. Cell. Longev.* **2013**, *2013*, 763257.
26. Wei, T.; Jiang, Y.-F.; Murugavel, P.; Mamadou, D. Role of Nrf2 in chronic liver disease. *World J. Gastroenterol.* **2014**, *20*, 13079–13087.
27. Pazdro, R.; Burgess, J.R. The role of vitamin E and oxidative stress in diabetes complications. *Mech. Ageing Dev.* **2010**, *131*, 276–286. [CrossRef] [PubMed]

28. Teixeira, T.M.; da Costa, D.C.; Resende, A.C.; Soulage, C.O.; Bezerra, F.F.; Daleprane, J.B. Activation of Nrf2-Antioxidant Signaling by 1,25-Dihydroxycholecalciferol Prevents Leptin-Induced Oxidative Stress and Inflammation in Human Endothelial Cells. *J. Nutr.* **2017**, *147*, 506–513. [CrossRef]
29. Chairatana, P.; Nolan, E.M. Defensins, lectins, mucins, and secretory immunoglobulin A: Microbe-binding biomolecules that contribute to mucosal immunity in the human gut. *Crit. Rev. Biochem. Mol.* **2017**, *52*, 45–56. [CrossRef]
30. Soroush, S.; Shahryar, K.; Ali, M.G.; Giovanna, C.M.; Mariangela, C. Heat stress abatement during the postpartum period: Effects on whole lactation milk yield, indicators of metabolic status, inflammatory cytokines, and biomarkers of the oxidative stress. *J. Anim. Sci.* **2019**, *97*, 122–132.
31. Schmautz, C.; Hillreiner, M.; Ballweg, I.; Pfaffl, M.W.; Kliem, H. Stimulated enrichment of Clostridium difficile specific IgA in mature cow's milk. *PLoS ONE* **2018**, *13*, e0195275. [CrossRef]
32. Franziska, R.; Eva, B.; Meike, E.; Rainer, M. Effect of different carbohydrates on in vitro fermentation activity and bacterial numbers of porcine inocula under osmotic stress conditions. *Arch. Anim. Nutr.* **2011**, *65*, 445–459.
33. Xu, H.; Zhang, Q.; Wang, L.; Zhang, C.; Li, Y.; Zhang, Y. Effects of 25-Hydroxyvitamin D3 and Oral Calcium Bolus on Lactation Performance, Ca Homeostasis, and Health of Multiparous Dairy Cows. *Animals* **2021**, *11*, 1576. [CrossRef]
34. Bivona, J.J.; Patel, S.; Vajdy, M. Induction of cellular and molecular Immunomodulatory pathways by vitamin E and vitamin C. *Expert Opin. Biol. Ther.* **2017**, *17*, 1539–1551. [CrossRef] [PubMed]
35. Li, M.; Li, D.; Tong, X.; Nan, X.; Ding, D.; Xu, B.; Wang, G. Nutritional strategies for alleviating the detrimental effects of heat stress in dairy cows: A review. *Int. J. Biometeorol.* **2019**, *63*, 1283–1302.
36. Corthesy, B. Multi-faceted functions of secretory IgA at mucosal surfaces. *Front. Immunol.* **2013**, *4*, 185. [CrossRef] [PubMed]
37. IlBin, K.; JaeHon, L.; SeonCheol, P. The Relationship between Stress, Inflammation, and Depression. *Biomedicines* **2022**, *10*, 1929.
38. Wang, L.; Liu, F.; Luo, Y.; Zhu, L.; Li, G. Effect of acute heat stress on adrenocorticotropic hormone, cortisol, interleukin-2, interleukin-12 and apoptosis gene expression in rats. *Biomed. Rep.* **2015**, *3*, 425–429. [CrossRef]
39. Rajendran, P.; Chen, Y.F.; Chen, Y.F.; Chung, L.C.; Tamilselvi, S.; Shen, C.Y.; Day, C.H.; Chen, R.J.; Viswanadha, V.P.; Kuo, W.W.; et al. The multifaceted link between inflammation and human diseases. *J. Cell. Physiol.* **2018**, *233*, 6458–6471. [CrossRef]
40. Chen, D.D.; Zhu, X.P. Vitamin A supplementation in the prevention and treatment of bronchopulmonary dysplasia in premature infants: A meta-analysis based on 6 randomized control trial. *Chin. J. Evid. Based Pediatr.* **2017**, *12*, 429–433.
41. D'Arcy, M.S. Cell death: A review of the major forms of apoptosis, necrosis and autophagy. *Cell Biol. Int.* **2019**, *43*, 582–592. [CrossRef]
42. Ke, H.; Wang, X.; Zhou, Z.; Ai, W.; Wu, Z.; Zhang, Y. Effect of Weimaining on Apoptosis and Caspase-3 Expression in a Breast Cancer Mouse Model. *J. Ethnopharmacol.* **2020**, *264*, 113363. [CrossRef] [PubMed]
43. Hockenbery, D.; Nuñez, G.; Milliman, C.; Schreiber, R.D.; Korsmeyer, S.J. Bcl-2 is an inner mitochondrial membrane protein that blocks programmed cell death. *Nature* **1990**, *348*, 334–336. [CrossRef] [PubMed]
44. Moldoveanu, T.; Czabotar, P.E. BAX, BAK, and BOK: A Coming of Age for the BCL-2 Family Effector Proteins. *Cold Spring Harb. Perspect. Biol.* **2020**, *12*, a036319. [CrossRef] [PubMed]
45. Liu, W.; Leng, J.; Hou, J.G.; Jiang, S.; Wang, Z.; Liu, Z.; Gong, X.J.; Chen, C.; Wang, Y.P.; Li, W. Saponins derived from the stems and leaves of Panax ginseng attenuate scrotal heat-induced spermatogenic damage via inhibiting the MAPK mediated oxidative stress and apoptosis in mice. *Phytother. Res. PTR* **2020**, *35*, 311–323. [CrossRef]
46. Lee, J.; Park, S.H. Tumor-suppressive activity of 1,25-dihydroxyvitamin D3 against kidney cancer cells via up-regulation of FOXO3. *Biosci. Biotechnol. Biochem.* **2016**, *80*, 1947–1953. [CrossRef] [PubMed]
47. Erdemli, M.E.; Zayman, E.; Erdemli, Z.; Gul, M.; Gul, S.; Gozukara, B.H. Protective effects of melatonin and vitamin E in acetamiprid-induced nephrotoxicity. *Environ. Sci. Pollut. Res. Int.* **2020**, *27*, 9202–9213. [CrossRef]

animals

MDPI

Article

Movements after Captive Bolt Stunning in Cattle and Possible Animal- and Process-Related Impact Factors—A Field Study

Anika Lücking [1,2,*], Helen Louton [2], Martin von Wenzlawowicz [1], Michael Erhard [3] and Karen von Holleben [1]

1 bsi Schwarzenbek—Training and Consultancy Institute for Animal Welfare at Transport and Slaughter, Grabauer Str 27A, 21493 Schwarzenbek, Germany; mvw@bsi-schwarzenbek.de (M.v.W.); info@bsi-schwarzenbek.de (K.v.H.)

2 Animal Health and Animal Welfare, Faculty of Agricultural and Environmental Sciences, University of Rostock, Justus-von-Liebig-Weg 6b, 18059 Rostock, Germany; helen.louton@uni-rostock.de

3 Animal Welfare, Animal Behavior, Animal Hygiene and Animal Husbandry, Department of Veterinary Sciences, Faculty of Veterinary Medicine, LMU Munich, 80539 Munich, Germany; m.erhard@tierhyg.vetmed.uni-muenchen.de

* Correspondence: al@bsi-schwarzenbek.de

Simple Summary: Regarding animal welfare during cattle slaughter, two issues emerge: firstly, consumers are becoming steadily alienated from meat production, and secondly, some media-released slaughterhouse footage is causing dissatisfaction among a public increasingly focused on animal welfare. Such footage often reveals cattle kicking during shackling and exsanguination, described by commentators as indicative of insufficient stunning effectiveness. Consequently, this undermines confidence in both the meat industry and the competence of supervising authorities. At slaughter, cattle movements after stunning affect occupational safety and often lead to a prolonged interval until bleeding occurs. The objective of this study was to comprehensively describe and analyse these movements in cattle (*Bos taurus*) in relation to stunning effectiveness, as well as to identify influencing factors. The results show that movements occurred in most cattle after captive bolt stunning. However, none of the movements observed were related to stunning effectiveness. Breed type and sex category, as well as the type and design of the captive bolt stunner used, influenced the movements of cattle after stunning. The results of this investigation underscore the importance of applying reliable indicators to assess stunning effectiveness.

Abstract: Movements in cattle after captive bolt stunning cause problems in the slaughter process and lead to uncertainties in assessing stunning effectiveness. The objective of this study was to categorize and quantify these movements and determine animal- and process-related impact factors, as well as connections to stunning effectiveness and shooting position. In total 2911 cows, heifers, and bulls (dairy, beef, and crossbreeds) were examined (mean age 3.02 years). Movements from landing until at least four minutes after sticking were recorded by action cams (Apeman® A100). Nine movement categories were defined ("kicking hind limb", "twitching", "bending and stretching hind limb", "lifting and bending forelimb", "body arching laterally", "body arching ventrally", and "arching backwards"). According to the movement severity, a score was assigned to each category. The scores were summed, either for certain process intervals, e.g., LANDING (ejection from the stunning box), HOISTING, or STICKING, or for the total time between LANDING and end of the FOURTH MINUTE OF BLEEDING (sum score). Statistical analysis (ANOVA) was performed on the scores. Only 6.6% of cattle showed no movement. Most movements occurred during STICKING and FIRST MINUTE OF BLEEDING, occurring rarely up to 8 min after sticking. While cows moved most at LANDING, bulls and heifers moved more if all process intervals were considered. The sum score was highest in German Angus, Charolais, and Limousin and lowest in Brown Swiss and Simmental. The score at LANDING was highest in German Angus and Black Holstein. The use of pneumatic stunners and an increase in bolt-exit length significantly reduced movements. No impact of stunning effectiveness on movements was found, but only 19 cattle showed reduced effectiveness.

Citation: Lücking, A.; Louton, H.; von Wenzlawowicz, M.; Erhard, M.; von Holleben, K. Movements after Captive Bolt Stunning in Cattle and Possible Animal- and Process-Related Impact Factors—A Field Study. *Animals* **2024**, *14*, 1112. https://doi.org/10.3390/ani14071112

Academic Editors: Temple Grandin and Elbert Lambooij

Received: 26 January 2024
Revised: 19 March 2024
Accepted: 2 April 2024
Published: 4 April 2024

Keywords: animal welfare; captive bolt; cattle; movements; slaughter; stunning effectiveness; unconsciousness

1. Introduction

In the European Union (EU), most cattle are slaughtered after captive bolt stunning using pneumatic or cartridge-driven devices [1]. After stunning, the collapsed cattle are released from the restraining box onto a landing grid, where one hind leg is shackled, and the cattle are hoisted for sticking (exsanguination). In Germany, bleeding is usually performed via chest stick after a skin incision and changing of the knife.

According to Regulation (EC) No. 1099/2009 [2] on the protection of animals at the time of killing, restraining equipment and facilities should be designed and used to optimise the application of the stunning method, prevent injury or contusions to the animals, and minimize struggling. If cattle are stunned using pneumatic captive bolt devices, restraining boxes must restrict both the lateral and vertical movement of the animal's head during stunning. The regulation requires effective stunning and comprehensive monitoring of stunning effectiveness to prevent the animal from being conscious during slaughter and subsequent exsanguination, experiencing pain, fear, and anxiety before finally dying due to blood loss. To ensure good stunning effectiveness, restraining, stunning, and bleeding must be carried out by persons holding a certificate of competence [2]. Each stunning method should adhere to defined key parameters, and the equipment utilized must be well maintained to ensure the effective stunning of all animals [3]. For penetrative captive bolt stunning, these key parameters include the position and direction of the shot, appropriate velocity, exit length and diameter of the bolt according to the animal's size and species, and the maximum stun-to-stick/kill interval. In Germany, the Animal Welfare at Slaughter Ordinance specifies a maximum interval between stunning and sticking of 60 s for cattle [4].

Before 2001, the pithing rod was regularly used in cattle after captive bolt stunning, both to meet the time limit for bleeding and to reduce the reflex kicking movements. Pithing and, thus, mechanical destruction of the brain and spinal cord prevented the animal from regaining consciousness during exsanguination. By reducing post-stun movements at the same time, pithing also led to a lower risk of injury for staff at hoisting and sticking. Due to the increasing incidence of bovine spongiform encephalopathy (BSE) in Europe in the mid-1990s, from 1 January 2001 onwards the European Commission banned the use of pithing rods in cattle (*Bos taurus*), sheep (*Ovis gmelina aries*), and goats (*Capra aegagrus hircus*) slaughtered for human consumption [5]. Following this ban, convulsions and reflex-like body movements significantly increased in cattle after captive bolt stunning [6]. Furthermore, more animals regained consciousness during bleeding [7–11]. This was mostly a result of deficiencies regarding the shot placement, the stunning equipment used, and/or the strength of the cartridge used [6]. Both the increased occurrence of movements and signs of regaining consciousness during bleeding were likely to be due to the lack of pithing. However, due to the increasing use of modern stun boxes with tight head restraints and the development of more powerful captive bolt devices (especially pneumatically powered devices), stunning effectiveness at European cattle abattoirs has improved significantly in recent years [1,12]. This also applies to other bovines [13].

Previous studies indicate that following the ban on pithing, more than half of all captive bolt-stunned cattle show movements during hoisting, sticking, and bleeding [14,15]. However, movements do not always occur in a consistent pattern. Martin et al. [16] already found that Holstein cattle showed more kicking movements than other breeds, especially when stunned using a device with a relatively longer exit length of the bolt. In studies by Terlouw et al. [17], all cattle showed at least one movement after stunning, and post-stun movements by cattle with the spinal cord severed after stunning did not differ from those by cattle on which this manipulation was not performed. Clonic seizures such as reflex-like paddling of the hind limb are signs of a correctly performed bolt stun in cattle [17,18]. This is caused by the lack of inhibition of medullary or spinal reflexes due to the trauma-induced failure of higher-level centres in the brain [19] and also explains the occurrence of the above-mentioned convulsions during a zero-line EEG [20,21].

Nevertheless, there are currently two problems in practice since pithing has been banned. While the sometimes-strong excitatory movements often impede quick and safe shackling and sticking [22], leading to an extended stun-to-stick interval, there is also inherent uncertainty in the assessment of stunning effectiveness regarding movements. This uncertainty is heightened by the growing prevalence of video surveillance in abattoirs, coupled with public reactions to unauthorized video footage entering a society that is increasingly concerned about animal welfare. Consequently, the attention of animal welfare activists and some veterinarians is repeatedly drawn to the sometimes very impressive movements, which are then interpreted as a sign of regained consciousness. In the literature, there are only a few detailed descriptions of these post-stun movements and the possible impact factors, further contributing to existing uncertainty.

The aim of this publication is to record (1) the occurrence of movements after captive bolt stunning in cattle and to describe them in detail and to identify (2) animal-related and (3) process-related impact factors, as well as (4) a possible connection to reduced stunning effectiveness. The process-related factors included in the analysis also cover the various key parameters of the bolt gun. We hypothesised that movements would regularly occur in well-stunned cattle and that cows, especially the Black Holstein breed, would show the most movements. Furthermore, we hypothesised an association between the velocity of the bolt and the amount of movement. This work aims to contribute to the correct assessment of stunning effectiveness in cattle following captive bolt stunning.

2. Materials and Methods

2.1. Experimental Procedure

The investigation took place between June 2020 and April 2021 at five different German abattoirs (A–E), on two to four days each during routine slaughter. On each examination day, approximately 200 cattle were continuously observed by two pre-trained veterinarian investigators and filmed with the help of three to four action cameras, starting at stunning and lasting for at least four minutes after sticking. One camera was mounted above the restraining device, focussed on the animal's head and body, to evaluate the effectiveness of the shot and detect second shots. The other cameras were positioned to continuously observe the head and body of the animals from landing (ejection from the stunning box) until at least four minutes after sticking. For each animal, animal-related and process-related factors were collected. The animal-related factors were sex category (bulls, heifers, and cows), breed, and carcass weight, as well as fat and conformation class, using the European Union's EUROP grid method of beef carcass classification. The traditional grid is commonly used by most beef plants. Conformation is assessed on an E to P basis (EUROP), with E being a convex and shapely carcass, R being an average shape or straight profile, and P being a plainer carcass with a concave profile. Fat is assessed on a 1 to 5 basis, with 1 being very lean and 5 being very fat. Regarding stunning equipment, the device model of the stunner, function type (cartridge or pneumatic), cartridge strength (resp., air pressure) used, exit length of the bolt, bolt diameter, bolt velocity and weight, and the resulting kinetic energy were monitored. In addition, the stunner operator, the stun-to-stick interval, and the start of further dressing procedures (duration of bleeding) were recorded.

2.2. Slaughter Facilities

The participating abattoirs were medium to large-sized slaughter facilities with a daily slaughter capacity of 300 to 1000 animals (Table 1).

Table 1. Features for restraining, stunning, and bleeding at the participating abattoirs.

Abattoir	Approximate Slaughter Capacity per Day (no. of Cattle)	Slaughter Speed per Hour (no. of Cattle/h)	Number of Staff Performing the Stunning	Stunning Device	Head Restraint/ Manufacturer	Mean Stun-to-Stick Interval (s)	Bleeding Time before further Processing (min)
A	400	50	2	EFA® VB 315 (Schmid & Wezel GmbH, Maulbronn, Germany); Schermer® KR, KS (Freund Maschinenfabrik GmbH & Co. KG, Paderborn, Germany)	Tight restraint [1], Allkon® (Allkon GmbH & Co. KG, Steinau an der Straße, Germany)	37.8	3.0
B	700	65	3	Jarvis® USSS-21 (Jarvis GmbH, Buchholz, Germany)	Tight restraint [1], MPS® (Marel, Lichtenvoorde, Netherlands)	47.5	7.0
C	1000	60–70	3	EFA® VB 315 (Schmid & Wezel GmbH, Maulbronn, Germany);	Tight restraint [1], BANSS® (JWE-BANSS GmbH; Biedenkopf, Germany)	49.6	6.0
D	500	55	4	EFA® VB 315 (Schmid & Wezel GmbH, Maulbronn, Germany);	Tight restraint [1], DGS® (DGS Processing Solution, Haaksbergen, Netherlands)	42.0	6.0
E	350	50–55	2	Schermer® KL, KS (Freund Maschinenfabrik GmbH & Co. KG, Paderborn, Germany)	Loose (passive), self-built	51.0	4.5

[1] Tight restraint: lateral and vertical restriction of movement of the head [2].

At all facilities, cattle were delivered on the day of slaughter and temporarily (20 min to 7 h) kept in lairage. Except for abattoir E, all facilities used a modern stun box with tight head restraint (active) for restraining before stunning (Figure 1). The staff performing the stunning were trained (certificate of competence) and were aware of the optimum shooting position, which is 1.0 cm above the intersection of two imaginary lines between the centre of the eye and the opposite centre of the horn base, with deviations of less than 2 cm, and a shooting direction perpendicular to the skull [1].

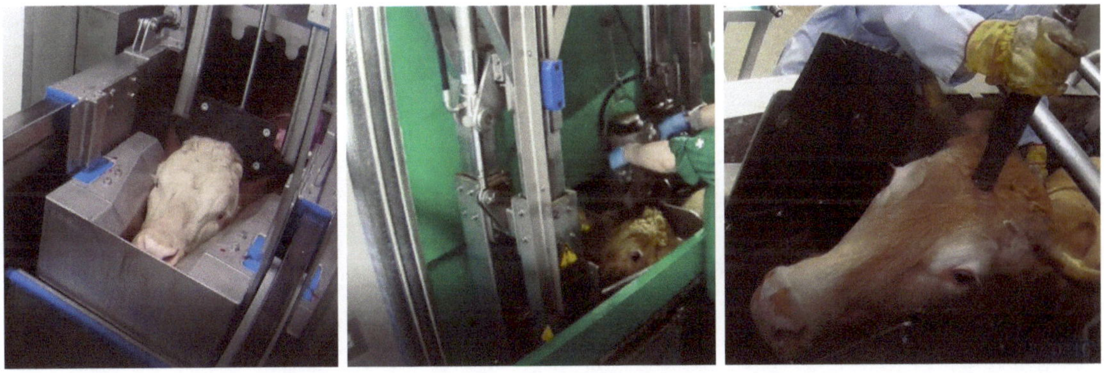

Figure 1. Different types of stun boxes to restrain cattle for captive bolt stunning; tight (active) head restraint DGS® (**left**) and Banss® (**middle**), and self-built loose (passive) head restraint (**right**).

2.3. Animals

A total of 2911 cattle, consisting of dairy, beef, and crossbreeds, were examined. Twenty animals were excluded from this study due to missing information, particularly due to incomplete video recordings. Further analyses thus included 2891 cattle. Of these, 31.4% were cows, 51.3% bulls, and 17.2% heifers. The distribution included 48.5% dairy cattle, 10.9% beef cattle, and 40.6% dual-purpose breeds. The most common breeds were Black Holstein (40.6%), Simmental (18.6%), and crossbreeds (14.7%). Most of the cows (79.0%) were Black Holstein. Among the bulls and heifers, the breeds Simmental (32.9%/8.0%), crossbreed (19.3%/24.3%), and Black Holstein (17.7%/39.0%) were predominant. The information on the breed, sex category, age, and carcass weight of the selected animals was gained from the cattle passports and slaughter lists.

2.4. Key Parameters and Features of Captive Bolt Devices

Different penetrating captive bolt devices, both pneumatic and cartridge-driven, and powered by different cartridge strength or air pressures, were used. For specifications, see Table 2.

Table 2. Specifications of captive bolt stunners used, including key parameters according to Reg. (EC) No. 1099/2009 [2].

Abattoir	Manufacturer, Type of Stunning Device	Function Type of Stunning Device	Pressure [1] or Cartridge Strength Used	Velocity in Air in (m/s) [4]	Extension Lengths in (mm) [4]	Diameter in (mm) [4]	Kinetic Energy in (Joule)	Position of the Person in Charge of Stunning
A	EFA® VB 315 (Schmid & Wezel GmbH, Maulbronn, Germany)	Pneumatic	13.5 bar	40.1	119.0	14.5	453	Left of the animal
A	Schermer® KR (Freund Maschinenfabrik GmbH & Co. KG, Paderborn, Germany)	Cartridge	Red [2]	58.3	88.0	12.0	389	Left of the animal
A	Schermer® KS (Freund Maschinenfabrik GmbH & Co. KG, Paderborn, Germany)	Cartridge	Red [2]	52.6	81.0	12.0	334	Left of the animal
B	Jarvis® USSS-21 (Jarvis GmbH, Buchholz, Germany)	Pneumatic	15.5 bar	43.5	88.0	12.0	488	Right of the animal
B	Jarvis® USSS-21 (Jarvis GmbH, Buchholz, Germany)	Pneumatic	15.5 bar	37.6	90.0	14.5	421	Right of the animal
C	EFA® VB 315 (Schmid & Wezel GmbH, Maulbronn, Germany)	Pneumatic	14.2 bar	41.6	119.0	14.5	488	Left of the animal
D	EFA® VB 315 (Schmid & Wezel GmbH, Maulbronn, Germany)	Pneumatic	14.2 bar	41.6	119.0	14.5	488	Front of the animal
E	Schermer® KL (Freund Maschinenfabrik GmbH & Co. KG, Paderborn, Germany)	Cartridge	Red [2]	58.4	121.0	12.0	433	Right of the animal
E	Schermer® KS (Freund Maschinenfabrik GmbH & Co. KG, Paderborn, Germany)	Cartridge	Blue [3]	50.3	79.0	12.0	306	Right of the animal

[1] Air pressure as shown by the manometer; [2] red cartridge strength corresponds to 4.9 grain or 320 mg; [3] blue cartridge strength corresponds to 4.8 grain or 315 mg; and [4] key parameters according to Reg. (EC) No. 1099/2009.

All the devices used were examined beforehand. The diameter of each bolt was determined using a digital calliper (Digital ABS AOS calliper, Mitutoyo Germany GmbH, Neuss, Germany), and the weight of the bolt was determined using a precision balance (Sartorius ENTRIS II precision balance, WHI-Wägetechnik für Handel und Industrie GmbH & Co. KG, Hamburg, Germany). In order to obtain comparable values for exit velocity and kinetic energy, all stunning devices included were checked using a velocity tester (AST-106; AST 107-111 Stun Tester, Jarvis Products Corporation, Middletown, CT, USA, https://jarvisproducts.com/ accessed on 1 November 2023). The kinetic energy was then calculated from the velocity and bolt weight.

$$\text{Ekin} = \frac{1}{2} \times \text{m} \times \text{v}^2$$

To determine the exit length, the stunning device used was shot at floral foam (ELES VIDA®, Bremen, Germany), and then the penetration depth was measured using a calliper. Measurements of the velocity and exit length were performed for all combinations of cartridge strengths or pneumatic pressure applied. To validate the method of exit length and velocity determination, a portion of the devices was rechecked in collaboration with a federal physics institute (PTB, Braunschweig, Germany). For this purpose, the shooting process was recorded using a high-speed camera (Fastcam 20,000 fps (frames per second), Photron, Reutlingen, Germany) and the recordings were then analysed using specific software ("Tracker"; https://physlets.org/tracker; open-source Physics, version number 5.1.5, accessed on 1 May 2020). The values determined by both measuring methods match well, although we could only perform the tests on 3–4 shots using a combination of cartridge and pressure.

2.5. Measurements

After the cattle had been stunned, an employee of the abattoir checked the state of consciousness in the stun box (collapse, relaxed eyelids, and ears) and at the landing grid (wide pupil, fixed eye, and no breathing; see Table 3) before they were shackled and hoisted. Bleeding was carried out via a chest stick after a previous skin incision (two-knife technique). Except at abattoir B, the chest stick procedure was performed using a single-edged slaughter knife. Abattoir B used a double-edged knife with a hollow handle and a peristaltic pump for blood collection. Individual animals were bled in a recumbent position if hoisting was delayed due to strong tonic and clonic seizures. The time to further dressing was between three and seven minutes. Of the two veterinarian investigators one had 30 years of experience and the other had trained for 10 months in advance to determine stunning effectiveness, respectively. An animal was considered properly stunned when it collapses immediately after stunning, exhibits tonic and clonic seizures, floppy ears, and no attempts to regain posture, both on the landing grid and during hoisting, sticking, and exsanguination. Additionally, apnoea starts immediately after the shot, corneal reflex on the landing grid is negative and throughout the entire observation period the eyes remain fixed with dilated pupils and no respiratory movements are noticed [23]. We distinguished between signs of sufficient stunning, doubtful stunning, and insufficient stunning (Table 3). Sufficiently stunned animals do not exhibit any signs of preserved brainstem activity, such as eye or eyelid movements and respiratory movements. The category insufficient stunning includes a high risk of regaining consciousness as well as remaining or regained consciousness [9].

Table 3. The evaluation scheme for stunning effectiveness is mainly based on EFSA AHAW Panel [23], von Holleben et al. [24], and von Wenzlawowicz et al. [9].

Body Part	Doubtful Effect of Stunning [1]		Insufficient Stunning Effect [2]	
	Parameter	Definition	Parameter	Definition
Whole body	Abnormal cramps	The animal does not show a typical tonic or tonic–clonic convulsive phase after the shot [10]	No collapse after shot, posture	Failure to collapse, attempts to regain posture after being shot
			Arched back righting reflex	Arching of the back and sustained backward lifting of the head, while the animal hangs on the rail [25]; the symptom can also be shown while an animal is lying in a horizontal position [26]
Ears			Ear tone	Ears do not hang down limply but are tense or straightened (visual and palpatory examination in case of suspicion)
Eyes	Eyeball rotation	The eyeball is not centred within 25 s after the shot; the eyeballs may be rotated to a great extent so that the pupils may not be visible [10]; the eyeball is rotated [27]	Corneal reflex	Repeated blinking response elicited by touching or tapping the cornea
	No pupil dilatation	The pupil is not fully dilated	Spontaneous blinking	Spontaneous closing of the eyelids without prior irritation of the eyelid or cornea
	Nystagmus	Spontaneous rapid side-to-side movements of the eyeball	Focused eye movements	Accommodation of the eye, the eyeball follows movements in the vicinity
	Corneal reflex	Single blinking response elicited by touching or tapping the cornea		
Respiratory system	Breathing movements (<4 times)	The animal shows up to 3 breathing movements after the shot, which can be recognised by movements of the flank, muzzle, or nostrils	Breathing movements ($\geq$4 times)	The animal shows more than 3 breathing movements after the shot, which can be recognised by movements of the flank, mouth, or nostrils
			Vocalisation	Vocalisation in the form of moaning, grunting, or mooing

[1] Transition zone between definitely unconscious/brain dead and definitely conscious but low risk of awakening; no pain and suffering but first indicators for a shallow depth of stunning [25]; [2] high risk of regaining consciousness (still in transition zone) as well as remaining or regained consciousness. Signs of a conscious state are no loss of posture, righting reflex, spontaneous blinking, and focused eye movements.

The first investigator stood next to the head of the animal at the landing grid and recorded the ear tag number, stunning efficiency during LANDING and HOISTING, and stun-to-stick interval (stopwatch; Delta E 100, Hanhart 1882 GmbH. Gütenbach, Germany). The second investigator checked stunning efficiency during bleeding and subsequently measured the angle and position of the shot hole using a geo-triangle (Westcott E-10132 00 Geodreieck, Acme United Europe GmbH, Solingen, Germany), a plastic stick (diameter 8.4 mm or 12.4 mm, depending on the bolt diameter; POM round bar, Nattmann GmbH, Willich, Germany; Figure 2), and a multi-angle ruler (Wohao, Shenzen, China; Figure 2). Deviations of at least 2 cm from the ideal shooting position [1] and at least 10° from the perpendicular were recorded.

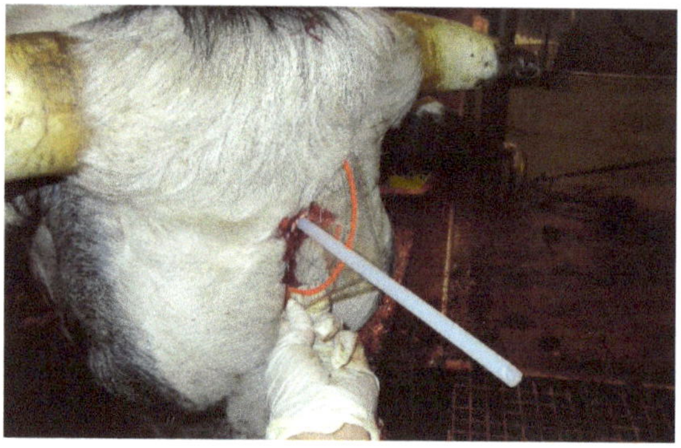

Figure 2. Recording possible deviations of shot position and angle using a plastic stick inserted in the shot hole and a geo-triangle at the end of the bleeding line.

On each examination day, three to four action cameras (Apeman A100S, Apeman, Shenzhen, China, https://de.apemans.com/collections/action-kamera, accessed on 1 November 2023) were installed, depending on the facility, to record each animal from the stun box until at least the end of the fourth minute after sticking. The observation period was divided into eight process intervals. The first interval, LANDING (1), started when the animal fell out of the stun box and touched the landing grid (approx. 7 s–20 s after the shot) and ended as soon as the shackle chain tightened during lifting (approx. 15 s–55 s after shot). The next process interval, HOISTING (2), was followed by SKIN INCISION (3). The latter started with the first contact between knife and skin (approx. 24 s–60 s after the shot) and ended before STICKING via the chest stick (4). This interval started when the knife was inserted into the chest entrance (26 s–78 s after the shot) and ended 3 s after the knife had been pulled out. STICKING was followed by the FIRST (5), SECOND (6), THIRD (7), and FOURTH MINUTE OF BLEEDING (8). These last four process intervals each lasted for one minute, the first starting four seconds after STICKING and the last ending 240 s later at the earliest (maximum 400 s). The subsequent analysis of the video material regarding movement category, frequency, and time of movement was always carried out by the same pre-trained investigator. For each animal, the movement categories were recorded based on process interval. Each movement category was counted only once per process interval. A score value was assigned to each movement category (0–3), reflecting the severity of the movement in terms of vigour, speed, frequency, and impact on the process. A score per animal and process interval was obtained by addition. The scores of all process intervals summed up resulted in the total score (sum score) for an individual animal. The definitions of movement categories and scores are shown in Table 4. Examples showing the different movement categories can be found as movies in the Supplementary Materials.

Table 4. Definitions of post-stun movement categories and scores (for examples, see also the Supplementary Materials).

Movement Category	Definition	Intensity [1]	Severity of Movement, Score Explanation	Score Value
Kicking hind limb	A single/free hind limb is moved away from the body and back repeatedly and rapidly.	Moderate [2]	Fast Vigorous Dangerous to workers	2.0
		Intense [3]	Very fast Vigorous Dangerous to workers	3.0
Twitching	More than one limb is moving, possibly together with the trunk and neck. The movement can be synchronous or asynchronous and may involve the hind limb and forelimb (e.g., forelimb paddling together with hind limb kicking).	Moderate [2]	Fast Vigorous Dangerous to workers	2.0
		Intense [3]	Very fast Vigorous Dangerous to workers	3.0
Bending forelimb	One or both forelimbs are bent towards the body. The movement can be synchronous or asynchronous, single or repeated (e.g., forelimb paddling with bent limbs).		Either fast or slow Predominantly vigorous May impede the work	1.0
Lifting forelimb	One or both forelimbs lift in an extended position. The movement can be synchronous or asynchronous, single or repeated (e.g., forelimb paddling with stretched limbs).		Either fast or slow Predominantly vigorous May impede the work	1.0
Body arching laterally	The longitudinal axis of the hoisted animal is bending to one side and the head and body are not hanging straight down; in most cases without repetition.		Either fast or slow Predominantly vigorous May impede the work	1.0
Body arching ventrally	The head and possibly the trunk of the hoisted animal are bending ventrally; in most cases without repetition.		Either fast or slow Predominantly vigorous May impede the work	1.0
Stretching hind limb	The free hind limb extends away from the body for at least 3 s.		Slow No repetition within 5 s No impact on work	0.5
Bending hind limb	The free hind leg is bent and pulled towards the body.		Slow No repetition within 5 s No impact on work	0.5
Arching backwards/righting reflex	Arching of the back and sustained backward lifting of the head, while the animal hangs on the rail [25].		Slow No repetition within 5 s No impact on work	0.5
No movement	Animal shows none of the movements listed above.			0

[1] A graduation of intensities is only used for kicking and twitching during LANDING and HOISTING: from the process interval SKIN INCISION onwards, no differentiation was made between moderate and intense; [2] Moderate: 1. frequency <1/s and duration <5 s; 2. frequency >1/s and duration <5 s; and 3. frequency <1/s and duration >5 s; [3] Intense: frequency >1/s and duration >5 s; example for the calculation of the sum score: An animal shows "kicking hind limb intense" (3.0) at LANDING, "body arching ventrally" (1.0) at STICKING, "kicking hind limb" (2.0) and "bending forelimb" (1.0) in the FIRST MINUTE OF BLEEDING and "stretching hind limb" (0.5) in the FOURTH MINUTE OF BLEEDING. The scores for the individual process intervals would then look as follows: LANDING 3.0, HOISTING 0, SKIN INCISION 0, STICKING 1.0, FIRST MINUTE OF BLEEDING 3.0 (2.0 + 1.0), SECOND MINUTE OF BLEEDING 0, THIRD MINUTE OF BLEEDING 0, and FOURTH MINUTE OF BLEEDING 0.5. This results in a total sum score of 3.0 + 0 + 0 + 1.0 + 3.0 + 0 + 0 + 0.5 = 7.5.

2.6. Statistical Analysis

Data were documented in MS Excel (Microsoft Corporation, (2018), Microsoft Excel, Redmond, WA, USA) and pivot tables were used for plotting and analysis. The software JMP v. 15 (SAS Institute Inc., Cary, NC, USA) was used for statistical analyses. The dependent variable was the sum score across all process intervals or the score of a specific process interval, calculated for each individual animal. The influence of possible animal- and process-related factors (independent variables) on the scores was investigated via analysis of variance (ANOVA model) and the Tukey–Kramer post hoc test to compare means. A result was considered statistically significant for $p < 0.05$. Independent variables included in the model were captive bolt parameters, model and function type of stunning device (pneumatic or cartridge powered), exit length, diameter, velocity, and kinetic energy of the bolt. Animal-related factors included were sex category (cow, heifer, and bull), breed, carcass weight, fat, and conformation class. Other variables in the model were stunning effectiveness, as well as the process-related variables stun-to-stick interval and deviation in position or angle of the shot. The relative frequencies of individual nominal parameters were described using contingency tables and tested for random distribution using Chi-square tests. In addition, decision trees (stepwise partitions) were used to determine combinations (of independent variables) that most strongly influence a certain expression of the sum score. In this process, the sample is split into subgroups in such a way that the means of the sum score of subgroups differ as much as possible until no further sensible differentiation can be made. Only effects leading to significant deviations from the mean are presented, and results are not shown when concerning an insignificantly small group (threshold: n < 146 ≙ 5%).

3. Results

3.1. Occurrence of Movements

Most cattle showed movements after captive bolt stunning (number of movements across all process intervals: minimum: 0; maximum: 15; mean: 3.5). Only 6.6% of the animals did not show any movement at any process interval (bulls 6.1%, heifers 4.0%, cows 8.7%). Most movements were recorded in the process interval FIRST MINUTE OF BLEEDING. Here, 61.7% of the animals showed at least one movement. In addition, at STICKING, more than half of the cattle (58.0%) showed at least one movement. In the SECOND (14.2%) and THIRD MINUTES (6.4%) OF BLEEDING, considerably fewer movements were observed in cattle compared with the FIRST MINUTE OF BLEEDING. In the FOURTH MINUTE OF BLEEDING, the proportion of animals moving increased slightly (12.2%). Late movements during bleeding also occurred in cattle that had previously been hanging on the rail completely motionless and with limp tails. In some cattle, movements could still be observed more than 8 min after STICKING. During one process interval, several movements of different categories could occur. In some animals, up to five different movement categories were observed within the same process interval.

The mean sum score, a measure of the frequency and intensity of an animal's movements across all process intervals, was 4.87 (minimum: 0; maximum: 20.5; median: 4.5). The interquartile range of the sum score was 2.5 to 7.0. The distribution of the sum score is shown in Figure 3. The mean score at LANDING was 0.79 (minimum: 0; maximum: 6.0; median: 0).

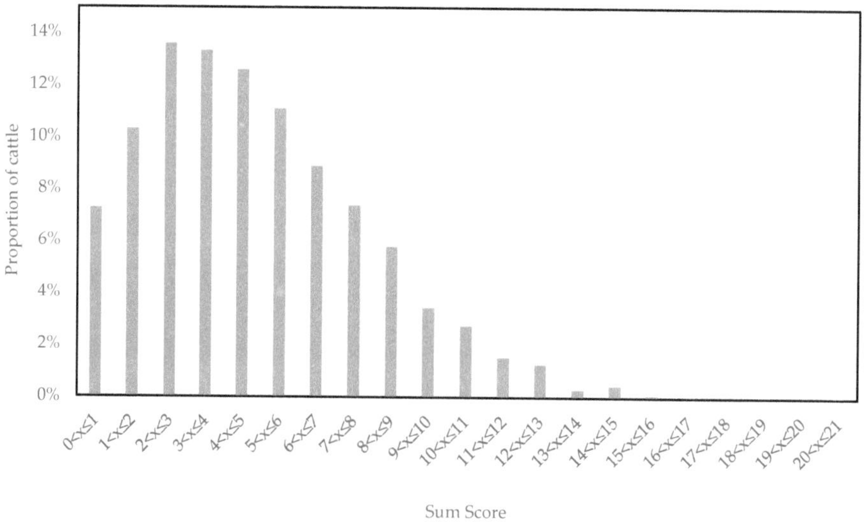

Figure 3. Distribution of sum score values (a measure of frequency and intensity of an animal's movements across all process intervals; mean: 4.87; minimum: 0; maximum: 20.5; and median: 4.5) in the study group (N = 2891) (relative frequency).

All movements occurred spontaneously as well as in connection with manipulation (grasping and lifting limbs, pulling of the shackle on the hind limb, sticking, and further dressing of the lower feet). Particularly strong and intense movements were observed at LANDING and during HOISTING. Movements were obvious during LANDING in 31.7%

of the cattle and during HOISTING in 43.7%. The most common movements during these early process intervals were "kicking hind limb moderate" (21.9%), "twitching moderate" (18.0%), and "twitching intense" (13.6%). In particular, "kicking hind limb intense" and "twitching intense" on the landing grid made it very difficult for the staff to start up shackling, which often led to an extended stun-to-stick interval. Occasionally, the "kicking hind limb" and "twitching" movements caused shackles to be thrown out of position. This was especially observed in Holstein dairy cows.

During the process intervals SKIN INCISION and STICKING, "kicking hind limb" (39.0%), "body arching ventrally" (25.5%), "lifting forelimb" (22.9%) and "bending forelimb" (22.4%) were observed most frequently. The majority of late movements towards the end of exsanguination were "stretching hind limb" (7.2%), "kicking hind limb" (3.7%), and "bending forelimb" (1.1%). In the late phases of bleeding, it was frequently noticed that the tail tension temporarily increased again. In some cases, there was also intensive tail flapping, even though the tail had already been hanging limply for minutes before. None of the animals that moved for more than four minutes after sticking showed signs of regaining consciousness.

Category of Movements during Different Process Intervals

The absolute number of individual movements related to the eight different process intervals is shown in Figure 4. The movement most frequently counted was "kicking hind limb", including "kicking hind limb moderate" (n = 2810). Kicking occurred in each process interval but most frequently between LANDING and the end of the FIRST MINUTE OF BLEEDING. In some cases, "kicking hind limb" was observed over 50 times, lasting several process intervals. During LANDING and HOISTING, 21.9% of the animals expressed "kicking hind limb moderate" and 10.0% expressed "kicking hind limb intense". During HOISTING, "kicking hind limb", either moderate or intense, was observed in 20.3% of the cattle (18.8% bulls, 27.7% heifers, 18.7% cows). "Kicking hind limb" during STICKING occurred in 41.6% of the animals (32.0% bulls, 30.0% heifers, 63.0% cows). The next frequent movements were "lifting forelimb" (n = 1761) and "bending forelimb" (n = 1655), which occurred in all process intervals from HOISTING onwards but most frequently during SKIN INCISION, STICKING, and the FIRST MINUTE OF BLEEDING. We observed that "bending forelimb" tended to occur later than "lifting forelimb". "Twitching", either moderate or intense, occurred almost exclusively during LANDING or HOISTING. "Stretching hind limb" was observed in each process interval, but most frequently in the FIRST MINUTE OF BLEEDING. The slight increase in movements at the FOURTH MINUTE OF BLEEDING was predominantly due to "stretching hind limb", often together with shivering of the same. The category "body arching ventrally" almost always occurred during SKIN INCISION or STICKING, and in individual cases, also during HOISTING or still in the FIRST MINUTE OF BLEEDING (see Figure 4). In the later course of bleeding (SECOND MINUTE OF BLEEDING and later), "body arching ventrally" was observed only four times. In particular, provoked by sticking, cattle moved their heads and necks vigorously in the ventral direction. Across all process intervals, "body arching ventrally" was observed in 649 cattle (22.4%), regularly accompanied by "kicking hind limb" or "lifting forelimb". "Body arching laterally" was detected in 291 animals overall (10.0%), mainly during HOISTING, SKIN INCISION, and STICKING. Only six times was this category of movement observed at a later process interval, i.e., four times during the SECOND MINUTE and twice during the THIRD MINUTE OF BLEEDING. The movement of "body arching laterally" as such could either be short or held for more than a minute. In this study, no "arching backwards" was noticed in any animal.

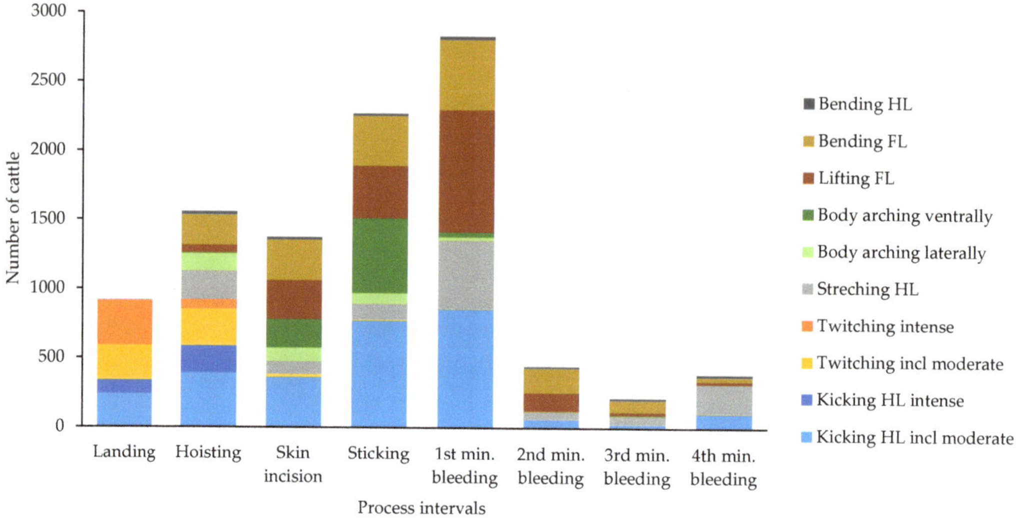

Figure 4. The absolute number of observed movement categories related to the eight different process intervals (more than one movement by animal and process interval possible); HL = hind limb; FL = forelimb.

3.2. The Effect of Animal-Related Factors on Movements

3.2.1. The Effect of Sex Category

Sex categories differed in terms of frequency of movement during the different process intervals. A total of 48.9% of cows, 37.2% of heifers, and 19.1% of bulls showed movements at LANDING. During HOISTING, the percentage of heifers and bulls with movements increased (52.4% and 39.7%, respectively), while the percentage of cows decreased (45.4%). During SKIN INCISION and STICKING, most heifers (78.3%) and bulls (70.3%) showed at least one movement, whereas this was the case for less than half of the cows (47.2%). During the four-minute bleeding interval, 78.3% of heifers, 70.3% of bulls, and 62.3% of cows still showed at least one movement. In addition, concerning the frequency of movement categories, differences between sex categories were determined for certain categories. Both "twitching moderate" and "twitching intense" were observed, especially in cows (see Figure 5). According to the ANOVA, "body arching laterally" and "body arching ventrally" occurred significantly more in bulls and less in cows ($p < 0.001$). Forelimb movements occurred significantly more in heifers ($p < 0.001$).

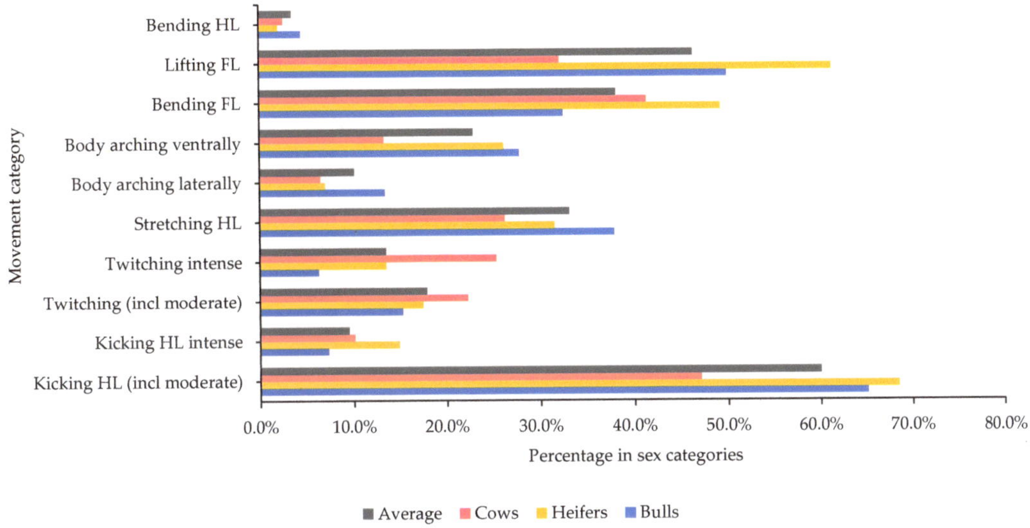

Figure 5. Relative frequency of the movement categories across all process intervals in relation to sex categories (cows, heifers, and bulls); HL = hind limb; FL = forelimb.

The results of the contingency test comparing the sex categories regarding movement frequencies by process intervals show that, especially at LANDING ($p < 0.001$), "twitching intense" occurred significantly more often in cows (22.8%) than in heifers (11.2%) or bulls (4.1%). At LANDING ($p < 0.001$) and HOISTING ($p < 0.001$), "kicking hind limb intense" was observed significantly more often in heifers (5.6%/11.2%) than in bulls (1.5%/5.9%). At STICKING ($p < 0.001$), "body arching ventrally" was recorded significantly more often for heifers (22.9%) and bulls (21.6%) than for cows (10.6%). "Lifting forelimb" in the FIRST MINUTE OF BLEEDING ($p < 0.001$) was observed significantly more often in heifers (42.4%) and bulls (33.2%) than in cows (18.7%).

Heifers had the highest average sum score. Despite the intensive movements at LANDING, across all process intervals cows had a lower average sum score (mean: 4.42) than heifers (mean: 6.06) or bulls (mean: 4.70). The ANOVA revealed a significant effect of sex category on the sum score ($p < 0.0001$), with least square means being significantly lower in cows (5.58) than in bulls (6.74) and heifers (6.90). However, in the ANOVAs used to check animal- and process-related impacts on scores, the variables examined explained only 10.0% to 12.0% of the score variability.

3.2.2. The Effect of Breed

Table 5 displays the frequency of movement categories for the most commonly represented breed types. Some breeds differed considerably regarding the category of movements and frequency of occurrence. "Twitching intense" during LANDING or HOISTING was shown especially by German Angus (GA: 21.4%, total: 13.5%), Black Holstein (20.6%), Red Holstein (19.4%), and Limousin (18.9%) but rarely by Brown Swiss (0.0%) and Simmental (3.5%). "Kicking hind limb intense" during LANDING and HOISTING was observed most frequently in Limousin (LIM: 16.7%, total: 9.6%) and Red Holstein (16.0%) and least frequently in Brown Swiss (2.0%). "Stretching hind limb" occurred particularly often in Limousin (LIM: 50.0%, total: 32.9%), and "body arching ventrally" occurred in German Angus and Charolais (see Table 5).

Table 5. Movement categories are shown by cattle following stunning across various breeds (relative frequency).

Breed Type	Kicking HL/Kicking HL Moderate	Kicking HL Intense	Twitching/ Twitching Moderate	Twitching Intense	Stretching HL	Body Arching Laterally	Body Arching Ventrally	Bending FL	Lifting FL	Bending HL
Brown Swiss	53.0%	2.0%	9.0%	0.0%	28.0%	7.0%	21.0%	21.0%	45.0%	5.0%
Charolais	71.1%	11.1%	15.6%	15.6%	42.2%	11.1%	37.8%	53.3%	62.2%	6.7%
German Angus	71.4%	7.1%	42.9%	21.4%	33.3%	19.0%	40.5%	40.5%	57.1%	2.4%
German Red Pied	76.1%	11.9%	14.9%	9.0%	38.8%	14.9%	25.4%	43.3%	56.7%	6.0%
Simmental	58.5%	4.8%	10.4%	3.5%	39.9%	13.4%	25.9%	23.8%	44.7%	4.7%
Limousin	64.4%	16.7%	13.3%	18.9%	50.0%	10.0%	35.6%	35.6%	66.7%	2.2%
Red Holstein	54.9%	16.0%	16.0%	19.4%	23.3%	7.8%	22.3%	46.6%	57.8%	2.9%
Black Holstein	57.6%	11.7%	20.9%	20.6%	26.0%	8.3%	15.2%	43.8%	35.2%	3.1%
CBB	69.6%	4.0%	16.0%	9.6%	46.4%	12.0%	28.0%	35.2%	69.6%	5.6%
CBD	65.3%	8.0%	23.1%	6.4%	39.6%	11.1%	32.1%	38.7%	57.5%	2.4%
Total	60.2%	9.6%	18.1%	13.5%	32.9%	10.2%	22.7%	38.1%	46.2%	3.5%

FL = forelimb; HL = hind limb; CBB = crossbreed beef × beef; CBD = crossbreed beef × dairy.

The highest average sum score was measured for German Angus (mean: 6.46), followed by Charolais (mean: 6.04) and Limousin (mean: 5.91). Breeds with low sum scores covered Brown Swiss (3.21) and Simmental (mean: 3.88). The score at LANDING was highest in German Angus (mean: 1.29), Black Holstein (mean: 1.08), and Red Holstein (mean: 1.01).

The ANOVA revealed a significant impact of the breed type ($p < 0.0001$) on the sum score. The breeds Brown Swiss and Simmental significantly decreased, while German Angus increased the sum score. Least square means for German Angus (7.84), Charolais (7.35), Limousin (6.75), Black Holstein (6.52), Crossbreed Beef × Beef (6.43), Crossbreed Beef × Dairy (6.39), and Red Holstein (6.22) were significantly higher than for Simmental (4.98) and Brown Swiss (4.21).

An impact of the class of conformation (E-P) on the extent of movements was found regarding the score at HOISTING ($p < 0.01$), which was higher in cattle classified as moderately conformed (R-O). Class of conformation also affected the score at STICKING ($p < 0.01$). Here, low conformation (O-P) decreased the score. The effect of slaughter weight was less significant ($p < 0.05$). In the model (ANOVA), an increase in slaughter weight by 1.0 kg each lowered the sum score by 0.003 (mean: 4.87).

3.3. *The Effect of Process-Related Factors on Movements*

Of the process-related factors, only those related to stunning devices are presented here. The ANOVA did not show an effect of the stun-to-stick interval (mean: 44.6 s, min: 26.0 s, and max: 78.0 s).

The Effect of Captive Bolt-Related Factors

Of 2891 animals examined, 731 (bulls n = 203, heifers n = 239, and cows n = 289) were stunned with cartridge-powered captive bolt devices (KS Schermer® n = 301, KR Schermer® n = 325, KL Schermer® n = 105). The other 2160 animals (bulls n = 1281, heifers n = 259, and cows n = 620) were stunned with a pneumatically powered device (USSS-21 Jarvis® n = 508, VB 315 EFA® n = 1652). Overall, more movements were observed in cattle stunned with a cartridge-powered bolt gun than in cattle stunned with a pneumatic device. "Kicking hind limb intense" during LANDING and HOISTING, for example, occurred considerably more often in animals stunned with a cartridge-driven captive bolt (17.9%) than in animals stunned with a pneumatically powered device (6.6%). When comparing the two function types regarding movement frequencies by process intervals by means of contingency tests, i.e., without taking other effects into account (see Table 6), the described effect (cartridge-powered > pneumatic powered) often became significant ($p < 0.001$) for most movement categories and for the process intervals from HOISTING up to and including the FIRST MINUTE OF BLEEDING. An exception was "twitching moderate" at LANDING and HOISTING, which was observed more frequently for pneumatic guns (Table 6). However, most of the other movement categories, such as "kicking hind limb intense" at LANDING and HOISTING, "kicking hind limb moderate" at LANDING and HOISTING, "kicking hind limb" in the FIRST MINUTE OF BLEEDING, and "lifting forelimb" from HOISTING to FIRST MINUTE OF BLEEDING, also occurred significantly more often in animals stunned with a cartridge-powered device.

Table 6. The results of the contingency test: relative frequency of movement categories by process interval and stunning device function type (pneumatic and cartridge).

Category of Movement	Landing		Hoisting		Skin Incision		Sticking		1st min. of Bleeding		2nd min. of Bleeding		3rd min. of Bleeding		4th min. of Bleeding	
	P	C	P	C	P	C	P	C	P	C	P	C	P	C	P	C
Kicking HL; incl. moderate	7.4 *	11.2 *	11.3 *	19.8 *	12.7	11.6	27.1	24.9	26.6 *	38.6 *	2.2	1.9	1.0	0.8	3.2	5.1
Kicking HL intense	2.4 *	6.0 *	4.4 *	13.3 *												
Twitching incl. moderate	9.9 *	5.5 *	10.5 *	5.6 *	0.6	0.7	0.2	0.3	0.1	0.0	0.0	0.0	0.0	0.0	0.0	0.0
Twitching intense	10.4	13.7	2.5	2.2												
Stretching HL	0.1	0.3	5.8 *	9.8 *	3.2	3.0	3.9	4.1	17.2	15.3	1.9	1.5	2.3	1.8	6.7	8.6
Body arching laterally	0.0	0.0	4.9	3.0	3.5	2.9	2.5	3.4	0.2 *	2.7 *	0.1	0.1	0.0	0.0	0.0	0.0
Body arching ventrally	0.0	0.0	0.2	0.0	8.7 *	2.7 *	17.8	20.0	0.9	2.1	0.0	0.0	0.1	0.0	0.0	0.0
Bending FL	0.0	0.0	7.0	8.6	9.2	12.3	10.7 *	17.2 *	16.0 *	22.0 *	5.9	6.4	3.2	1.6	1.3	0.4
Lifting FL	0.0	0.0	1.5 *	3.1 *	8.3 *	14.0 *	11.6 *	17.6 *	28.2 *	36.1 *	4.4	4.8	1.0	0.0	0.7	0.5
Bending HL	0.0	0.0	0.7	1.0	0.7	0.3	0.3	0.8	0.6	0.7	0.4	0.1	0.5	0.3	0.5	0.4

P = pneumatically powered device; C = cartridge-powered device; FL = forelimb; HL = hind limb; * high statistical difference between function types ($p < 0.001$).

The average sum score was 5.68 for the cartridge-powered devices and 4.60 for the pneumatically powered devices. In particular, the greatest difference occurred between the pneumatically powered device VB-315 (mean: 4.34) and the two cartridge-powered devices KL (mean: 6.52) and KS (mean: 6.06).

The ANOVA revealed a significant impact of the function type ($p < 0.001$) on the sum score. In the model, the use of cartridge-powered stunners increased the sum score while the use of pneumatically powered stunners decreased it by 4.1. The exit length of the bolt had a decreasing impact on the frequency and intensity of movements (ANOVA, $p < 0.001$). In the statistical model, an increase in exit length by 1 mm reduced the sum score (mean: 4.87) by 0.044, and the score at LANDING (mean: 0.78) by 0.009. The effect of the bolt diameter was less clear ($p < 0.05$). In the model (ANOVA), increasing the bolt diameter by 1 mm increased the sum score (mean: 4.87) by 0.38 and the score at LANDING (mean: 0.78) by 0.26. The exit velocity of the bolt had an increasing effect on movements ($p < 0.001$). In the model (ANOVA), an increase in bolt velocity by 1 m/s increased the sum score by 0.07 and the score at LANDING by 0.03. This effect was only found if the function type (pneumatic or cartridge) and type of stunning device were not included in the analysis, as the effect was otherwise attributed to the function or type of device. A clear effect of bolt kinetic energy could not be shown by the ANOVA. An analysis of the decision trees revealed a lower sum score (mean: 4.43) for stunning devices with kinetic energy above 453 J (see Table 2) compared with devices with kinetic energy below 453 J (mean: 5.75). A similar effect was also found for the score at LANDING. Bulls stunned using devices of at least 488 J had a lower sum score at LANDING (mean: 0.32) than those stunned using devices with kinetic energy below 488 J (mean: 0.72).

3.4. *The Effect of Stunning Quality on Movements*

A total of 99.4% of the animals examined (n = 2911) were described as "sufficiently stunned" (bulls 99.5%, heifers 99.0%, and cows 99.3%). The stunning effect was stated as "doubtful" for 0.3% (n = 9: 3 bulls, 3 heifers, and 3 cows) and "insufficient" for 0.3% (n = 10: 5 bulls, 2 heifers, and 3 cows), of which only four cattle stayed conscious for a few seconds. Signs observed in animals called "doubtful" were twofold breathing movements (recognised on muzzle or nostrils), nystagmus at the landing grid, and returning tension/movement of the tongue. Animals classified as "insufficiently" stunned showed repeated respiratory movements (n > 4) at the muzzle or nostrils, spontaneous eyelid closure, or lack of collapse following the shot together with focused eye movements. In one animal showing multiple respiratory movements of muzzle and jaw, as well as a tense eyelid during bleeding, ear tension was regained three minutes after sticking. In two other animals that showed repeated respiratory movements and/or tongue tone, persistent "body arching laterally" was observed as well, lasting until the THIRD MINUTE OF BLEEDING. These two animals were the only ones that still showed "body arching laterally" after the end of the SECOND MINUTE OF BLEEDING. A total of 17 animals were re-stunned by the staff. All animals classified as "insufficiently" stunned (n = 10) and most of the animals classified as "doubtful" were immediately re-stunned on the spot by the staff. After case-by-case analysis, the following potential causes were considered. In 6 of 19 animals with "doubtful" or "insufficient" stunning effects, angular and/or positional deviation of the shot position of more than 20° or more than 3 cm were recorded. In two cattle, slight angular deviations (≤15°) were found together with slight deviations in shot position (≤2 cm). One animal had an abnormal swelling of the os frontale. In four other cases, it was suspected that the stunning device used (Schermer KS, Freund Maschinenfabrik GmbH & Co. KG, Paderborn, Germany) might not have been strong enough in relation to body weight. For the remaining six cases, no possible reason for reduced effectiveness could be identified. Overall (n = 2911), deviations in shooting position (≥2 cm from the ideal position) and/or angle (≥10°) were found in 21.1% (n = 614) of the cattle and we obtained indications for thereby increasing effects on movements. The sum score increased with increasing distance from the ideal shot position (<2 cm: mean 4.81; ≥2 cm: mean 5.46; ≥3 cm: mean 5.39; and ≥4 cm: mean 6.0). In the model (ANOVA), increasing the shot deviation by 1 cm increased the score at HOISTING (mean: 0.90) by 0.11 ($p < 0.001$). An increase in the sum score was also seen with angular deviations of more than 20° (mean: 5.35) compared with animals with angular deviations of less than 10° (mean: 4.85).

Due to the very small number of animals with the stunning effect stated as "doubtful" or "insufficient" (n = 19), no statistics are possible regarding the impact of stunning effectiveness on the sum score or the occurrence of certain movement categories. Proceeding purely descriptively, the sum score of the cattle categorised as "doubtful" (mean: 6.44, minimum: 3.00, and maximum: 10.00) or "insufficient" (mean: 5.90, minimum: 1.00, and maximum: 12.50) is comparable to the entire study group (mean: 4.87, minimum: 0.00, and maximum: 20.50). All movements observed during this investigation occurred in both well-stunned cattle and cattle with reduced stunning effectiveness.

4. Discussion

4.1. Occurrence of Movements

In this study, the analyses of movements show that after captive bolt stunning, there were hardly any cattle that did not show any movements at all. Overall, movements were observed in 93.4% of all cattle during at least one process interval. Most movements occurred during sticking (58.0%) and during the first minute of bleeding (61.7%). This could be attributed to the continued seizure activity after the shot [10,27] and the simultaneous reduction of inhibitory effects by higher brain centres already damaged by the effect of the captive bolt stun [28], while major manipulations such as hoisting and sticking could act as external triggers. This is in line with the results of Hilsenbeck [14] and those of von Holleben and von Wenzlawowicz [15]. Both studies report the proportion of cattle with movements during hoisting, sticking, and bleeding between 51.0% and 65.0%. In our study, a significant proportion of animals (12.2%; between 9.1% (bulls) and 14.4% (cows)) were found to still show movements in the fourth minute of bleeding. As well as described by Hilsenbeck [14] we observed that "stretching hind limb", often in combination with shivering, was the most frequent movement during advanced exsanguination. Before these late movements occurred, cattle were usually already hanging completely relaxed with their tails hanging limply. This was also the case in one animal that was moving eight minutes after sticking. Our results show that all movements, including neck and back movements such as "body arching ventrally" (22.4%) and "body arching laterally" (10.0%), regularly occur in unconscious cattle after captive bolt stunning. In practice, however, the latter movements are often equated with "arching backwards". Nevertheless, only "arching backwards" is a form of righting reflex [25,29], an active attempt of the animal to bring the head and body into a normal position, requiring a functional medulla and midbrain [30] and thus indicating the return of consciousness. "Body arching ventrally" in this study and in that by Terlouw et al. [17] was observed predominantly during sticking. In almost all cases, it occurred as a reflex-like response to the cutting of skin, muscles, and blood vessels. The fact that this movement was only very rarely observed without previous manipulation, e.g., during exsanguination, again confirms the trigger effect of manipulation. In this study, "body arching ventrally" was observed only four times during the second to third minute of bleeding, mainly at abattoir B, which was the only facility using a sticking knife with a hollow handle for blood collection. The late occurrence of "body arching ventrally" at this abattoir during bleeding could, therefore, be a consequence of the manipulation when removing the sticking knife. "Body arching laterally" in our study was observed predominantly during hoisting and in connection with the tonic phase, a typical sign for well-stunned animals after captive bolt stunning, during which the muscles of the back and legs are rigid and the hind limbs are flexed [27]. These clonic–tonic seizures typically occur following captive bolt stunning and may vary in duration [31]. Due to the use of modern stun boxes, nowadays cattle are often already shackled and hoisted during the tonic phase, which fosters the occurrence of "body arching laterally" [15]. In only two animals was "body arching laterally" observed during the third minute of bleeding.

4.2. Animal-Related Impact Factors on Movements

Our analyses demonstrate that the frequency of movements varies with regard to both sex category and breed. In particular, German Angus, Black Holstein, and Red Holstein exhibited a high movement score during landing, which may impede shackling. We confirmed the field experience that cows show especially strong movements such as "kicking hind limb intense" at landing. However, across all process intervals, cows moved less (mean sum score: 5.58) than heifers or bulls (mean sum score: 6.90 and 6.74, respectively), as from hoisting onwards they already showed less frequent movements. Our results are similar to those of von Holleben and von Wenzlawowicz [15], who recorded movements in 55.0% of bulls, 61.0% of heifers, and 51.0% of cows. Differences between sex categories regarding the frequency of certain movements were another result of our investigation, e.g., 30.6% of the bulls showed "body arching ventrally", while this movement was only observed in 22.7% of cows. The latter, on the other hand, showed "twitching intense" significantly more often (23.2%) than did heifers (13.7%) or bulls (10.4%). By contrast, Terlouw et al. [17] did not find any clear differences between sex categories regarding type and frequency of movements, possibly because of the significantly smaller sample size (n = 40) in their study. Our results also differ compared with those of Hilsenbeck [14] regarding the frequency of "kicking hind limb" during hoisting, which was observed in 55.7% of cattle, while our results show a prevalence of 20.3% in cattle (18.8% bulls, 27.7% heifers, 18.7% cows). This could be attributed to the different groups of animals investigated or the exclusive use of cartridge-powered stunners in the study by Hilsenbeck [14]. However, the results are similar with regard to movements at sticking, with 34.3% of the cattle not showing any movements in Hilsenbeck [14], compared with 41.6% in the present study. Furthermore, we observed an impact of breed type on the sum score, as well as on the occurrence of certain movement categories. During hoisting, "kicking hind limb" was observed significantly more often in dairy cattle (e.g., Black Holstein) than in beef or cross breeds. Similar results were obtained by Martin et al. [16]. Results by Kline et al. [22] and von Holleben and von Wenzlawowicz [15] show that, especially during landing, strong movements like "twitching intense" and "kicking hind limb intense" occur more often in dairy breeds (Black and Red Holstein). Despite the high score for dairy breeds during hoisting and especially landing, we calculated a higher sum score for breeds like German Angus than for Black or Red Holstein. This is due to the finding that movements in German Angus lasted longer and, therefore, occurred over more process intervals than those in dairy breeds. A lower sum score than that for German Angus (mean: 6.46), Limousin (mean: 5.91), or Charolais (mean: 6.04) was obtained for Brown Swiss cattle (mean: 3.21). Skull anatomy features may have influenced the extent of movements, but our data do not provide conclusive evidence. For instance, Simmental and German Angus bulls share comparable skull anatomy but their sum scores differ significantly. A possible impact factor could be the animals' temperament or the state of excitement/stress level just before stunning. Both physical exertion and pre-slaughter stress cause an increased muscle metabolism shortly after death [32] and could, therefore, affect the expression of spinal cord reflexes and automatisms. According to Grandin and Deesing [33], certain "Common Continental European breeds", such as Limousin and Charolais, are known to be nervous and flighty. The observations made during our study indicate that temperament and stress impact could be important, without these two parameters explicitly being monitored.

4.3. Captive Bolt-Related Impact Factors on Movements

The effects of features of the stunning devices on movements cannot always be completely separated from each other. Our research showed that cattle stunned using pneumatically powered devices (mean sum score: 4.60) moved less than those stunned using cartridge-powered guns (mean sum score: 5.68). One possible reason for this could be the higher bolt velocity of cartridge-powered stunners. Pneumatically powered devices also differ from cartridge-powered devices in terms of their kinetic energy. The mass of the bolt is significantly higher for pneumatically powered guns, which, despite a lower bolt speed, results in higher kinetic energy than for cartridge-powered stunners. According to our study (analyses of decision trees), for a higher kinetic energy (>453 J), the average sum score—a measure of the frequency and intensity of movements—was lower. It is possible that, due to their higher kinetic energy, pneumatically powered devices caused more damage to the brain, especially the brainstem and upper spinal cord, thus reducing the occurrence of medullary and spinal reflexes. However, another reason for the more extensive damage in pneumatically stunned animals could be the use of a closer head restraint, which is usually practised when using pneumatic guns. As there is hardly any space left for the head to move while being shot, energy is transferred directly to the skull and brain without loss. By contrast, if hand-held cartridge-powered stunners are used, the heads of the animals are often not restrained as tightly, and kinetic energy may be lost if the head swerves on impact. The assumption that more pronounced damage in deep brain structures may reduce movements is also supported by the fact that in our analyses, an increasing exit length of the bolt was associated with a significantly lower sum score or score at hoisting. A longer bolt penetrates deeper into the brain, thus causing more extensive damage to deeper brain structures. However, our results regarding exit length are not in line with those by Martin et al. [16], who observed more "kicking hind limb" with increasing penetration depth in Holstein cattle. Martin et al. [16], however, considered just "kicking hind limb" and only in the period from hoisting to sticking, which may explain the deviating results, along with differences in the group of animals studied, captive bolt devices used, and the lack of head restraint when using a centre track conveyor restrainer system.

4.4. Stunning Quality

In this study, a "doubtful" or "insufficient" stunning effectiveness was found in only 0.6% of the cattle. This is a remarkable improvement compared with previous studies. For the period 2003 to 2012, rates of failed stunning were reported at 4.0% to 9.2% [7–11,34]. Dörfler [12] estimated the proportion of inadequately stunned cattle to be only between 0.9% and 1.9%, rising to 5.7% in exceptional cases. This trend is presumably due to the further development and improved maintenance of captive bolt stunning devices [35], the establishment of modern stun boxes with tight head restraints, and increasing animal welfare monitoring at abattoirs. The presence of the investigators may also have positively affected the performance of the staff performing the stunning. When using well-maintained modern captive bolt devices, slight deviations in position or angle do not necessarily lead to a reduced stunning effect [36,37], which is also confirmed by our results, as we found deviations in position or angle in 614 (21.1%) heads but only 19 out of 2911 showed signs of a reduced stunning effect. However, our results, as well as those of Ilgert [38] and Kaegi [39], indicate that, independently of stunning effectiveness, an increased deviation in shooting position or angle may lead to more movements.

Our results demonstrate that movements occur regularly in cattle that simultaneously show no signs of an active brainstem such as eye or respiratory movements. This is in line with previous studies, such as the work of Fricker and Riek [20], who confirmed that convulsive activity still occurs in cattle with absent brain function (isoelectric EEG). Convulsions result from the failure of higher-level motor control centres in the brain, thus being incompatible with a simultaneously maintained consciousness [18,28,40]. Therefore, the mere presence of movements is not suitable to distinguish between consciousness and unconsciousness [29]. Concerning "arching backwards", which in cattle always indicates an insufficient stunning effect [25], studies by McKinstry and Anil [41] and Grandin [42] show that cattle when regaining consciousness first show a resumption of respiration or a positive corneal reflex before exhibiting righting reflex and attempts to raise their heads.

In this study, none of the animals showed "arching backwards". With regard to "body arching laterally", another movement often mistaken for the righting reflex, our investigation revealed that this movement occurs in 10.0% (n = 291) of all animals, mainly during hoisting and sticking, but in nearly all cases within two minutes after sticking. At this time, the animal may still show tonic–clonic seizures following the stun. In only two animals were "body arching laterally" still observed during the third minute of bleeding. These two animals also expressed signs of reduced stunning effectiveness, i.e., respiratory movements and/or tongue tone. However, based on these two single cases, we would not recommend late "body arching laterally" as a sole indicator of an inadequate stunning effect, but the occurrence of this movement after the end of the second minute of bleeding should lead to intensive monitoring for signs of an active brainstem.

An association between stunning effectiveness and movements cannot be statistically proven in this study due to the small number of animals with reduced stunning effectiveness. However, when looking at the movements monitored and scores calculated in cattle showing signs of shallow depth of stunning, these are comparable to those of properly stunned cattle when analysed on a case-by-case analysis.

Based on these results, the authors would like to emphasise the importance of paying attention to reliable indicators such as eye movements and/or resumption of breathing to evaluate stunning effectiveness, and not to be distracted by, e.g., kicking movements of the limbs or lateral body arching. The indicators have to be assessed in context and not just considered individually. The results show that movements occur regularly in properly stunned cattle.

4.5. Limitations

Due to technical reasons, we could not precisely measure the efficiency of the exsanguination, which could have had an effect either on stunning effectiveness or on movements. However, all staff involved were experienced and promptly repeated sticking if reduced blood flow was suspected. Another possible limitation in this study is that the values for the key parameters of bolt exit length and bolt velocity could only be approached and thus were not measured for every single shot. Currently, there is no technical solution to measure bolt velocity during the shot. Regarding the bolt exit length, examining a section of the skull of every animal would have been a possible solution, but this could not be included because of limitations in resources. As stated above, due to the small number of animals showing signs of reduced stunning effectiveness, only case-by-case analysis was possible concerning an association between stunning effectiveness and movements. Nevertheless, considering the current state of scientific knowledge, we assume that indicators other than movement are far more important for evaluating the effectiveness of stunning in cattle after captive bolt stunning.

4.6. Animal Welfare Implications

This description and analysis of movements can contribute to assessing the stunning effectiveness of cattle after captive bolt stunning more reliably. According to our results, most movements observed are not suitable to determine an insufficient stunning effect. When assessing stunning efficiency, we recommend looking for signs at the head such as breathing, eye movements, or recurrent ear tension since other movements like leg and body movements can still occur a few minutes after sticking in well-stunned cattle. The identification of the factors influencing these movements can contribute to understanding the movements, optimising the stunning process, and the further development of stunning devices. In order to correctly assess movements, the staff responsible for slaughter should be able to distinguish between conscious righting attempts and unconscious body or limb movements. Impressive movements after stunning should not divert the attention of those responsible for detecting the pertinent signs of reduced stunning effectiveness, which are comparatively inconspicuous. On the other hand, movements can have an impact on animal welfare when they impede prompt shackling and sticking, thereby posing challenges for employees in efficiently and safely performing these tasks. This aspect must be considered, as staff play a key role in maintaining high animal welfare standards during slaughter. Therefore, it is necessary to continue looking for ways to reduce movements.

5. Conclusions

For the first time, movements in cattle after captive bolt stunning were categorised and systematically described from landing up to at least the end of the fourth minute of bleeding. A total of 93.4% of all cattle examined showed at least one movement within the observation period. Although most cattle movements ended one minute after sticking, there were individual animals that showed movements for even longer, sometimes up to eight minutes after sticking. Factors affecting the category and frequency of movements could be identified, in relation to both the animals and the features of the captive bolt devices used. Breed, sex category, and the exit length and function type of the stunning device were the main impact factors found. The common field experience of Black and Red Holstein cows showing especially strong movements during landing and hoisting was confirmed. In total, 48.9% of all cows showed movements on the landing grid, whereas related to the whole observation period, bulls and heifers had a higher movement activity (sum score) than cows. Factors related to the stunning device also had an impact on movements. Both the use of pneumatically powered devices and the use of captive bolt guns with an increased exit length resulted in a significantly lower sum score. However, the variables investigated explain only 12% of the variance; thus, the effects of unknown confounding variables cannot be excluded.

Excitatory movements at landing hinder the ability of employees to quickly and safely perform hoisting and sticking. As the occurrence and intensity of movements are only explained to a limited extent by the identified process-related factors, it is necessary to continue looking for ways to reduce movements from landing to sticking.

Out of 2911 cattle, 99.4% showed no signs of reduced stunning effectiveness. This positive development in enhanced stunning effectiveness is attributed to, among other things, the use of modern stunning devices (esp. pneumatic-driven stunners) and stun boxes with tight head restraint. Thus, slight angular or positional shooting deviations do not necessarily lead to a reduced stunning effect. Movements of limbs or the tail, as well as lateral or ventral movements of the head and/or trunk, do not indicate limited stunning effectiveness. Only "body arching laterally" still being expressed in the third minute of bleeding might be an indication for insufficient stunning and bleeding, but it should only be considered together with respiratory movements or preserved eye reflexes. It is assumed that after successful captive bolt stunning in cattle, movements commonly occur for several minutes after the shot and, with the exception of "arching backwards", are not a suitable sole indicator of reduced stunning effectiveness. In terms of animal welfare, it is crucial to accurately recognize signs of reduced stunning effectiveness. Staff, veterinarians, and external inspectors should prioritize signs of an active brainstem over limb, body, and tail movements. Ensuring the correct identification of signs of remaining or regaining consciousness is a key factor in guaranteeing animal welfare at slaughter.

Supplementary Materials: With regard only to the definition of movements, the following supporting information (video examples) can be downloaded at: https://doi.org/10.5281/zenodo.10572807 (accessed on 19 March 2024), Video S1: 1 Kicking hind limb moderate landing unconscious cow; Video S2: 2 Twitching intense landing unconscious cow; Video S3: 3 Kicking hind limb intense landing unconscious cow; Video S4: 4 Twitching moderate hoisting unconscious bull; Video S5: 5 Twitching intense hoisting unconscious bull; Video S6: 6 Body arching laterally hoisting unconscious heifer; Video S7: 7 Stretching hind limb hoisting lifting forelimb sticking unconscious bull; Video S8: 8 Body arching ventrally kicking hind limb sticking 1st min of bleeding unconscious bull; Video S9: 9 Body arching laterally 1st min of bleeding unconscious bull; Video S10: 10 Body arching ventrally bending hind and forelimbs 1st min of bleeding unconscious heifer; Video S11: 11 Lifting forelimb kicking hind limb 1st min of bleeding unconscious bull; Video S12: 12 Bending hind limb 3rd minute of bleeding unconscious bull; Video S13: 13 Bending forelimb bending hind limb 4th minute of bleeding unconscious cow; Video S14: 14 Stretching hind limb with shivering sixth min of bleeding unconscious cow; Figure S1: 15 Arching backwards.

Author Contributions: Conceptualization, K.v.H. and M.v.W.; methodology, A.L. and K.v.H.; software, K.v.H. and M.v.W.; validation, A.L., K.v.H., H.L. and M.v.W.; formal analysis, A.L.; investigation, A.L., K.v.H. and M.v.W.; resources, K.v.H. and M.v.W.; data curation, A.L.; writing—original draft preparation, A.L.; writing—review and editing, K.v.H. and H.L.; visualization, A.L. and K.v.H.; supervision, K.v.H., H.L. and M.E.; project administration, K.v.H.; funding acquisition, K.v.H. All authors have read and agreed to the published version of the manuscript.

Funding: This research was funded by the Federal Ministry of Food and Agriculture (Germany) (BMEL) based on a decision of the Parliament of the Federal Republic of Germany via the Federal Office for Agriculture and Food (BLE), project code 2817801X18.

Institutional Review Board Statement: Ethical review and approval were waived for this study due to reason This study did not involve procedures on animals requiring formal approval from an animal ethics committee. All data collection occurred during and after routine slaughter, with all animals intended for use in the food chain. Thus, this study did not constitute an animal experiment in terms of the German Animal Welfare Act [43]. The observations were conducted without any interference with the animals. The video material was produced in a written agreement with the respective abattoirs and the competent authorities.

Informed Consent Statement: Informed consent was obtained from all subjects involved in this study.

Data Availability Statement: The data are not publicly accessible due to security concerns for the companies involved. Further enquiries can be directed to the author.

Acknowledgments: We particularly thank the participating slaughter companies for their great cooperation, Jarvis GmbH, Buchholz in der Nordheide, and PTB Braunschweig for providing the measuring instruments and technical assistance, as well as the Max Rubner Institute for statistical support.

Conflicts of Interest: The funders had no role in the design of this study; in the collection, analyses, or interpretation of data; in the writing of the manuscript; or in the decision to publish the results. The authors declare no conflicts of interest.

References

1. EFSA. *Welfare of Cattle at Slaughter*; 1831-4732; EFSA Panel on Animal Health Welfare: Parma, Italy, 2020; p. e06275.
2. Council Regulation (EC) No 1099/2009 of 24 September 2009 on the protection of animals at the time of killing. 2009, 1–30. Available online: https://eur-lex.europa.eu/legal-content/EN/ALL/?uri=CELEX:32009R1099 (accessed on 3 April 2024).
3. Grist, A.; Lines, J.A.; Bock, R.; Knowles, T.G.; Wotton, S.B. An Examination of the Performance of Blank Cartridges Used in Captive Bolt Devices for the Pre-Slaughter Stunning and Euthanasia of Animals. *Animals* **2019**, *9*, 552. [CrossRef]
4. Tierschutz-Schlachtverordnung. TierSchlV-Verordnung zum Schutz von Tieren im Zusammenhang mit der Schlachtung oder Tötung zur Durchführung der Verordnung (EG) Nr. 1099/2009 des Rates. *BGBl* **2012**, *1*, 2982. Available online: https://www.gesetze-im-internet.de/tierschlv_2013/ (accessed on 1 November 2023).
5. Commission Regulation (EC) No 1326/2001 of June 2001 laying down transitional measures to permit the changeover to the Regulation of the European Parliament and of the Council (EC) No 999/2001 laying down rules for the prevention, control and eradication of certain transmissible spongiform encephalopathies, and amending Annexes VII and XI to that Regulation. *(2000/418/EG)*. 2001, pp. 60–67. Available online: https://eur-lex.europa.eu/legal-content/DE/ALL/?uri=CELEX:32001R1326 (accessed on 1 November 2023).
6. Von Holleben, K.; Schütte, A.; von Wenzlawowicz, M.; Bostelmann, N. Call for veterinary action in slaughterhouses-deficient welfare at CO_2-stunning of pigs and captive bolt stunning of cattle. *Fleischwirtsch. Int.* **2002**, *3*, 8–11.
7. Gregory, N.G.; Lee, C.J.; Widdicombe, J.P. Depth of concussion in cattle shot by penetrating captive bolt. *Meat Sci.* **2007**, *77*, 499–503. [CrossRef] [PubMed]
8. Troeger, K.; Moje, M. Zero percent stunning failure rate? New equipment for captive bolt stunning of cattle shows convincing effect. *Fleischwirtschaft* **2012**, *92*, 26–29.
9. Von Wenzlawowicz, M.; von Holleben, K.; Eser, E. Identifying reasons for stun failures in slaughterhouses for cattle and pigs: A field study. *Anim. Welf.* **2012**, *21*, 51–60. [CrossRef]
10. EFSA. *Opinion of the Scientific Panel on Animal Health and Welfare on a Request from the Commission Related to Welfare Aspects of the Main Systems of Stunning and Killing the Main Commercial Species of Animals*; Question N° EFSA-Q-2003-093; EFSA: Parma, Italy, 2004; pp. 1–29.
11. Endres, J.M. Effectiveness of Concussion Stunning in Comparison to Captive Bolt Stunning in Routine Slaughtering of Cattle. Ph.D. Thesis, Ludwig-Maximilians-Universität München, Munich, Germany, 2005.
12. Dörfler, K. Assessment of the Efficiency of Different Captive Bolt Stunning Devices in Cattle and Their Influence on the Degree of Bleeding. Ph.D. Thesis, Universität Leipzig, Leipzig, Germany, 2015.
13. Grandin, T.; Velarde, A.; Strappini, A.; Gerritzen, M.; Ghezzi, M.; Martinez-Burnes, J.; Hernandez-Avalos, I.; Dominguez-Oliva, A.; Casas-Alvarado, A.; Mota-Rojas, D. Slaughtering of Water Buffalo (*Bubalus bubalis*) with and without Stunning: A Focus on the Neurobiology of Pain, Hyperalgesia, and Sensitization. *Animals* **2023**, *13*, 2406. [CrossRef]
14. Hilsenbeck, E.M. Examinations of Bleeding-Time in Cattle after Capitive Bolt Stunning. Ph.D. Thesis, Ludwig-Maximilians-Universität München, Munich, Germany, 2007.
15. Von Holleben, K.; von Wenzlawowicz, M. Are movements after captive bolt stunning of cattle a sign of regaining consciousness. In Proceedings of the 65th ICoMst 2019, Berlin, Germany, 4 August 2019; pp. 1–4.
16. Martin, M.S.; Kline, H.C.; Wagner, D.R.; Alexander, L.R.; Edwards-Callaway, L.N.; Grandin, T. Evaluation of different captive bolt lengths and breed influence upon post-stun hind limb and forelimb activity in fed cattle at a commercial slaughter facility. *Meat Sci.* **2018**, *143*, 159–164. [CrossRef] [PubMed]
17. Terlouw, E.M.; Bourguet, C.; Deiss, V.; Mallet, C. Origins of movements following stunning and during bleeding in cattle. *Meat Sci.* **2015**, *110*, 135–144. [CrossRef]
18. Verhoeven, M.T.W. Assessing Unconsciousness in Livestock at Slaughter. Ph.D. Thesis, Wageningen University, Wageningen, The Netherlands, 2016.
19. Grillner, S.; Wallén, P.; Saitoh, K.; Kozlov, A.; Robertson, B. Neural bases of goal-directed locomotion in vertebrates—An overview. *Brain Res. Rev.* **2008**, *57*, 2–12. [CrossRef]
20. Fricker, C.; Riek, W. Using a captive bolt pistol to stun cattle before slaughter. *Fleischwirtschaft* **1981**, *61*, 124–127.
21. Gibson, T.J.; Oliveira, S.E.O.; Costa, F.A.D.; Gregory, N.G. Electroencephalographic assessment of pneumatically powered penetrating and non-penetrating captive-bolt stunning of bulls. *Meat Sci.* **2019**, *151*, 54–59. [CrossRef] [PubMed]
22. Kline, H.C.; Wagner, D.R.; Edwards-Callaway, L.N.; Alexander, L.R.; Grandin, T. Effect of captive bolt gun length on brain trauma and post-stunning hind limb activity in finished cattle Bos taurus. *Meat Sci.* **2019**, *155*, 69–73. [CrossRef]
23. EFSA AHAW Panel. Scientific Opinion on monitoring procedures at slaughterhouses for bovines. *EFSA J.* **2013**, *11*, 3460. [CrossRef]
24. Von Holleben, K.; Luecking, A.; von Wenzlawowicz, M. Signs of reduced stunning effectiveness in cattle and presentation of open access animations. In Proceedings of the 4th Annual Meeting of the European Veterinary Congress of Behavioural Medicine and Animal Welfare Palma, Mallorca, Spain, 28 September 2022.
25. Grandin, T.; Cockram, M. *The Slaughter of Farmed Animals: Practical Ways of Enhancing Animal Welfare*; Grandin, T., Cockram, M., Eds.; CABI International: Wallingford, UK, 2020.
26. Welfare Quality®. *Assessment Protocol for Cattle*; Welfare Quality®: Lelystad, The Netherlands, 2009.

27. Gregory, N.G. Physiology of Stress, Distress, Stunning and Slaughter. In *Animal Welfare and Meat Science*; Gregory, N.G., Ed.; CAB International: Wallingford, UK, 1998; pp. 64–92.
28. Terlouw, C.; Bourguet, C.; Deiss, V. Consciousness, unconsciousness and death in the context of slaughter. Part I. Neurobiological mechanisms underlying stunning and killing. *Meat Sci.* **2016**, *118*, 133–146. [CrossRef] [PubMed]
29. Terlouw, C.; Bourguet, C.; Deiss, V. Consciousness, unconsciousness and death in the context of slaughter. Part II. Evaluation methods. *Meat Sci.* **2016**, *118*, 147–156. [CrossRef] [PubMed]
30. Magnus, R. Some results of studies in the physiology of posture II. *Lancet* **1926**, *2*, 585–588.
31. EFSA. *Scientific Opinion—Guidance on the Assessment Criteria for Studies Evaluating the Effectiveness of Stunning Interventions Regarding Animal Protection at the Time of Killing*; EFSA Panel on Animal Health and Welfare (AHAW); European Food Safety Authority (EFSA): Parma, Italy, 2013; p. 41.
32. Terlouw, E.M.C.; Bourguet, C.; Deiss, V. Stress at slaughter in cattle: Role of reactivity profile and environmental factors. *Anim. Welf.* **2012**, *21*, 43–49. [CrossRef]
33. Grandin, T.; Deesing, M. Genetics and Learned Behavior. In *Humane Livestock Handling*; Boyd-Owens, R., Burns, D., Guare, S., Eds.; Storey Publishing: North Adams, MA, USA, 2008; pp. 16–30.
34. Marzin, V.; Collobert, J.F.; Jaunet, L.; Marrec, L. Critères pratiques de mesure de l'efficacité et de la qualité de l'étourdissement par tige perforante chez le bovin. *Rev. Med. Vet.* **2008**, *159*, 423–430.
35. Gibson, T.J.; Mason, C.W.; Spence, J.Y.; Barker, H.; Gregory, N.G. Factors affecting penetrating captive bolt gun performance. *J. Appl. Anim. Welf. Sci.* **2015**, *18*, 222–238. [CrossRef]
36. Algers, B.; Atkinson, S. Stun quality in relation to cattle size, gun type and brain hemorrhages. In Proceedings of the 13th International Congress in Animal Hygiene, Tartu, Estonia, 17 June 2021; pp. 1028–1031.
37. Atkinson, S.; Velarde, A.; Algers, B. Assessment of stun quality at commercial slaughter in cattle shot with captive bolt. *Anim. Welf.* **2013**, *22*, 473–481. [CrossRef]
38. Ilgert, H. Effizienz der Bolzenschußbetäubung beim Rind mit Berücksichtigung der Einschußstelle und der Eindringrichtung des Bolzens unter Praxisbedingungen. Ph.D. Thesis, Freie Universität Berlin, Berlin, Germany, 1985.
39. Kaegi, B. Untersuchung zur Bolzenschussbetäubung beim Rind. Ph.D. Thesis, Universität Zuerich, Zuerich, Switzerland, 1988.
40. Maennl, M. Betäubung und Entblutung von Schwein und Rind. *Mitteilungsblatt-Bundesanst. Für Fleischforsch. Kulmb.* **1994**, *1*, 347–347.
41. McKinstry, J.L.; Anil, M.H. The effect of repeat application of electrical stunning on the welfare of pigs. *Meat Sci.* **2004**, *67*, 121–128. [CrossRef] [PubMed]
42. Grandin, T. How to determine insensibility (unconsciousness) in cattle, pigs, and sheep in slaughter plants. 2015, revised June 2017. Available online: http://www.grandin.com/humane/insensibility.html (accessed on 15 March 2024).
43. Animal Protection Act. Federal Act on the Protection of Animals (Animal Protection Act–TSchG). 2020. Available online: https://www.gesetze-im-internet.de/tierschg/BJNR012770972.html (accessed on 3 April 2024).

animals

MDPI

Article

Analysis of the Broiler Chicken Dead-on-Arrival (DOA) Rate in Relation to Normal Transport Conditions in Practice in Germany

Julia Gickel [1,*], Christian Visscher [1,2], Nicole Kemper [1,3] and Birgit Spindler [3]

1 WING (Science and Innovation for Sustainable Poultry Production), University of Veterinary Medicine Hannover, Foundation, 49456 Bakum, Germany; christian.visscher@tiho-hannover.de (C.V.); nicole.kemper@tiho-hannover.de (N.K.)
2 Institute for Animal Nutrition, University of Veterinary Medicine, Foundation, 30173 Hannover, Germany
3 Institute for Animal Hygiene, Animal Welfare and Farm Animal Behavior, University of Veterinary Medicine Hannover, Foundation, 30173 Hannover, Germany; birgit.spindler@tiho-hannover.de
* Correspondence: julia.gickel@tiho-hannover.de

Simple Summary: The dead-on-arrival (DOA) rate is a good indicator of the quality of broiler chicken transport. This evaluation investigated factors that might have an influence on the DOA rate. Therefore, data from the transport of broilers in Germany between January 2022 and May 2023 were analyzed. The results showed no influence on the DOA rate of the amount of animals per transport, the duration and the distance of the transport, the stocking density, and the average daily temperature. Transport between 11:00 and 17:00 and in fall tended to have a higher DOA rate.

Abstract: In total, around 631 million broilers were slaughtered in Germany in 2022. This evaluation included data of approx. 198 million broilers of different ages and breeds that were transported in Germany in 2022 (31% of all cases of broiler chicken transport in 2022). The aim of this study was to analyze German broiler chicken transport (n = 14,054) to the slaughterhouse between January 2022 and May 2023 with regard to the dead-on-arrival (DOA) rate and the possible influencing factors. Therefore, the relation between the total amount of animals per transport, the duration and distance of the transport, the planned stocking density in the transport cages, the average daily temperature and time of day and season of the transport as well as the DOA rate were statistically evaluated. The results showed a mean DOA rate of 0.09% (SD 0.09). Transport conducted at midday (11:00 to 17:00) showed higher DOA rates ($p < 0.05$) than transport at other times (day split into 6 h intervals). The highest mean DOA rate (0.10%) was found in the fall, followed by the winter, while transport in the spring and summer resulted in the lowest DOA rate ($p < 0.05$). All in all, the relatively low DOA rate (%) in Germany indicates the good standard of their broiler transport compared to available data from research in other European countries.

Keywords: slaughterhouse; broiler; transport; DOA; influencing factors; duration; distance; temperature

Citation: Gickel, J.; Visscher, C.; Kemper, N.; Spindler, B. Analysis of the Broiler Chicken Dead-on-Arrival (DOA) Rate in Relation to Normal Transport Conditions in Practice in Germany. *Animals* **2024**, *14*, 1947. https://doi.org/10.3390/ani14131947

Academic Editor: Elbert Lambooij

Received: 12 June 2024
Revised: 27 June 2024
Accepted: 30 June 2024
Published: 30 June 2024

1. Introduction

In 2022, about 631 million broilers were slaughtered in Germany, and the preliminary data for 2023 indicate a similar level [1]. It can be assumed that approximately the same number of animals were transported to the slaughterhouse in both years. The transport of animals is regulated in accordance with European Union (Council Directive (EC) 1/2005) and national law (Animal Welfare Transport Ordinance—TierSchTrV, 2009) [2,3]. These regulations deal with requirements concerning the stocking density and duration of transport and the conditions of transportation cages, for instance. It is known that the catching and loading of the animals for transport, and also the transport itself, may cause stress for the animals [4,5]. The amount of dead animals (dead on arrival, DOA) is used as a first indication to identify pre-slaughter welfare issues [6].

The European Food Safety Authority (EFSA) stated in its scientific opinion on 'Welfare of domestic birds [...] transported in containers' that DOA rates above 0.1% should be investigated [7]. In Germany, the Board of Trustees for Technology and Construction in Agriculture e.V. (KTBL) published a target range of 0.05% and an alarm range of 0.3% [8]. Both should be used as indicators for farmers to evaluate the actual status of animal welfare on their farms [8]. The implementing provision of the national law for Lower Saxony (2014) published a target DOA rate of 0.5%. If a specific transport exceeds this threshold, the slaughterhouse is obliged to report it to the veterinary authorities, which are allowed to initiate follow-up controls of this transport [9].

Comprehensive and recent data about DOA rates are partly available from different European countries, and some of them describe factors influencing the DOA rate. In general, the published mean DOA rates range from 0.126% (in the United Kingdom, 2005 [10]) to 0.46% (data from 2000 and 2001, focusing on Dutch and German flocks slaughtered in the Netherlands [11]).

In the United Kingdom, higher DOA rates were detected between June and August [10]. In the Netherlands, a longer transport time and a higher stocking density led to increased DOA values [11]. A study with data from 1997 to 2004 from the Czech Republic noticed that a shorter distance of transport resulted in fewer dead animals [12]. In the latter data, the highest DOA values were found between June and August and between December and February. A large Italian study with 1226 million broiler chickens (data from August 2001 to July 2005, mean DOA rate: 0.35%) confirmed higher DOA rates during summer [13]. A study from France [14] published a DOA rate of 0.18%, showing the influence of the density in crates (more space allowance being associated with less mortality) and climatic conditions (rain and wind being associated with more DOA). More recent European studies were published with data from Belgium from 2013/2014 (DOA: 0.3% [6]) and with data from Spain from 2015/2016 (DOA: 0.187% [15]). In the Spanish study, a maximum temperature above 21.5 °C, a high distance of transport, a long time of transport, and a high average weight were rather related to a high level of DOA [15].

In 2006, a Belgian publication investigated possible reasons for DOA, finding that caution in loading and the general good condition of the animals is the best way to prevent dead animals during transport [16]. A more recent study also stated that lung congestion is the most frequent reason for dead animals during transit [17].

In Germany, comprehensive data on the transport of broiler chickens to the slaughterhouse are missing, as these were previously only used for internal control purposes and have not been collected from the companies for research purposes. This evaluation has succeeded in collecting and harmonizing these data in order to gain a meaningful insight into the German transportation of broilers. The aim of this study was to analyze data on the transport of broiler chickens under usual transport conditions in practice in Germany conducted between January 2022 and May 2023 with regard to the DOA rate and to identify possible influencing factors. For this purpose, the relationship between the number of transported animals, the distance and duration of the transport, the stocking density in the truck, the average daily temperature and the season of the transport, the time of transport and loading as well as the origin of the transport and the DOA rate was investigated. In addition, it was investigated whether transports with a short/long duration, a specific season or a specific average daily temperature were postponed to a specific time in order to avoid high DOA rates, as indicated by the results of previous smaller studies.

2. Materials and Methods

2.1. Description of the Data Set

In total, 14,054 broiler chicken transports between January 2022 and May 2023 (3 January 2022 to 9 May 2023), with about 210 million broilers, were included. These data were provided by several companies with access to databases containing data from already conducted transports. For each transport, the data set included the DOA rate, defined as the percentage of animals per transport unit that were dead on arrival at the slaughterhouse. In this case, the transport unit was not the single truck but all the animals loaded at this destination. The other included parameters were information about the number of transported broiler chickens per transport unit, the actual date of the transport, planned start and end time of the loading and transport, intended duration (min) and distance (km) of the transport, and the zip code (for transports with a German origin only) or the country of origin (Figure 1). About 85% of the data on conducted transports also included the planned stocking density inside the transport cages in the truck (kg per cm^2, calculated with the average live weight at loading and the size of the cages). The loading times and the durations and distances of the transports were provided as planned data for the respective transport, while in contrast, the arrival at the slaughterhouse, date of transport, number of broilers, and origin of transport were real data (Figure 2). Possible changes, for example, because of a lot of traffic, delays during the loading process, or general rearrangements were not considered. Transports with known unplanned incidents, such as failure of technology on the truck, were excluded from this data set. If either the distance or the duration of a specific transport was missing, a mean speed of 60 km per hour was assumed to calculate the missing value.

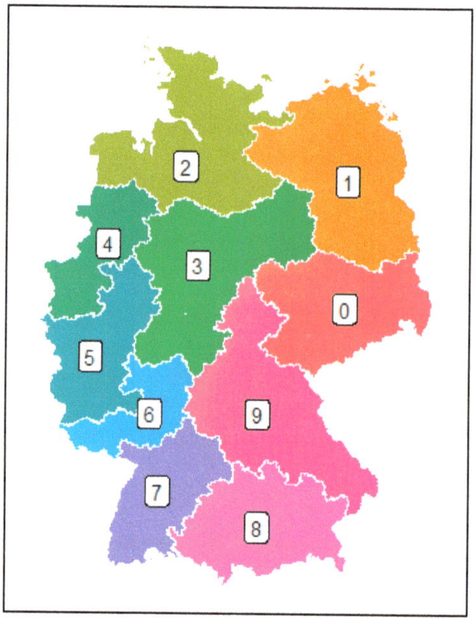

Figure 1. Overview of Germany, with areas classified based on the first digit of the zip code.

The date of each individual transport was used to find out the average daily temperature recorded in Northern Germany (historic data from Bremen, Germany) by the German Meteorological Service (DWD).

The date was also used to categorize each transport into a certain season. Therefore, all the transports in March to May were considered as "spring", all the transports from June to August as "summer", all the transports from September to November as "fall", and all the transports between December and February were classified as "winter". To distinguish the times of loading and transport, for each of them, the end time was used to classify them into "night" (23:00 to 05:00), "morning" (05:00 to 11:00), "midday" (11:00 to 17:00), or "evening" (17:00 to 23:00).

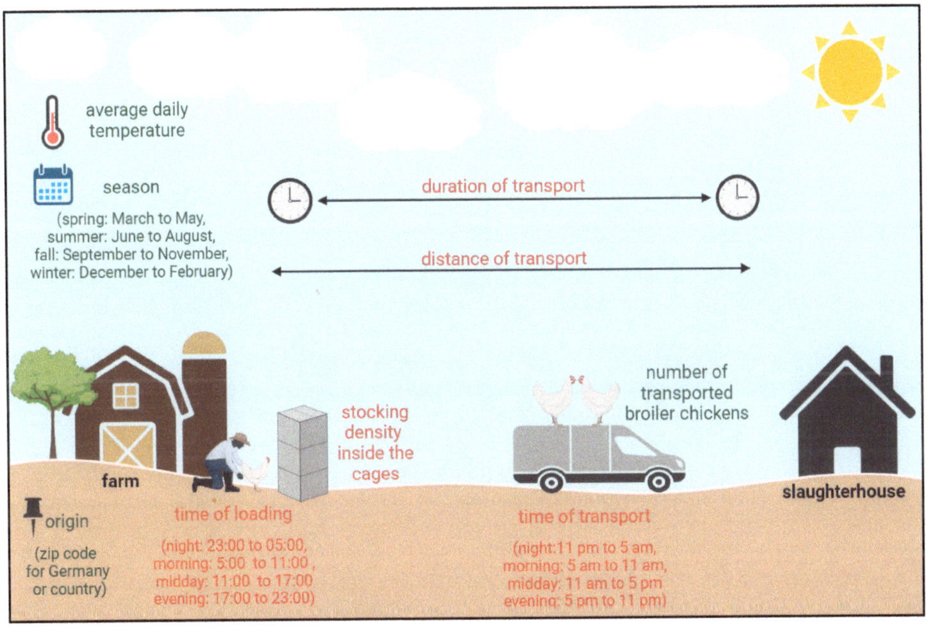

Figure 2. Overview of the evaluated factors influencing the dead-on-arrival (DOA) rate; planned data displayed in red, real data displayed in green (created with BioRender.com).

All the parameters were classified as planned or real data (Figure 2).

2.2. Statistics

The influence of the number of animals per transport unit, the duration and distance of the transport, the mean daily temperature, and the stocking density on the DOA rate was evaluated using a correlation analysis (according to Pearson). Cohen's guideline (1988) was used to assess the correlation as follows: <0.1 = no relation, ≥0.1 and <0.3 = small relation, ≥0.3 and <0.5 = medium relation, and ≥0.5 = high relation [18]. Differences in the season, loading time, time of transport, first digit of the zip code or country on the level of the DOA rate were evaluated using an ANOVA with a post hoc Tukey's test with a significance level of 0.05. If boxplots are shown, they include the first and third quartile as the frame of the box, the mean marked with a line in between the box, and values below the first quartile and above the third quartile marked with dots. To describe the distribution of the data, the 95% confidence interval and/or the standard deviation (SD) were calculated. A possible influence of different factors on the loading time and the time of transports was furthermore studied by focusing on the percentage distribution.

All the calculations were performed using R Statistical Software (v4.3.0 [19]).

3. Results

3.1. Distribution of the Level of the DOA Rate (%)

Most transports showed a low DOA rate (Figure 3). The lowest DOA rate (%) was 0.00 and the highest 3.36, with a mean DOA rate (%) of 0.09. In total, 0.09% of all the transports reached the slaughterhouse without any dead broilers. In 50% of all the transports, the DOA rate (%) was below 0.07; in 75% of all the transports, it was below 0.12; and in 90% of all the transports, it was below 0.19. Only 0.06% of all the transports were above the rate of 1%. A DOA rate of below 0.5% was reached by 99.4% of all the transports (Table 1).

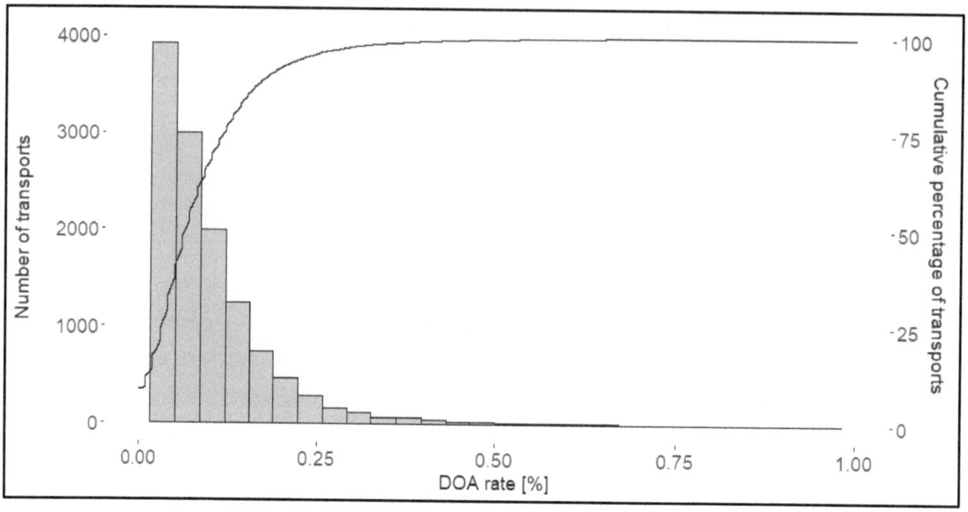

Figure 3. Number of transports (bars) and cumulative percentage (line) of transports per dead-on-arrival (DOA) rate [%]) (n = 14,054 broiler transports).

Table 1. Percentage of transports and number of transported animals within the data set below published target values [3,7,8] (KTBL = Board of Trustees for Technology and Construction in Agriculture e.V., EFSA = European Food Safety Authority, TierSchTrV = Animal Welfare Transport Ordinance—national law).

Origin of Target Value	Value	Proportion of Transports within This Target Area (%)	Number of Animals within These Transports
KTBL—target range	0.05%	37.3	70 million
EFSA	0.1%	67.9	137 million
KTBL—alarm range	0.3%	97.3	205 million
TierSchTrV	0.5%	99.4	209 million

3.2. *Factors Influencing the DOA Rate (%)*

The following sections describe the results for each influencing factor. In each figure, the y-axis is set to the boundaries of 0% at the lower and 1.5% at the upper end. Outliers above this level (very few; see Section 3.1) are therefore not visible in the figures.

3.2.1. Relation between the Number of Transported Animals and the DOA Rate (%)

The number of transported broilers per transport unit ranged between 242 and 62,377 animals (Figure 4), with a mean of 14,936 broilers (SD: 7260). The confidence interval was 706 to 29,165. With a correlation coefficient of 0.06, no relation was found between the number of broilers per transport unit and the DOA rate.

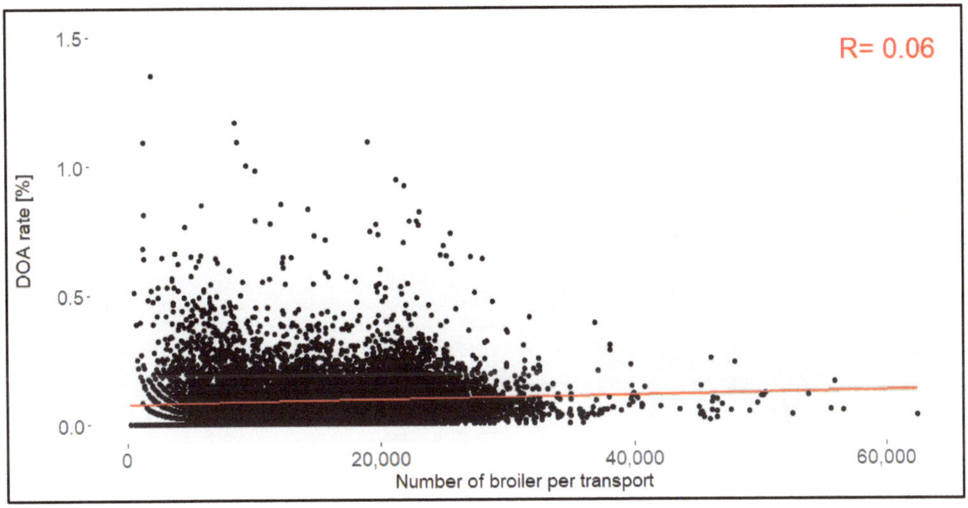

Figure 4. Relation between the number of broilers per transport unit and the dead-on-arrival (DOA) rate [%], including the correlation coefficient (R) (n = 14,054 broiler transports).

3.2.2. Relation between the Distance or Duration of Transport and the DOA Rate (%)

The transport distance between the loading location and the slaughterhouse ranged between 4.0 km and 450.0 km (Figure 5), with a mean of 95.3 km (SD: 81.0 km). This led to a duration of between 1 min and 465 min and a mean of 92 min (SD: 78 min), respectively (Figure 6). With correlation coefficients of 0.08 and 0.09, no relation was found between the distance or the duration and the DOA rate. The confidence intervals included all the data below 254 km and below 245 min.

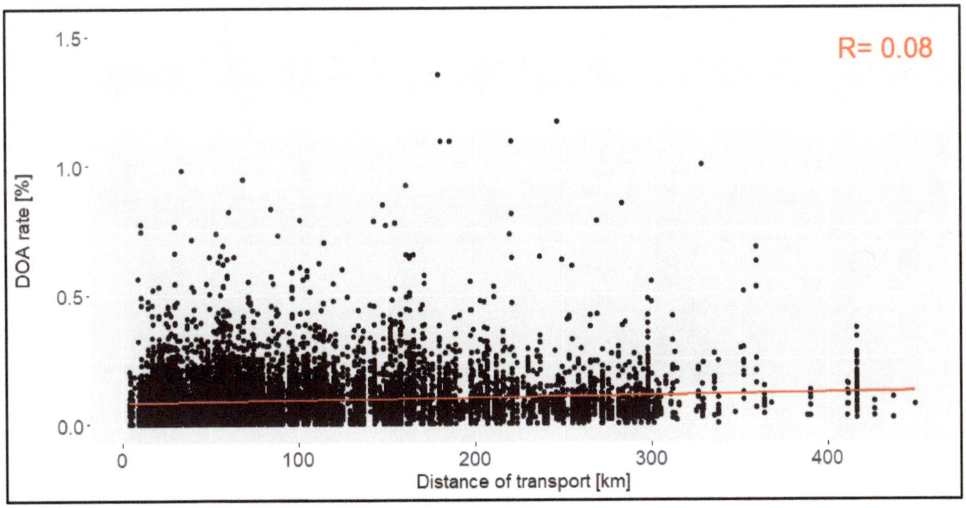

Figure 5. Relation between the distance of the transport [km] and the dead-on-arrival (DOA) rate [%], including the correlation coefficient (R) (n = 14,054 broiler transports).

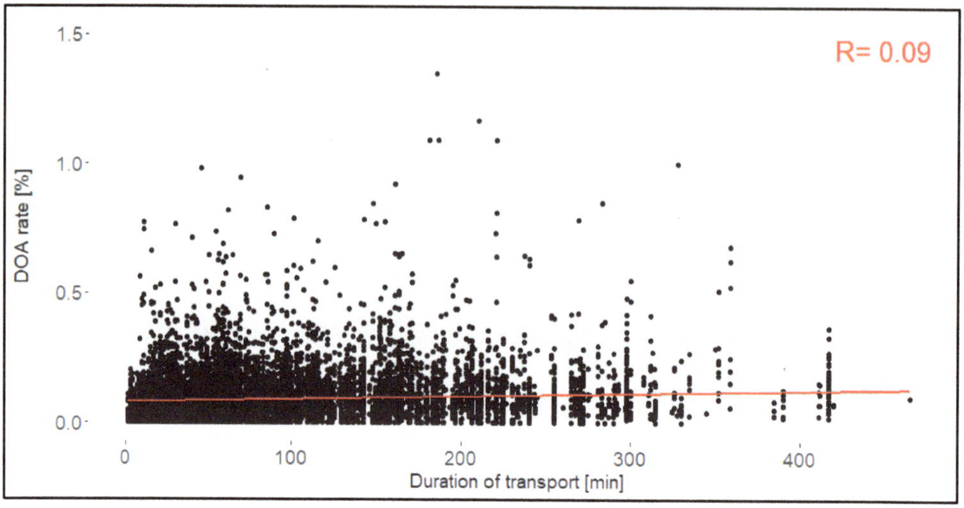

Figure 6. Relation between the duration of the transport [min] and the dead-on-arrival (DOA) rate [%], including the correlation coefficient (R) (n = 14,054 broiler transports).

3.2.3. Relation between the Stocking Density inside the Truck and the DOA Rate (%)

The stocking density in all 11,966 transports with data available concerning this factor differed between 0.0004 kg per cm^2 and 0.023 kg per cm^2, with a mean value of 0.005 kg per cm^2 (Figure 7). The correlation coefficient was 0.08, indicating that there was no relation between the stocking density and the DOA rate. The confidence interval was 0.0036 to 0.0068 kg per cm^2.

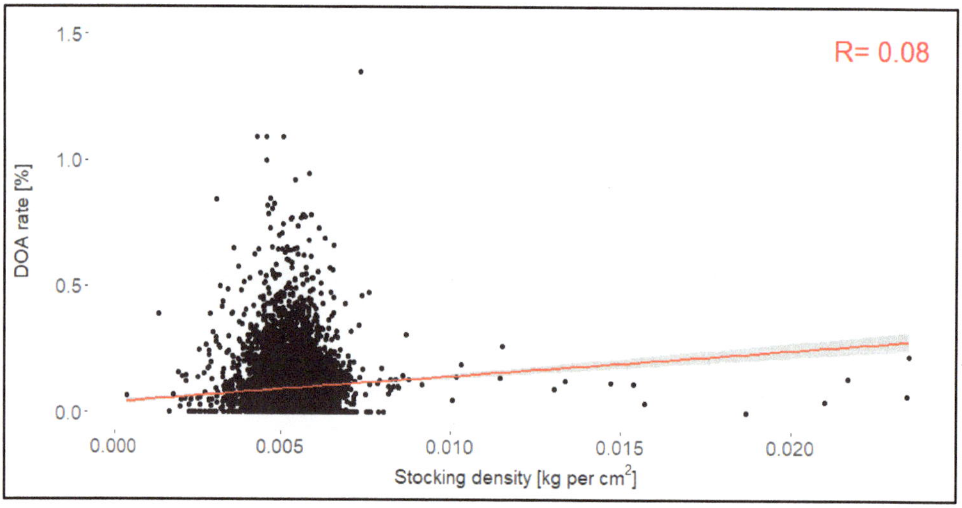

Figure 7. Relation between the stocking density [kg per cm^2] and the dead-on-arrival (DOA) rate [%], including the correlation coefficient (R) (n = 11,966 broiler transports).

3.2.4. Relation between the Average Daily Temperature in Northern Germany and the DOA Rate (%)

The average daily temperature in Northern Germany for all the transports differed from -6.6 °C to 28.1 °C, showing a mean value of 11.0 °C (Figure 8). The confidence interval ranged from -1.8 °C to 24.3 °C. The correlation was 0.03, indicating that there was no relation between the average daily temperature in Northern Germany and the DOA rate. Transports below 0 °C (3.4% of all the transports) and above 25 °C (0.8% of all the transports) were rare.

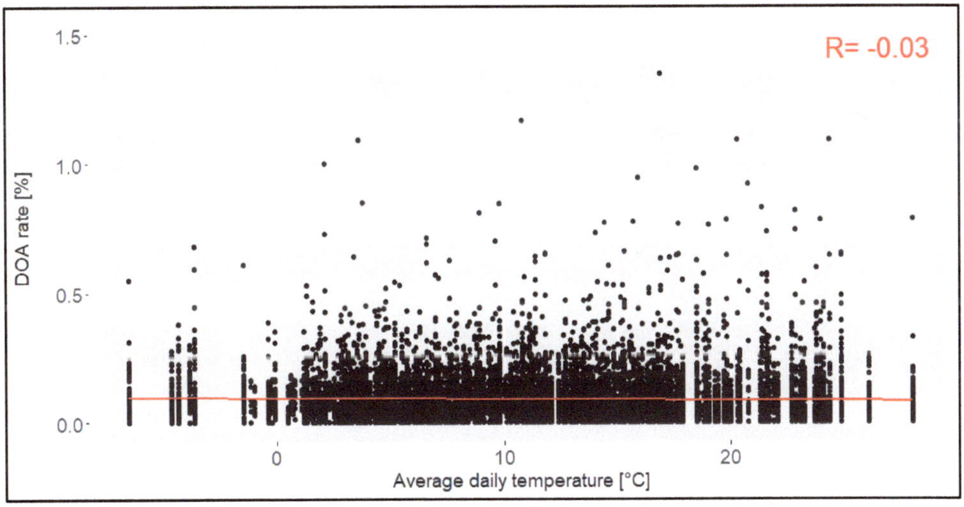

Figure 8. Relation between the average daily temperature [°C] and the dead-on-arrival (DOA) rate [%], including the correlation coefficient (R) (n = 14,054 broiler transports).

3.2.5. Relation between the Season of the Transport and the DOA Rate (%)

In general, the numbers of transports per season were similar (Table 2). In all the seasons, the minimum DOA rate was 0.00%, while the highest values differed from 1.00% (spring) to 3.36% (summer). The ANOVA results showed that the DOA rate was the highest in fall (significant), followed by the DOA rate in winter (Figure 9). The DOA rates in spring and summer were similar to each other. The boxplots, or rather the location of the first and third quartiles, indicated that most of the transports in all the seasons showed a DOA rate above 0.02% and below 0.13%.

Table 2. Measures of the position and dispersion of the dead-on-arrival (DOA) rate [%] per season (n = 14,054 broiler transports).

Measures of Location	Spring	Summer	Fall	Winter
Minimum	0.00	0.00	0.00	0.00
First Quartile	0.03	0.02	0.04	0.04
Median	0.06	0.06	0.07	0.07
Mean	0.08	0.08	0.10	0.09
Third Quartile	0.11	0.11	0.13	0.12
Maximum	1.00	3.36	1.17	1.73
n	3688	3657	3437	3272

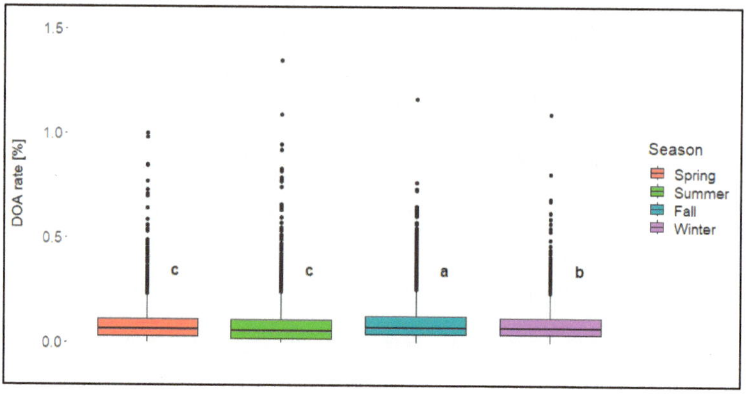

Figure 9. Boxplots of the dead-on-arrival (DOA) rate [%] per season; significant differences ($p < 0.05$) are marked with different letters (n = 14,054 broiler transports).

3.2.6. Relation between the Loading Time or the Time of Transport and the DOA Rate (%)

The highest percentage of all the loadings (approx. 37%) and the highest percentage of all the transports (approx. 42%) were conducted in the morning (Tables 3 and 4), while the lowest percentage was conducted in the evening (7% of all the loadings, 8% of all the transports). The significant ($p < 0.05$) highest mean DOA rate was found concerning the loadings at midday and concerning the transports at midday or in the evening (Figures 10 and 11). For both influencing factors, all the other time categories were similar to each other.

Table 3. Measures of the position and dispersion of the dead-on-arrival (DOA) rate [%] per loading time (categories: night [23:00 to 05:00], morning [05:00 to 11:00], midday [11:00 to 17:00], evening [17:00 to 23:00] (n = 14,054 broiler transports).

Measures of Location	Night	Morning	Midday	Evening
Minimum	0.00	0.00	0.00	0.00
First Quartile	0.03	0.03	0.04	0.03
Median	0.06	0.07	0.07	0.06
Mean	0.08	0.09	0.10	0.09
Third Quartile	0.10	0.12	0.13	0.11
Maximum	1.73	0.94	3.36	1.09
n	5099	5142	2801	1012

Table 4. Measures of the position and dispersion of the dead-on-arrival (DOA) rate [%] per time of transport (categories: night [23:00 to 05:00], morning [05:00 to 11:00], midday [11:00 to 17:00], evening [17:00 to 23:00]) (n = 14,054 broiler transports).

Measures of Location	Night	Morning	Midday	Evening
Minimum	0.00	0.00	0.00	0.00
First Quartile	0.03	0.03	0.04	0.03
Median	0.06	0.06	0.08	0.07
Mean	0.08	0.08	0.10	0.09
Third Quartile	0.11	0.11	0.14	0.12
Maximum	1.73	1.00	3.36	1.09
n	3809	5872	3203	1170

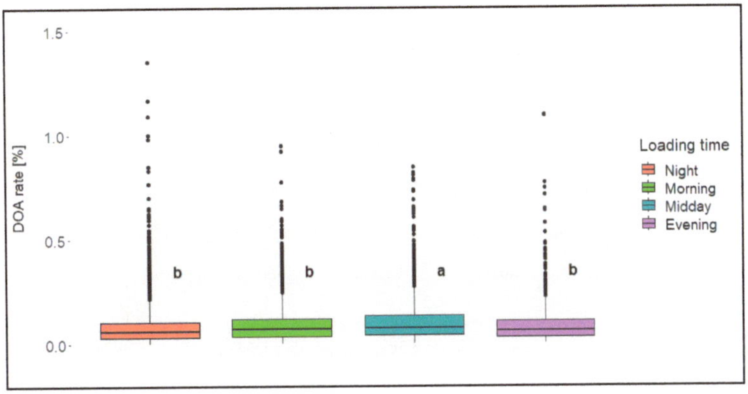

Figure 10. Boxplots of the dead-on-arrival (DOA) rate [%])per loading time (categories: night [23:00 to 05:00], morning [05:00 to 11:00], midday [11:00 to17:00], evening [17:00 to 23:00]); significant differences ($p < 0.05$) are marked with different letters (n = 14,054 broiler transports).

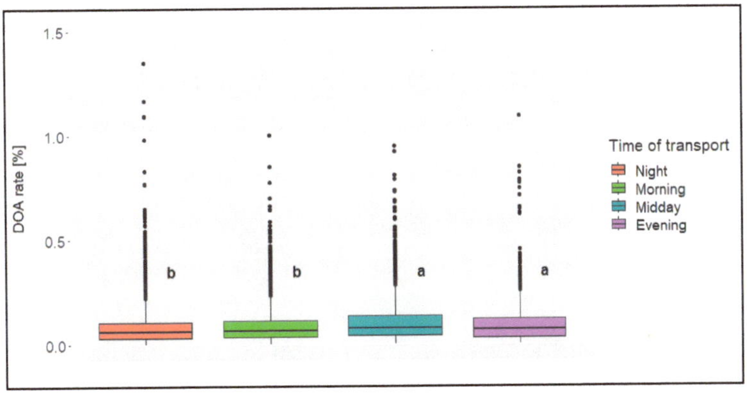

Figure 11. Boxplots of the dead-on-arrival (DOA) rate [%] per time of transport (categories: night [23:00 to 05:00], morning [05:00 to 11:00], midday [11:00 to 17:00], evening [17:00 to 23:00]); significant differences are marked with different letters (n = 14,054 broiler transports).

3.2.7. Relation between the Origin of the Transports and the DOA Rate (%)

Nearly all the transports (99.9%) set off from barns in Germany; only 13 transports came from the Netherlands, while 5 transports originated in Belgium. The ANOVA results indicated no influence of the country of origin on the DOA rate. However, these results need to be classified as not expressive due to the distinctly different amount of data per country.

Concerning the transports from Germany only, no data were given for zip codes starting with the digit seven or eight, which mainly include the south of Germany (Table 5, Figure 1). It is also striking that the majority (approx. 50%) came from the region with zip codes starting with the digit four, covering an area in the west of Germany. This region was followed by the region with zip codes starting with the digit two (approx. 30%), covering an area in the north and north-west of Germany. Approx. 14% of all the transports originated from the region with a zip code starting with the digit three (central Germany), while approx. 5% of all the transports could be assigned to the area with zip codes starting with the digit five (west Germany, south of west Germany). The other regions only accounted for approx. 1% of all the transports.

Table 5. Measures of the position and dispersion of the dead-on-arrival (DOA) rate [%] per first digits of zip code (including transports from Germany only) (n = 14,037 broiler transports).

Measures of Location	0	1	2	3	4	5	6	9
Minimum	0.01	0.02	0.00	0.00	0.00	0.00	0.01	0.02
First Quartile	0.05	0.10	0.04	0.03	0.03	0.04	0.03	0.03
Median	0.08	0.15	0.07	0.06	0.06	0.07	0.04	0.05
Mean	0.15	0.16	0.10	0.08	0.08	0.08	0.05	0.18
Third Quartile	0.15	0.20	0.13	0.09	0.11	0.11	0.07	0.21
Maximum	1.35	0.51	1.09	0.98	3.36	0.42	0.14	0.68
n	40	57	4160	2004	7059	668	32	16

The regions with zero, one or nine at the beginning of the zip code showed the highest mean values, with significant differences from the other groups (Figure 12). These included the eastern part of Germany. However, since all these regions included low frequencies, the influence of one single transport was higher than in other regions, which can contribute to distorting the results. When focusing on the three regions with larger amounts (zip code starting with the digit two, three, or four), the highest mean value could be found for the region starting with the digit two, followed by the region starting with the digit four, while the region starting with three showed the lowest mean value of these three regions. All of these differences were significant.

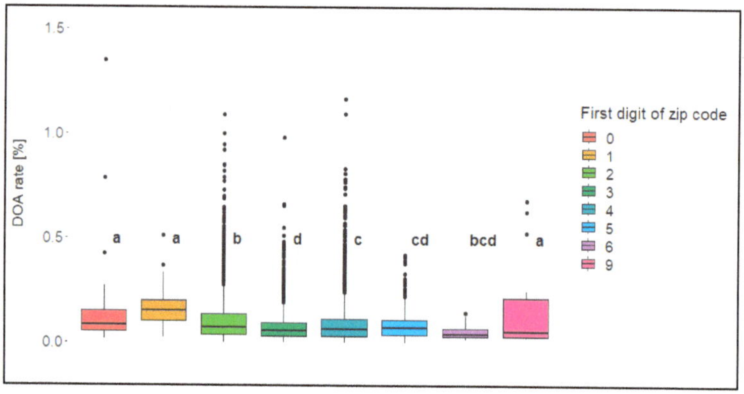

Figure 12. Boxplots of the dead-on-arrival (DOA) rate [%] per the first digits of zip codes, including transports from Germany only; significant differences are marked with different letters (n = 14,037 broiler transports).

3.3. *Influence of Factors on the Decision Regarding a Loading Time and a Time of Transport*

3.3.1. Influence of the Duration of Transport, Season, and Average Daily Temperature on the Loading Time

The loadings for transports of between 100 min and 400 min were mostly conducted during the night, while most loadings for short transports (<100 min) were conducted in the morning (Table 6). For a transport duration of 400 min or more, the highest percentage of loadings was found in the evening. However, concerning transports with less than 200 min of duration, loadings in the evening had the smallest percentage. Above 200 min, loadings at midday had the smallest percentage.

Table 6. Distribution of the percentage frequency of different loading times depending on the duration of transport, including the total number of transports per category transport (categories: night [23:00 to 05:00], morning [05:00 to 11:00], midday [11:00 to 17:00], evening [17:00 to 23:00]) (n = 14,054 broiler transports).

Duration of Transport	Night	Morning	Midday	Evening	Total (Absolute)
<100 min	30%	40%	23%	7%	9497
100–200 min	44%	32%	18%	5%	2672
200–300 min	57%	27%	5%	11%	1719
300–400 min	49%	30%	8%	13%	119
≥400 min	23%	30%	6%	40%	47
Total (absolute)	3809	5872	3203	1170	14,054

Regardless of the season, loadings in the evening had the lowest percentage (Table 7). The highest percentage of loadings could be found during the night or in the morning, irrespective of the season. From all the season loadings, those in the fall seemed to differ the most from the other seasons, showing a slight trend toward more loadings during the night and fewer loadings at midday.

Table 7. Distribution of the percentage frequency of different loading times depending on the season of transport, including the total number of transports per category transport (categories: night [23:00 to 05:00], morning [05:00 to 11:00], midday [11:00 to 17:00], evening [17:00 to 23:00) (n = 14,054 broiler transports).

Season	Night	Morning	Midday	Evening	Total (Absolute)
Spring	35%	37%	20%	8%	3688
Summer	36%	36%	20%	8%	3657
Fall	38%	36%	19%	7%	3437
Winter	36%	37%	21%	7%	3272
Total (absolute)	3809	5872	3203	1170	14,054

Concerning the average daily temperature, loadings below 0 °C and above 25 °C were rare (Table 8). In all the categories of average daily temperature, the lowest percentage of loadings occurred in the evening and the highest percentage was found in the morning (excluding 10 °C to 15 °C).

Table 8. Distribution of the percentage frequency of different loading times depending on the average daily temperature of the transport, including the total number of transport per range transport (categories: night [23:00 to 05:00], morning [05:00 to 11:00], midday [11:00 to 17:00], evening [17:00 to 23:00]) (n = 14,054 broiler transports).

Average Daily Temperature	Night	Morning	Midday	Evening	Total (Absolute)
<0 °C	35%	40%	19%	7%	481
0 to <5 °C	37%	38%	19%	7%	2088
5 to <10 °C	36%	36%	21%	7%	3796
10 to <15 °C	37%	36%	19%	7%	3259

Table 8. *Cont.*

Average Daily Temperature	Night	Morning	Midday	Evening	Total (Absolute)
15 to <20 °C	36%	36%	20%	8%	2946
20 to <25 °C	36%	36%	20%	8%	1375
≥25 °C	36%	43%	15%	6%	109
Total (absolute)	3809	5872	3203	1170	14,054

3.3.2. Influence of the Duration of Transport, Season and Average Daily Temperature on the Time of Transport

All the transports below a duration of 400 min were mostly conducted in the morning (Table 9). Between 200 min and 400 min, more than half of the transports may be assigned to this period. Transports equal and above 400 min (n = 47) were mostly conducted during the night. Excluding this small group of transports, it seems that short transports (<100 min) tended to be conducted at midday, whereas longer transports were conducted in the evening. In general, transports below 100 min were the largest group (approx. 68%).

Table 9. Distribution of the percentage frequency of different times of transport depending on the duration of the transport, including the total number of transport per category (categories: night [23:00 to 05:00], morning [05:00 to 11:00], midday [11:00 to 17:00], evening [17:00 to 23:00]) (n = 14,054 broiler transports).

Duration of Transport	Night	Morning	Midday	Evening	Total (Absolute)
<100 min	25%	41%	25%	10%	9497
100–200 min	32%	39%	21%	7%	2672
200–300 min	31%	52%	14%	3%	1719
300–400 min	21%	51%	21%	7%	119
≥400 min	38%	28%	26%	9%	47
Total (absolute)	3809	5872	3203	1170	14,054

Regardless of the season, transports in the morning had the highest percentage and transports in the evening had the lowest percentage (Table 10). Both percentages were nearly stable for all the seasons. When considering the proportion of the remaining percentage, it seems that transports at midday were rarer in the fall than in the winter, while those at night were more frequent.

Table 10. Distribution of the percentage frequency of different times of transport depending on the season of the transport, including the total number of transports per category (categories: night [23:00 to 05:00], morning [05:00 to 11:00], midday [11:00 to 17:00], evening [17:00 to 23:00]) (n = 14,054 broiler transports).

Season	Night	Morning	Midday	Evening	Total (Absolute)
Spring	27%	41%	23%	8%	3688
Summer	27%	42%	22%	9%	3657
Fall	28%	42%	22%	8%	3437
Winter	26%	42%	24%	8%	3272
Total (absolute)	3809	5872	3203	1170	14,054

Concerning the average daily temperature, transports below 0 °C and above 25 °C were rare (Table 11). In all the categories of average daily temperature, the lowest percentage of loadings occurred in the evening, whereas the highest percentage was found in the morning.

Table 11. Distribution of the percentage frequency of different times of transport depending on the average daily temperature of the transport, including the total number of transports category (categories: night [23:00 to 05:00], morning [05:00 to 11:00], midday [11:00 to 17:00], evening [17:00 to 23:00]) (n = 14,054 broiler transports).

Average Daily Temperature	Night	Morning	Midday	Evening	Total
<0 °C	27%	44%	22%	7%	481
0 to <5 °C	27%	42%	24%	7%	2088
5 to <10 °C	26%	42%	24%	9%	3796
10 to <15 °C	29%	41%	22%	8%	3259
15 to <20 °C	27%	42%	23%	9%	2946
20 to <25 °C	27%	41%	22%	10%	1375
≥25 °C	24%	48%	22%	6%	109
Total (absolute)	3809	5872	3203	1170	14,054

4. Discussion

The available data for the year 2022, with a total of approx. 198 million broilers, amount to around 31% [1] of all the broiler chicken transports to slaughterhouses carried out in Germany in that year and thus constitute a comprehensive sample. However, by excluding some transports with special incidents, the data set does not represent the entire range of all the transports in Germany. In this evaluation, there was a focus on transports representing a typical "normal transport" without issues. It is therefore to be expected that the overall number of outliers is somewhat higher. According to the EFSA, the DOA rate may serve as an "iceberg indicator for transport conditions" [7] for the welfare of animals during these specific transports, although it is known that the DOA might also be influenced by other factors. The mean DOA rate of 0.09% in this evaluation represents a value below previous European studies with a range from 0.13% (United Kingdom, 2005 [10]) to 0.46% (data from 2000 and 2001, focusing on Dutch and German flocks slaughtered in the Netherlands [11]). This leads to the assumption that broiler chicken transports in Germany have a comparable low DOA rate. On the other hand, it may also be assumed that the excluded transports had a higher DOA rate. It could be possible that unplanned incidents such as the failure of technology on the truck (left out in this evaluation) could have resulted in a higher DOA rate, thus raising the overall mean. This leads to the assumption that this overall mean is not representative of broiler chicken transport in Germany in general but of broiler chicken transports usual in practice that experience no major difficulties. However, in view of the large number of transports already included in the evaluation, it can be assumed that these data would not have had a serious impact on the results of this evaluation regarding the influencing factors.

More than 97% of all the observed transports in this evaluation complied with the target values of TierSchTrV (0.5%) [9] and the alarm value of the KTBL (0.3%) [8] and may therefore be classified as having no severe issues. The target value of the EFSA (0.1%) [7] was only complied with by 67.9% of all the transports, leading to the assumption that this is not a generally feasible standard and cannot be accomplished without adjustments. Even more adjustments in the transport of broiler chickens are probably needed to comply with the target range stipulated by the KTBL [7]. This target DOA rate of 0.05% was achieved by only 37.3% of all the transports within this evaluation.

In our findings, the number of animals per transport unit (i.e., all the animals loaded at any one destination) did not affect the DOA rate. It is important to mention that this value did not reflect the number of animals per truck but the number of animals within a transport and can therefore be equated with a slaughter batch. Neither small nor large loads led to a different DOA rate. As large loads mostly entailed longer loading times and a higher number of loaded animals per worker, it may be assumed that these factors did not affect the DOA rate. Since to the best of our knowledge, no other study has examined this factor, a comparison is not possible.

The distance of transports with a mean of 95 km and the duration of transports with a mean of 92 min did not affect the DOA rate in this study. Prior studies, for example, with data from transports in the Czech Republic (categories from <50 km to >300 km) [12] and with Spanish data (8 km to 119 km, 53 min to 1184 min) [15], however, indicated a clear relation, with longer and long-distance transports leading to higher DOA rates. Although the general ranges of the duration and distance in this evaluation were comparable with those in the two studies above, no reason for this distinction may be stated. It may be assumed that either technology has improved in recent years so that longer and long-distance transports are no longer a problem, or that transports in Germany did not have an issue with the duration and distance of the transports at all.

In this evaluation, the stocking density (range between first and third quantile: 0.0046 and 0.0057 kg per cm^2) did not affect the DOA rate. However, a previous study from 2004 described that higher stocking densities led to higher DOA rates [11]. As the stocking density in this study was described using birds per compartment without defining the size of a compartment, a direct comparison is not possible. A later study from 2011 published a range between the first and third quartile of between 0.0053 kg per cm^2 and 0.0062 kg per cm^2 for their data [14], which is higher overall than the stocking density in our evaluation. It may therefore be assumed that the general mean of the stocking density might have been too low to see the effects that occurred in previous evaluations. Within the framework of the national regulations for the stocking density in the cages, no effects on the DOA rate (%) were to be expected for broiler transports in Germany. To evaluate a possible effect on high or low levels, more data and especially data on the actual (not planned) stocking density should be studied in further evaluations. The average daily temperature in northern Germany (Bremen) did not affect the DOA rate in this evaluation. Concerning this, it is important to note that these temperatures represent mean values for the entire day and therefore do not necessarily reflect the transport phase and certainly not the temperature on the truck itself. In addition, the actual region of transport is not represented in these temperature values, as they are all weather data from a station located in Northern Germany (Bremen). However, they proved to be a good compromise when considering the regions of origin of most of the transports throughout the zip codes. In the Spanish evaluation from 2016, the authors postulated that a maximum temperature above 21.5 °C led to a higher DOA rate [15]. As that study used the maximum daily temperature, while our evaluation used the average daily temperature, these factors are not directly comparable. In this evaluation, transports at a high outside temperature were rare, so it might be assumed that transports were postponed to a colder day if possible. It was also found in this evaluation that the most loadings and transports were conducted during the night or in the morning, meaning that they were not affected by the maximum daily temperature. To study the effects of the temperature on the DOA rate, it is important to link the transport date with the actual temperature at the specific time in the specific region and on the truck to assess all the temperature influences on the animals. However, earlier conditions before loading the animals (e.g., mortality on the farm or the duration of feed withdrawal before loading) also showed a relation with the rate DOA in prior studies [20,21], with indications that these periods should also be considered in an evaluation of DOA rates in broiler chickens. Ritz et al. (2005) also found that the transport itself did not result in higher environmental temperatures for the animals but that temperatures may become higher during the loading due to crowding, delays, or technical issues [21].

No conclusion may be drawn regarding the influence of the country of origin on the DOA rate due to the rarity of data from countries besides Germany. Concerning the region of origin in Germany (zip codes), it may be concluded that regions with a lot of broiler production (with four as the first digit of the zip code) did not show the highest but also not the lowest DOA rate when compared to regions with fewer broilers. Comparisons of North and South Germany are impossible because of missing data from South Germany. Similarly, comparisons of West (with two, four, five, six as the first digits of the zip code) and East (with zero, one, and nine as the first digit of the zip code) would not be meaningful due to the widely different number of data (West: 11,919 transports, East: 113 transports). The region of North-West Germany is known to have more farms on the same area than the region of South-East Germany has [22]. Therefore, it may be assumed that transports from areas with more farms per square kilometer do not result in a higher DOA rate than from regions with fewer farms.

With regard to the seasons, it could not be concluded in our study that the loading and transport times are adapted to these conditions. Fewer transports were generally carried out at night in winter, which may be due to the lower temperature or unsafe road conditions. In general, the mean DOA rate was highest in fall (0.13%) and second highest in winter (0.12%). Both seasons differed significantly from each other as well as from the other two seasons (0.11%). One explanation for this may be the increased incidence of wet weather with storms and rain in fall and winter, which makes it necessary to use tarpaulins on the trucks to protect the animals from external influences. By limiting the exchange of air with the outside, this may also have a negative impact on the air circulation inside the trucks, which could lead to problems in the animals' cardiovascular system. However, as no data are available that show the climate in the animal area of the trucks, no statement can be made in this regard.

The average DOA rate differed by only 0.01 to 0.02% between the individual times of day. However, it can be ascertained from the available data that loadings at midday led to a significantly higher mean DOA rate when compared to the other times of day, while at all the other times of day, the DOA rates were at a similar level. In addition to the possibly higher temperatures at midday (not covered due to the use of average daily temperature in this evaluation), one reason for this could be the presence of daylight during loading, which cannot always be completely avoided despite management-related factors such as darkening the barn. This could possibly lead to more restless or nervous animals during loading [23]. Shorter transports tended to take place in the morning rather than longer transports. This might be a result of management decisions based on experience values. The same reason might apply to transports in fall (showing the highest DOA rate of all the seasons in this evaluation), where fewer transports were conducted at midday (showing the highest DOA rate of all the time categories in this evaluation).

Even though a wide range of influencing factors were considered in this evaluation, there are still many other possible influencing factors that could also have had a greater or lesser impact on the DOA rate. Factors such as the temperature on the truck, the general behavior of the flocks during fattening (more active or calm), the weight of the animals, the loading process (also with regard to the use of different systems and methods), and the waiting time at the slaughterhouse, etc. were, for example, not studied in this evaluation. However, it may be assumed that these factors were considered for the transports in this evaluation due to the generally low DOA rate. As an earlier study published findings on the relation between the weight of broiler chickens and the DOA rate [15], it would also be interesting to study different rearing systems and breeds. To gain a closer understanding of the factors affecting the DOA rate, it is important to generate more detailed and specific data (for example, concerning the real actual temperature) for as many transports as possible.

5. Conclusions

The evaluated data showed no influence of the number of animals, the duration and the distance of the transport, the stocking density, and the average daily temperature on the DOA rate (%). Transports conducted at midday (11:00 to 17:00) were associated with higher rates of DOA than transports at other times. The highest DOA rate was found in the fall, followed by the winter, while transports in the spring and summer resulted in the lowest DOA rate. All in all, the relatively low DOA rate (%) in Germany indicates the good standard of the conducted broiler transports. To gather more knowledge on the welfare of the animals, the evaluation of more factors should be considered. Therefore, further research should focus on more detailed data on different transports, including the weights, rearing systems and breeds, as well as on the actual transport conditions inside the truck and the recording of other animal welfare indicators.

Author Contributions: Conceptualization, J.G. and B.S.; methodology, J.G. and B.S.; formal analysis, J.G.; data curation, J.G.; writing—original draft preparation, J.G.; writing—review and editing, J.G., C.V., N.K. and B.S.; visualization, J.G.; supervision, C.V. and N.K. All authors have read and agreed to the published version of the manuscript.

Funding: We acknowledge the financial support from the Open Access Publication Fund of the University of Veterinary Medicine Hannover, Foundation.

Institutional Review Board Statement: Ethical review and approval were waived for this study due to the fact that the all evaluated data were recorded during regular operation.

Informed Consent Statement: Not applicable.

Data Availability Statement: Restrictions apply to the availability of these data. Data were obtained from companies with access to data concerning the transport of broilers and are only available from the authors with the permission of these companies.

Acknowledgments: We wish to thank Frances Sherwood-Brock for proofreading the manuscript to ensure correct English.

Conflicts of Interest: The authors declare no conflicts of interest.

References

1. Statistisches Bundesamt Geflügelschlachtereien, Geschlachtete Tiere, Schlachtmenge. Available online: https://www-genesis.destatis.de/genesis/online?operation=abruftabelleBearbeiten&levelindex=1&levelid=1701889796394&auswahloperation=abruftabelleAuspraegungAuswaehlen&auswahlverzeichnis=ordnungsstruktur&auswahlziel=werteabruf&code=41322-0001&auswahltext=&w (accessed on 10 June 2024).
2. Council Regulation (EC). Pages 1–44 in No 1/2005 of 22 December 2004 on the Protection of Animals during Transport and Related Operations and Amending Directives 64/432/EEC and 93/119/EC and Regulation (EC) No 1255/97, OJ L. 2004. Available online: https://www.legislation.gov.uk/eur/2005/1 (accessed on 10 June 2024).
3. German Order on the Protection of Animals and the Keeping of Production Animals. 2006. Amended and Promulgated on August 22nd, 2006, Last Changed on January 29th, 2021. Available online: https://www.gesetze-im-internet.de/tierschnutztv (accessed on 10 June 2024).
4. Mitchell, M.A.; Kettlewell, P.J. Physiological Stress and Welfare of Broiler Chickens in Transit: Solutions Not Problems! *Poult. Sci.* **1998**, *77*, 1803–1814. [CrossRef] [PubMed]
5. Petracci, M.; Bianchi, M.; Cavani, C. Pre-Slaughter Handling and Slaughtering Factors Influencing Poultry Product Quality. *World's Poult. Sci. J.* **2010**, *66*, 17–26. [CrossRef]
6. Jacobs, L.; Delezie, E.; Duchateau, L.; Goethals, K.; Tuyttens, F.A.M. Broiler Chickens Dead on Arrival: Associated Risk Factors and Welfare Indicators. *Poult. Sci.* **2017**, *96*, 259–265. [CrossRef] [PubMed]
7. Nielsen, S.S.; Alvarez, J.; Bicout, D.J.; Calistri, P.; Canali, E.; Drewe, J.A.; Garin-Bastuji, B.; Gonzales Rojas, J.L.; Gortázar Schmidt, C.; Herskin, M.; et al. Welfare of Domestic Birds and Rabbits Transported in Containers. *EFSA J.* **2022**, *20*, e07441. [CrossRef] [PubMed]
8. Knierim, U.; Gieseke, D.; Michaelis, S. Tierschutzindikatoren für Masthühner: Vorschläge zu Ziel- und Alarmwerten für die Betriebliche Eigenkontrolle. 2020. Available online: https://www.researchgate.net/publication/350487620_Tierschutzindikatoren_fur_Masthuhner_Vorschlage_zu_Ziel-_und_Alarmwerten_fur_die_betriebliche_Eigenkontrolle (accessed on 10 June 2024).
9. Implementing Provisions for the Implementation of Section 20 (4) and (5) TierSchNutztV—Keeping Chickens for Fattening. 2014. Page 902 in Ministerial Gazette of Lower Saxony No 46 of 11 December 2014. Available online: https://www.niedersachsen.de/politik_staat/gesetze_verordnungen_und_sonstige_vorschriften/verkundungsblatter_vorjahre/niedersaechsisches-ministerialblatt-2014-130327.html (accessed on 10 June 2024).
10. Warriss, P.D.; Pagazaurtundua, A.; Brown, S.N. Relationship between Maximum Daily Temperature and Mortality of Broiler Chickens during Transport and Lairage. *Br. Poult. Sci.* **2005**, *46*, 647–651. [CrossRef] [PubMed]
11. Nijdam, E.; Arens, P.; Lambooij, E.; Decuypere, E.; Stegeman, J.A. Factors Influencing Bruises and Mortality of Broilers during Catching, Transport, and Lairage. *Poult. Sci.* **2004**, *83*, 1610–1615. [CrossRef] [PubMed]

12. Vecerek, V.; Grbalova, S.; Voslarova, E.; Janackova, B.; Malena, M. Effects of Travel Distance and the Season of the Year on Death Rates of Broilers Transported to Poultry Processing Plants. *Poult. Sci.* **2006**, *85*, 1881–1884. [CrossRef] [PubMed]
13. Petracci, M.; Bianchi, M.; Cavani, C.; Gaspari, P.; Lavazza, A. Preslaughter Mortality in Broiler Chickens, Turkeys, and Spent Hens under Commercial Slaughtering. *Poult. Sci.* **2006**, *85*, 1660–1664. [CrossRef] [PubMed]
14. Chauvin, C.; Hillion, S.; Balaine, L.; Michel, V.; Peraste, J.; Petetin, I.; Lupo, C.; Le Bouquin, S. Factors Associated with Mortality of Broilers during Transport to Slaughterhouse. *Animal* **2011**, *5*, 287–293. [CrossRef] [PubMed]
15. Villarroel, M.; Pomares, F.; Ibáñez, M.A.; Lage, A.; Martínez-Guijarro, P.; Méndez, J.; de Blas, C. Rearing, Bird Type and Pre-Slaughter Transport Conditions I. Effect on Dead on Arrival. *Span. J. Agric. Res.* **2018**, *16*, e0503. [CrossRef]
16. Nijdam, E.; Zailan, A.R.M.; Van Eck, J.H.H.; Decuypere, E.; Stegeman, J.A. Pathological Features in Dead on Arrival Broilers with Special Reference to Heart Disorders. *Poult. Sci.* **2006**, *85*, 1303–1308. [CrossRef] [PubMed]
17. Kittelsen, K.E.; Granquist, E.G.; Kolbjørnsen, O.; Nafstad, O.; Moe, R.O. A Comparison of Post-Mortem Findings in Broilers Dead-on-Farm and Broilers Dead-on-Arrival at the Abattoir. *Poult. Sci.* **2015**, *94*, 2622–2629. [CrossRef] [PubMed]
18. Cohen, J. *Statistical Power Analysis for the Behavioral Sciences*, 2nd ed.; Routledge: Hillsdale, NJ, USA, 1988.
19. R Core Team R. *A Language and Environment for Statistical Computing*; R Foundation for Statistical Computing: Vienna, Austria, 2023.
20. Cockram, M.S.; Dulal, K.J.; Mohamed, R.A.; Revie, C.W. Risk Factors for Bruising and Mortality of Broilers during Manual Handling, Module Loading, Transport, and Lairage. *Can. J. Anim. Sci.* **2019**, *99*, 50–65. [CrossRef]
21. Ritz, C.W.; Webster, A.B.; Czarick, M. Evaluation of Hot Weather Thermal Environment and Incidence of Mortality Associated with Broiler Live Haul. *J. Appl. Poult. Res.* **2005**, *14*, 594–602. [CrossRef]
22. Windhorst, H.W. Regional Patterns of Livestock and Poultry Production in Europe. In *Livestock Production and Society*; Wageningen Academic Publishers: Wageningen, The Netherlands, 2006; pp. 19–34. ISBN 9789076998893.
23. Mönch, J.; Rauch, E.; Hartmannsgruber, S.; Erhard, M.; Wolff, I.; Schmidt, P.; Schug, A.R.; Louton, H. The Welfare Impacts of Mechanical and Manual Broiler Catching and of Circumstances at Loading under Field Conditions. *Poult. Sci.* **2020**, *99*, 5233–5251. [CrossRef] [PubMed]

animals

MDPI

Article

Effects of Stocking Density and Illuminance in Lairage of Fattening Pigs in Different Temperatures

Dong-Cheol Song [1,†], Ji-Hwan Lee [2,†], Won Yun [3], Se-Yeon Chang [1], Se-Hyun Park [1], Kyeong-Ho Jeon [1], Hyuck Kim [1] and Jin-Ho Cho [1,*]

1 Department of Animal Sciences, Chungbuk National University, Cheongju 28644, Republic of Korea; paul741@daum.net (D.-C.S.); angella2425@naver.com (S.-Y.C.); jeonkh1222@gmail.com (K.-H.J.); harrck85@naver.com (H.K.)
2 Swine Science Division, National Institute of Animal Science, Rural Development Administration, Cheonan 31000, Republic of Korea; junenet123@naver.com
3 Central Research Institute, Woosung Feed Co., Ltd., Daejeon 34379, Republic of Korea; wyoun@woosung.kr
* Correspondence: jinhcho@cbnu.ac.kr; Tel.: +82-041-550-3652; Fax: +82-041-550-3604
† These authors contributed equally to this work.

Simple Summary: Concerns about animal welfare practices have increased as consumer awareness and regulatory standards have evolved. This study investigated the effects of lairage conditions on welfare and meat quality of pigs during lairage periods. A total of 3070 finishing pigs were assigned to one of six groups arranged in two trials in a 2 × 3 factorial design according to the illuminance (under 40 lux (LX), over 40 lux (HX)) and stocking density (low density (LD), higher than 0.83 m^2/100 kg; normal density (ND), 0.50–0.83 m^2/100 kg; high density (HD), lower than 0.50 m^2/100 kg) with high temperature (HT), higher than 24 °C; low temperature (LT), lower than 10 °C. Based on the obtained results, stocking of too-high (lower than 0.50 m^2/100 kg) density at HT and stocking of too-low (higher than 0.83 m^2/100 kg) density at LT are generally not good for meat quality and animal welfare.

Abstract: This study investigated the effects of lairage conditions on the welfare and meat quality of pigs during lairage periods. A total of 3070 finishing pigs were assigned to one of six groups arranged in two trials in a 2 × 3 factorial design according to the illuminance (under 40 lux (LX), over 40 lux (HX)) and stocking density (low density (LD), higher than 0.83 m^2/100 kg; normal density (ND), 0.50–0.83 m^2/100 kg; high density (HD), lower than 0.50 m^2/100 kg) with high temperature (HT), higher than 24 °C; low temperature (LT), lower than 10 °C. Pigs stocked with HD showed lower aggression behavior and overlap behavior than those stocked with LD at LT. Pigs stocked with HD showed higher standing, sitting, and aggression behavior than those stocked with LD at HT. Pigs stocked with HD showed higher pH than those stocked with LD at LT. At HT, pigs stocked with LD showed higher pH, WHC, DL, and CL than those stocked with HD. At LT, pigs stocked with LD showed higher cortisol levels than those stocked with HD. However, pigs stocked with LD showed lower cortisol levels than those stocked with HD at HT. Based on the obtained results, stocking of too-high (lower than 0.50 m^2/100 kg) density at HT and stocking of too-low (higher than 0.83 m^2/100 kg) density at LT are generally not good for meat quality and animal welfare.

Keywords: stocking density; illuminance; lairage; pig; welfare

Citation: Song, D.-C.; Lee, J.-H.; Yun, W.; Chang, S.-Y.; Park, S.-H.; Jeon, K.-H.; Kim, H.; Cho, J.-H. Effects of Stocking Density and Illuminance in Lairage of Fattening Pigs in Different Temperatures. *Animals* **2024**, *14*, 2145. https://doi.org/10.3390/ani14152145

Academic Editor: Elbert Lambooij

Received: 6 June 2024
Revised: 20 July 2024
Accepted: 22 July 2024
Published: 23 July 2024

1. Introduction

Generally, pigs are housed in pens before being slaughtered. Keeping familiar pigs in groups and ensuring adequate space in lairage pens are essential prerequisites for ensuring rest and minimizing aversive behavior in lairage. Lairage conditions, including stocking density, illuminance, and ambient temperature, can significantly influence the welfare, physiological responses, and meat quality of fattening pigs during the preslaughter phase [1–3]. Concerns about animal welfare practices have increased as consumer awareness and regulatory standards have evolved [4].

Stocking density, defined as the space allowance per pig within a holding area, directly affects animal stress levels and behavior [5]. High density often lead to increased aggression, stress, and subsequent physiological strain, which can compromise immune function and meat quality [6,7].

Illuminance, or the intensity of light exposure, also plays a crucial role in the behavior and stress management of livestock [8]. Proper lighting can reduce fear and injuries by improving visibility, whereas inappropriate lighting can lead to heightened stress and disruptive behaviors [9].

Temperature is a fundamental environmental factor that can affect animal comfort and physiological stability [10]. Extremes of temperature, be it high or low, can exacerbate the negative impacts of unsuitable stocking densities or inappropriate lighting conditions [11]. Thermoregulatory behavior of pigs can be significantly influenced by the thermal environment, which in turn affects their overall welfare and the quality of the pork produced.

Therefore, the objective of this study was to investigate physiological responses and meat quality changes according to lairage stocking density and illuminance in winter and summer.

2. Materials and Methods

2.1. Ethics

The protocol for this study was reviewed and approved by the Institutional Animal Care and Use Committee of Chungbuk National University, Cheongju, Republic of Korea (approval no. CBNUA-2185-23-02).

2.2. Animals, Preslaughter Conditions and Treatments

Between January 2023 and December 2023, a total of 3070 crossbred pigs of mixed sex with the same genetics ((Yorkshire × Landrace) × Duroc) were transported from the commercial finishing farms to the commercial slaughterhouse. A total of 3070 finishing pigs (ibw: 112.76 ± 4.33 kg) were randomly assigned to one of six groups arranged in a 2 × 3 factorial design according to the illuminance by [12] (under 40 lux (LX), over 40 lux (HX)), and stocking density by [13] (low density (LD), higher than 0.83 $m^2/100$ kg; normal density (ND), 0.50–0.83 $m^2/100$ kg; high density (HD), lower than 0.50 $m^2/100$ kg). The experiment was carried out at two different temperatures (high temperature (HT), higher than 24 °C in summer; low temperature (LT), lower than 10 °C in winter). This design was proposed emphasizing the control of all the factors associated with the experimental treatment (genotype, fasting, handling, bedding, distance, and lairage) in order to compare only the effects of lairage density and illuminance.

2.3. Meat Quality Measurements

Pig carcasses were graded with the Korean Pig Carcass Grade System. The conductor grades are as follows: 1+ grade (carcass weight: 83 to 93 kg, backfat thickness: 17 to 25 mm), 1 grade (carcass weight: 80 to 98 kg, backfat thickness: 15 to 28 mm), 2 grade (ranges of carcass weight and backfat thickness that do not correspond to 1, 1+ grade). The hot carcass weight was measured on an electronic scale 45 min postmortem and expressed in integer kg units. The left half carcass was used to measure the backfat thickness. The backfat thickness between the last thoracic vertebra and the first lumbar vertebra and that between the 11th and 12th thoracic vertebrae were measured with a ruler. Hot carcass weight and backfat thickness were measured and calculated as follows: (backfat thickness (mm)/hot carcass weight (kg)). Pig losses were measured by observing and classifying fractures and bruises after the pigs were unloaded after transport.

The moisture, protein, and fat content (%) was determined according to the Association of Official Analytical Chemists [14]. The pH was measured after adding 50 mL of distilled water to 5 g of the left carcass loin. All samples were homogenized for 30 s using a homogenizer (Stomacher® 400 Circulator, Seward, UK), and then measured with a pH meter (Orion Star™ A211 pH Benchtop Meter, Thermo scientific™, Cleveland, OH, USA) calibrated in phosphate buffer at pH of 4, 7, and 10. For meat color, the left carcass loin was measured with a Spectro Colorimeter (Model JX-777, Color Techno. System Co., Tokyo, Japan) standardized on a white plate (L*, 89.39; a*, 0.13; b*, −0.51). At this time, the light source used was a white fluorescent lamp (D65). Color values were expressed as L*(lightness), a*(redness), and b*(yellowness). Drip loss (DL) was assessed using the filter paper wetness (FPW) test. Cooking loss (CL) was determined with Oliveira et al.'s (2014) [15] methodology. CL value was measured as the ratio (%) of the weight of the initial sample to the weight after heating the sample. Sensory color was evaluated by 5 trained panelists. The sensory color was as follows: score 1 (pale), score 2 (grayish pink), score 3 (reddish pink), score 4 (purplish red), score 5 (dark). Marbling was evaluated by 5 panelists according to the detailed criteria for grading of livestock products. Marbling score was as follows: score 1 (practically devoid), score 2 (slight), score 3 (modest), score 4 (slightly abundant), score 5 (abundant).

2.4. Pork Quality Classes Measurements

The intra-measurement coefficients of variation for meat quality parameters were below 10%. Pork quality classes (pale, soft, and exudative—PSE; red, soft, and exudative -RSE; red, firm, and nonexudative—RFN; pale, firm, and nonexudative—PFN; dark, firm, and dry—DFD) were determined using pH values measured 24 h postmortem, DL variations, and light reflectance (L*), according to [16].

2.5. Behavioral and Status Observations and Blood Profile

During lairage, behaviors were continuously recorded using cameras (Intelbras VMH 1010 D HD 720p, Intelbras SA, São José, Brazil), installed on the ceiling of the lairage. During lairage, the number of pigs in each posture (lying, standing, sitting, aggression, and overlap) was recorded. Cortisol, lactate, and glucose samples were taken after 4 h arrival at the lairage. Blood samples were collected from 10 pigs in each group for the determination of concentration levels of cortisol, glucose, and lactate. At least 3 mL of blood samples were taken from the jugular vein. After collection, serum samples were centrifuged at 3000× *g* for 20 min at 4 °C. Thereafter, the blood sample tubes were stored in a −20 °C refrigerator until analysis. The cortisol values were measured using radioimmunoassay Coat-A-Count cortisol kits (Catalog number-TKCO5, Siemens Medical Solution Diagnostics, Los Anglos, CA, USA). Serum glucose was analyzed using an automatic Konelab analyser (Thermo Clinical Labsystems Oy, Vantaa, Finland) according to the manufacturer's instructions. Lactate levels were measured using a GM7 Analox analyzer (Analox Instruments, London, UK).

2.6. Skin Lesion Scoring

Skin lesion scoring was performed for each pig on the ears, shoulders, tails, and total body. Only pigs without skin lesions during the farming and transporting were selected and marked, and skin lesions were measured in lairage by a four-point scale from 0 to 3 [17]. Score 0 was given for a body part if no scratches were found. Score 1 was given if fewer than five superficial scratches were observed, and Score 2 was given if five to ten superficial scratches or fewer than five deep scratches were detected. Score 3 described a body area with more than ten superficial scratches or more than five deep scratches.

2.7. Statistical Analysis

Data generated were subjected to a two-way analysis of variance using JMP pro 16.0 software (Statistical Analysis System Software, 2012). Statistics for each factor were analyzed using general linear model (GLM) procedures of SAS. Significantly ($p < 0.05$) different means among the variables were separated using Tukey's multiple range test, as contained in the same statistical package. Pork quality classes proportion data were analyzed by the nonparametric Kruskal–Wallis test as the data were not normally distributed.

3. Results

3.1. Behavior Observation

Table 1 shows the effects of lairage stocking density and illuminance on behavior at LT. Pigs stocked with HD showed lower aggression behavior and overlap behavior than those stocked with LD at LT. Also, pigs stocked with LX tended to exhibit higher lying behavior than those stocked with HX at LT. Table 2 shows the effects of lairage stocking density and illuminance on behavior at HT. Pigs stocked with HD showed higher standing, sitting, and aggression behavior than those stocked with LD at HT. However, pigs stocked with LD showed higher lying behavior than those stocked with HD at HT. There was no interaction between stocking density and illuminance.

Table 1. Effects of lairage stocking density and illuminance on behaviors at low temperature.

Items	Main Effect					SE	p-Value		
	D			IL					
	HD	ND	LD	LX	HX		D	IL	D × IL
Standing	38.90	36.56	34.73	34.34	39.12	3.495	0.485	0.122	0.077
Sitting	6.43	5.55	6.20	5.61	6.51	0.068	0.289	0.129	0.095
Lying	14.67	17.89	19.07	20.05	14.38	3.612	0.397	0.078	0.179
Aggression	3.19 b	3.59 ab	4.05 a	3.6	3.62	0.212	0.002	0.919	0.583
Overlap	1.26 b	1.55 ab	2.30 a	1.63	1.77	0.348	0.037	0.652	0.696

Abbreviation: D, density; IL, illuminance; D × IL, density × illuminance; SE, standard error. HD (lower than 0.50 m^2/100 kg); ND (0.50 m^2~0.83 m^2/100 kg); LD (higher than 0.83 m^2/100 kg). LX, low illuminance; HX, high illuminance. Standing, the act of standing still without any other action, with the forelimbs and hind legs stretched perpendicularly to the floor or similar behavior; sitting, two front legs straight to the floor, two rear legs and hips sitting in contact with the floor or similar behavior; lying, the act of lying in the most comfortable position with the head, front legs, back legs, and abdomen touching the floor or similar behavior; aggression, pushing, biting, or beating another pig with the head, lifting the pigs by pushing the head under the body or similar behavior; overlap, the act of placing both forelimbs on the back of another pig or similar behavior. a, b: means in the same row with different superscripts differ ($p < 0.05$).

Table 2. Effects of lairage stocking density and illuminance on behaviors at high temperature.

Items	Main Effect					SE	p-Value		
	D			IL					
	HD	ND	LD	LX	HX		D	IL	D × IL
Standing	36.32 a	32.09 ab	28.83 b	31.17	33.65	2.558	0.044	0.253	0.478
Sitting	5.79 a	5.19 ab	4.33 b	4.99	5.21	0.484	0.034	0.573	0.559
Lying	17.89 b	22.73 ab	26.85 a	23.84	21.14	2.467	0.011	0.197	0.552
Aggression	0.87 a	0.72 ab	0.60 b	0.75	0.71	0.062	0.002	0.373	0.351
Overlap	0.53	0.56	0.59	0.52	0.6	0.064	0.707	0.138	0.278

Abbreviation: D, density; IL, illuminance; D × IL, density × illuminance; SE, standard error. HD (lower than 0.50 m^2/100 kg); ND (0.50 m^2~0.83 m^2/100 kg); LD (higher than 0.83 m^2/100 kg). LX, low illuminance; HX, high illuminance. Standing, the act of standing still without any other action, with the forelimbs and hind legs stretched perpendicularly to the floor or similar behavior; sitting, two front legs straight to the floor, two rear legs and hips sitting in contact with the floor or similar behavior; lying, the act of lying in the most comfortable position with the head, front legs, back legs, and abdomen touching the floor or similar behavior; aggression, pushing, biting, or beating another pig with the head, lifting the pigs by pushing the head under the body or similar behavior; overlap, the act of placing both forelimbs on the back of another pig or similar behavior. a, b: Means in the same row with different superscripts differ ($p < 0.05$).

3.2. Carcass Grade and Carcass Composition

Tables 3 and 4 show the effects of lairage stocking density and illuminance on carcass grade and composition at LT and HT, respectively. At HT, pigs stocked with LD showed lower backfat thickness than those stocked with HD. Also, pigs stocked at LD and HD showed numerically higher 1+ grade than those stocked at HD and LD at HT and LT, respectively. There was no interaction between stocking density and illuminance.

Table 3. Effects of lairage stocking density and illuminance on carcass trait at low temperature.

Items	Main Effect					SE	*p*-Value		
	D			IL					
	HD	ND	LD	LX	HX		D	IL	D × IL
Carcass composition traits									
Hot carcass weight, kg	87.58	86.21	85.87	86.42	86.68	0.79	0.11	0.69	0.19
Backfat thickness, mm	24.03	23.61	23.00	23.87	23.23	0.64	0.31	0.23	0.24
Carcass grade score									
Grade 1+, %	50.00	40.00	20.00	27.60	32.00	-	0.433	0.588	-
Grade 1, %	26.00	26.90	36.70	29.70	49.20	-	0.769	0.234	-
Grade 2, %	24.00	33.10	43.30	42.70	18.80	-	0.396	0.221	-

Abbreviation: D, density; IL, illuminance; D × IL, density × illuminance; SE, standard error. HD (lower than 0.50 $m^2/100$ kg); ND (0.50 m^2~0.83 $m^2/100$ kg); LD (higher than 0.83 $m^2/100$ kg). LX, low illuminance; HX, high illuminance; carcass grade score was measured by Korean Ministry of Agriculture, Food and Rural Affairs Notification. Grade 1+, 83 kg~93 kg of carcass weight with 17–25 mm of backfat thickness; Grade 1, 80 kg~98 kg of carcass weight with 15–28 mm of backfat thickness; Grade 2, pigs do not belong to Grade 1+ and Grade 1.

Table 4. Effects of lairage stocking density and illuminance on carcass trait at high temperature.

Items	Main Effect					SEM	*p*-Value		
	D			IL					
	HD	ND	HD	LX	HX		D	IL	D × IL
Carcass composition traits									
Hot carcass weight, kg	88.20	89.05	88.17	88.31	88.64	0.45	0.11	0.38	0.35
Backfat thickness, mm	24.70 a	24.45 ab	23.53 b	24.14	24.31	0.42	0.02	0.62	0.82
Carcass grade score									
Grade 1+, %	31.00	39.00	41.00	36.00	36.00	-	0.336	0.883	-
Grade 1, %	31.00	36.00	36.00	29.00	37.00	-	0.267	0.644	-
Grade 2, %	38.00	25.00	23.00	35.00	37.00	-	0.459	0.311	-

Abbreviation: D, density; IL, illuminance; D × IL, density × illuminance; SEM, standard error of means. HD (lower than 0.50 $m^2/100$ kg); ND (0.50 m^2~0.83 $m^2/100$ kg); LD (higher than 0.83 $m^2/100$ kg). LX, low illuminance; HX, high illuminance; carcass grade score was measured by Korean Ministry of Agriculture, Food and Rural Affairs Notification. Grade 1+, 83 kg~93 kg of carcass weight with 17–25 mm of backfat thickness; Grade 1, 80 kg~98 kg of carcass weight with 15–28 mm of backfat thickness; Grade 2, pigs do not belong to Grade 1+ and Grade 1. a, b: means in the same row with different superscripts differ ($p < 0.05$).

3.3. Pork Quality

Tables 5 and 6 show the effects of lairage stocking density and illuminance on pork quality at LT and HT, respectively. Pigs stocked with HD showed higher pH than those stocked with LD at LT. At HT, pigs stocked with LD showed higher pH, WHC, DL, and CL than those stocked with HD. There was no interaction between stocking density and illuminance.

Table 5. Effects of lairage stocking density and illuminance on pork quality at low temperature.

Items	Main Effect					SE	*p*-Value		
	D			IL			D	IL	D × IL
	HD	ND	LD	LX	HX				
Pork composition, %									
Moisture	73.15	73.49	73.87	73.76	73.25	0.597	0.508	0.312	0.684
Crude protein	22.82	23.08	21.81	22.26	22.88	1.072	0.483	0.495	0.716
Crude fat	2.24	2.56	1.63	2.18	2.11	0.645	0.372	0.892	0.898

Table 5. *Cont.*

Items	Main Effect					SE	*p*-Value		
	D			IL			D	IL	D × IL
	HD	ND	LD	LX	HX				
Pork quality parameters									
pH	5.59 a	5.50 b	5.50 b	5.52	5.54	0.034	0.031	0.520	0.72
WHC, %	63.87	60.26	60.9	60.46	62.89	3.696	0.595	0.437	0.613
DL, %	4.44	4.53	4.74	4.58	4.57	0.19	0.302	0.928	0.941
CL, %	24.43	23.19	26.22	25.11	24.11	2.621	0.527	0.649	0.924
L* value	49.25	49.77	50.62	50.54	49.22	2.543	0.864	0.536	0.715
a* value	7.65	6.48	5.42	6.71	6.32	0.863	0.071	0.583	0.402
b* value	5.89	5.62	4.8	5.63	5.24	0.443	0.072	0.299	0.358
sensory color [1)]	2.83	2.67	2.5	2.72	2.61	0.366	0.67	0.717	0.409
Marbling [2)]	3.15	3.15	3.23	3.12	3.23	0.71	0.991	0.851	0.889
Pork quality classes (%)									
PSE pork	10	15	30	20	16.67	-	0.215	0.321	-
RSE pork	10	10	10	10	10	-	0.684	0.745	-
RFN pork	40	10	50	43.33	43.33	-	0.573	0.198	-
PFN pork	40	35	10	26.67	30	-	0.742	0.138	-
DFD pork	0	0	0	0	0	-	1	1	-

Abbreviation: D, density; IL, illuminance; D × IL, density × illuminance; SE, standard error. HD (lower than 0.50 m^2/100 kg); ND (0.50 m^2~0.83 m^2/100 kg); LD (higher than 0.83 m^2/100 kg). LX, low illuminance; HX, high illuminance. [1)] Color score ranged from 1 (pale color) to 5 (dark color). [2)] Marbling score ranged from 1 (practically devoid) to 5 (abundant). WHC, water-holding capacity; DL, drip loss; CL, cooking loss; PSE, pale, soft, exudative; RSE, reddish–pink, soft, exudative; RFN, red, firm, nonexudative; PFN, pale, firm, nonexudative; DFD, dark, firm, dry.

Table 6. Effects of lairage stocking density and illuminance on pork quality at high temperature.

Items	Main Effect					SE	*p*-Value		
	D			IL			D	IL	D× IL
	HD	ND	LD	LX	HX				
Pork composition, %									
Moisture	72.41 b	73.38 ab	74.07 a	73.26	73.30	0.489	0.017	0.921	0.443
Crude protein	23.02	22.80	22.49	23.13	22.42	0.919	0.847	0.365	0.899
Crude fat	2.13	2.07	2.01	2.33	2.23	0.660	0.553	0.852	0.723
Pork quality parameters									
pH	5.46 b	5.55 ab	5.69 a	5.57	5.56	0.070	0.021	0.859	0.469
WHC, %	63.66 b	67.02 ab	69.70 a	67.37	66.22	1.821	0.020	0.456	0.855
DL, %	5.00 a	4.80 ab	4.42 b	4.78	4.71	0.164	0.013	0.599	0.452
CL, %	25.26 a	23.25 b	22.98 b	23.57	24.09	0.716	0.015	0.389	0.995
L* value	51.58	51.21	49.57	50.02	51.55	1.987	0.577	0.365	0.608
a* value	7.08	6.72	5.78	6.50	6.56	0.795	0.280	0.923	0.395
b* value	6.04	5.32	4.65	5.33	5.35	0.629	0.128	0.976	0.160
sensory color	2.83	2.75	2.92	2.78	2.89	0.495	0.945	0.788	0.517
Marbling	3.08	3.22	3.15	3.30	3.00	0.300	0.907	0.244	0.843
Pork quality classes (%)									
PSE pork	50.00	16.67	0.00	22.22	22.22		0.368	0.121	
RSE pork	16.67	16.67	0.00	11.11	11.11		0.684	0.105	
RFN pork	0.00	33.33	83.33	55.56	22.22		0.073	0.098	
PFN pork	33.33	33.33	16.67	11.11	44.45		0.842	0.138	
DFD pork	0.00	0.00	0.00	0.00	0.00		1	1	

Abbreviation: D, density; IL, illuminance; D × IL, density × illuminance; SE, standard error. HD (lower than 0.50 m^2/100 kg); ND (0.50 m^2~0.83 m^2/100 kg); LD (higher than 0.83 m^2/100 kg). LX, low illuminance; HX, high illuminance. [1)] Color score ranged from 1 (pale color) to 5 (dark color). [2)] Marbling score ranged from 1 (practically devoid) to 5 (abundant). WHC, water-holding capacity; DL, drip loss; CL, cooking loss; PSE, pale, soft, exudative; RSE, reddish–pink, soft, exudative; RFN, red, firm, nonexudative; PFN, pale, firm, nonexudative; DFD, dark, firm, dry. a, b: means in the same row with different superscripts differ ($p < 0.05$).

3.4. Blood Profile

Tables 7 and 8 show the effects of lairage stocking density and illuminance on blood profile at LT and HT, respectively. At LT, pigs stocked with LD showed higher cortisol level than those stocked with HD. However, pigs stocked with LD showed lower cortisol level than those stocked with HD at HT. There was no interaction between stocking density and illuminance.

Table 7. Effects of lairage stocking density and illuminance on blood profile at low temperature.

Items	Main Effect					SE	*p*-Value		
	D			IL					
	HD	ND	LD	LX	HX		D	IL	D × IL
Cortisol (ug/dL)	3.86 b	3.95 ab	4.05 a	3.93	3.98	0.08	0.048	0.399	0.645
Lactate (mmol/L)	4.66	4.71	4.71	4.68	4.71	0.13	0.892	0.776	0.225
Glucose (mg/dL)	81.60	82.15	82.00	81.73	82.10	1.14	0.883	0.695	0.495

Abbreviation: D, density; IL, illuminance; D × IL, density × illuminance; SE, standard error. HD (lower than 0.50 m^2/100 kg); ND (0.50 m^2~0.83 m^2/100 kg); LD (higher than 0.83 m^2/100 kg). LX, low illuminance; HX, high illuminance. a, b: means in the same row with different superscripts differ ($p < 0.05$).

Table 8. Effects of lairage stocking density and illuminance on blood profile at high temperature.

Items	Main Effect					SE	p-Value		
	D			IL					
	HD	ND	LD	LX	HX		D	IL	D × IL
Cortisol (ug/dL)	4.02 a	3.94 ab	3.84 b	3.94	3.93	0.06	0.021	0.876	0.544
Lactate (mmol/L)	4.61	4.51	4.44	4.55	4.49	0.10	0.255	0.499	0.361
Glucose (mg/dL)	78.70	78.80	78.00	78.20	78.80	1.19	0.765	0.539	0.765

Abbreviation: D, density; IL, illuminance; D × IL, density × illuminance; SE, standard error. HD (lower than 0.50 m^2/100 kg); ND (0.50 m^2~0.83 m^2/100 kg); LD (higher than 0.83 m^2/100 kg). LX, low illuminance; HX, high illuminance. a, b: means in the same row with different superscripts differ ($p < 0.05$).

3.5. Skin Lesion Score

Tables 9 and 10 show the effects of lairage stocking density and illuminance on skin lesion score at LT and HT, respectively. At LT, pigs stocked at LD showed numerically higher skin lesion (shoulder and body) than those stocked at HD. Under HT condition, pigs stocked at HD showed numerically higher skin lesion (ear and body) than those stocked at LD.

Table 9. Effects of lairage stocking density and illuminance on skin lesion score at low temperature.

Items	Main Effect				
	D			IL	
	HD	ND	LD	LX	HX
Skin Lesion Score					
Ear					
1	82.22	84.44	80.00	80.00	90.00
2	15.56	11.11	13.33	16.67	10.00
3	2.22	4.44	6.67	3.33	0.00
4	0.00	0.00	0.00	0.00	0.00
Tail					
1	77.78	86.67	76.67	83.33	86.67
2	17.78	11.11	23.33	10.00	10.00
3	4.44	2.22	0.00	6.67	3.33
4	0.00	0.00	0.00	0.00	0.00

Table 9. *Cont.*

Items	Main Effect				
	D			IL	
	HD	ND	LD	LX	HX
Shoulder					
1	77.78	88.89	80.00	83.33	86.67
2	15.56	4.44	6.67	10.00	13.33
3	6.67	6.67	13.33	6.67	0.00
4	0.00	0.00	0.00	0.00	0.00
Total body					
1	88.89	88.89	86.67	86.67	93.33
2	8.89	4.44	3.33	10.00	6.67
3	2.22	6.67	10.00	3.33	0.00
4	0.00	0.00	0.00	0.00	0.00

Abbreviation: D, density; IL, illuminance; D × IL, density × illuminance; HD (lower than 0.50 $m^2/100$ kg); ND (0.50 m^2~0.83 $m^2/100$ kg); LD (higher than 0.83 $m^2/100$ kg LX, low illuminance; HX, high illuminance; For lesion scoring, a four-point scale from 0 to 3 was used [17]. Score 0 given for a body part if no scratches were found; Score 1 given if fewer than five superficial scratches; Score 2 given if five to ten superficial scratches or fewer than five deep scratches; Score 3 describes a body area with more than ten superficial scratches or more than five deep scratches.

Table 10. Effects of lairage stocking density and illuminance on skin lesion score at high temperature.

Items	Main Effect				
	D			IL	
	HD	ND	LD	LX	HX
Skin Lesion Score					
Ear					
1	76.67	86.67	90.00	84.44	84.44
2	13.33	10.00	10.00	13.33	8.89
3	10.00	3.33	0.00	2.22	6.67
4	0.00	0.00	0.00	0.00	0.00
Tail					
1	73.33	76.67	80.00	77.78	75.56
2	20.00	20.00	20.00	17.78	22.22
3	6.67	3.33	0.00	4.44	2.22
4	0.00	0.00	0.00	0.00	0.00
Shoulder					
1	86.67	86.67	90.00	88.89	86.67
2	6.67	6.67	10.00	6.67	8.89
3	6.67	6.67	0.00	4.44	4.44
4	0.00	0.00	0.00	0.00	0.00
Total body					
1	80.00	86.67	90.00	86.67	84.44
2	10.00	10.00	10.00	11.11	8.89
3	10.00	3.33	0.00	2.22	6.67
4	0.00	0.00	0.00	0.00	0.00

Abbreviation: D, density; IL, illuminance; D × IL, density × illuminance; HD (lower than 0.50 $m^2/100$ kg); ND (0.50 m^2~0.83 $m^2/100$ kg); LD (higher than 0.83 $m^2/100$ kg). LX, low illuminance; HX, high illuminance. For lesion scoring, a four-point scale from 0 to 3 was used [17]. Score 0 given for a body part if no scratches were found; Score 1 given if fewer than five superficial scratches; Score 2 given if five to ten superficial scratches or fewer than five deep scratches; Score 3 describes a body area with more than ten superficial scratches or more than five deep scratches.

4. Discussion

Observed behavioral differences between pigs stocked at LD and HD indicated that pig behavior was sensitive to environmental conditions. In the LT condition, HD led to reduced aggression and overlap behaviors, indicating a potential decrease in stress levels among pigs. According to [18], low stocking density in lairage can lead to cold stress. Conversely, in the HT condition, a contrasting behavioral response was observed, with HD-stocked pigs displaying higher levels of standing, sitting, and aggression behavior than LD-stocked pigs. Increased activity and aggression among HD-stocked pigs at HT might be attributed to heat stress (HS) and heightened competition for resources, such as space and ventilation [19]. The HS can exacerbate behavioral responses, leading to restlessness and aggression as pigs seek relief from thermal discomfort [20]. Observed increases of standing and sitting behaviors might reflect efforts to dissipate heat and regulate body temperature, while heightened aggression may stem from increased frustration and irritability in crowded conditions [21]. Lying behavior serves as a crucial indicator of pig welfare as it reflects the animals' ability to rest and recuperate in lairage environment [7]. Interestingly, pigs stocked at LD at HT exhibited higher levels of lying behavior than HD-stocked pigs. Illuminance is another crucial factor that can significantly impact pig behavior, welfare, and overall productivity. Pigs are known to be sensitive to changes in light conditions, which can influence their physiological and psychological states. Appropriate lighting conditions can help reduce stress and aggression among pigs, promote more regular feeding and resting patterns, and enhance overall wellbeing [22]. In this study, pigs stocked with LX showed high tendency of lying behavior than those stocked with HX at LT. According to these findings, stocking density impacts behavior differently depending on temperature conditions, emphasizing the importance of adaptive management practices.

In HT conditions, pigs stocked at LD exhibited lower backfat thickness than those stocked at HD. This finding suggests that lower stocking densities may contribute to leaner carcasses, potentially due to increased physical activity and reduced adipose tissue deposition in less crowded environments [23,24]. The 1+ grade typically indicates superior meat quality characteristics, such as higher marbling and tenderness, which are desirable attributes in the pork industry [25,26]. The numerically higher prevalence of 1+ grade in LD-stocked pigs suggests potential benefits associated with lower stocking densities, including improved meat quality and carcass characteristics. This finding suggests that HS might influence adipose tissue deposition, potentially impacting carcass composition and yield [27]. However, further research is needed to elucidate the underlying physiological mechanisms driving these differences and to assess their implications for overall meat quality.

pH is a critical indicator of meat quality. Higher pH values are often associated with reduced acidity and improved meat tenderness and juiciness [28]. The observed increase in pH among HD-stocked pigs suggests potential differences in muscle metabolism and postmortem processes related to stocking density. Conversely, in HT conditions, pigs stocked at LD displayed superior pork quality characteristics to HD-stocked pigs. LD-stocked pigs exhibited higher pH, WHC, DL, and CL, indicating improved water retention and reduced cooking losses. Improved WHC, DL, and CL reflect better moisture retention and cooking properties, which are desirable traits of pork products [29]. These temperature-dependent effects underscore the importance of considering environmental conditions when evaluating pork quality parameters and implementing management strategies to optimize product attributes.

In LT conditions, pigs stocked at LD exhibited higher cortisol levels than those stocked at HD. Elevated cortisol levels indicate increased stress and arousal in pigs, which may result from factors such as crowding, competition for resources, and environmental discomfort [30]. Higher cortisol levels observed in LD-stocked pigs suggest that higher stocking densities may exacerbate stress levels and welfare concerns in pigs exposed to cooler temperatures during lairage. Conversely, in HT conditions, a contrasting pattern was observed, with LD-stocked pigs displaying lower cortisol levels than HD-stocked pigs. This unexpected finding suggests that lower stocking densities may mitigate stress responses in pigs exposed to higher temperatures during lairage. HS can exacerbate physiological stress responses, including cortisol secretion, as animals struggle to thermoregulate and cope with thermal discomfort [20]. Lower cortisol levels observed in LD-stocked pigs at HT may indicate reduced stress and improved welfare under less crowded conditions.

Skin damage can act as an animal welfare indicator and reflect the quality of the animal's social and physical environment [31]. Unadapted social environment can cause excessive fighting (e.g., after mixing), ultimately resulting in skin damage [32]. The higher prevalence of skin lesions observed in LD-stocked pigs at LT might be attributed to increased competition for resources and space within the lairage environment. Lower stocking densities may afford pigs more room to move and rest comfortably, potentially reducing the likelihood of skin abrasions and injuries [11]. In contrast, numerically higher skin lesions observed in HD-stocked pigs at HT suggest that HS and overcrowding may contribute to increased skin injuries in these animals. High ambient temperatures can exacerbate stress levels and discomfort among pigs, leading to heightened agitation and aggression, which may result in skin abrasions and lesions [33]. Additionally, reduced space availability in high-density stocking scenarios may exacerbate social stress and competition, further increasing the likelihood of skin injuries.

Overall, the findings of this study underscore the complexity of interactions between lairage stocking density, illuminance, and temperature conditions, and their effects on pig behavior, carcass quality, pork quality, blood profile, and skin lesion scores. Adopting integrated management approaches that consider these interactions is essential for promoting both animal welfare and meat quality in pig production systems. Further research is needed to elucidate the underlying mechanisms driving these effects and develop targeted interventions to optimize pig welfare and product quality in diverse production environments.

5. Conclusions

Pigs exposed to HD (lower than 0.5 $m^2/100$ kg) at HT during preslaughter caused acute stress. Also, pigs exposed to LD (lower than 0.5 $m^2/100$ kg) at LT during preslaughter caused acute stress. Based on the obtained results, stocking of too-high (lower than 0.50 $m^2/100$ kg) density at HT and stocking of too-low (higher than 0.83 $m^2/100$ kg) density at LT are generally not good for meat quality and animal welfare. Therefore, higher stocking density (lower than 0.50 $m^2/100$ kg) at LT (lower than 10 °C in winter) and lower stocking (higher than 0.83 $m^2/100$ kg) density at HT (higher than 24 °C in summer) are recommended.

Author Contributions: D.-C.S. and J.-H.L. conducted the experiment and wrote the manuscript. W.Y., S.-Y.C., S.-H.P., K.-H.J. and H.K. helped to conduct animal trials and laboratory work and helped to revise the manuscript. J.-H.C. was the principal investigator and wrote the last version of the manuscript. All authors have read and agreed to the published version of the manuscript.

Funding: This work was carried out with the support of "Cooperative Research program for Agriculture Science & Technology Development (Project No. PJ016214)" Rural Development Administration, Republic of Korea.

Institutional Review Board Statement: The experimental protocol was approved (CBNUA-2185-23-02) by the Institutional Animal Care and Use Committee of Chungbuk National University, Cheongju, Republic of Korea.

Informed Consent Statement: Not applicable.

Data Availability Statement: The original contributions presented in the study are included in the article, further inquiries can be directed to the corresponding author.

Conflicts of Interest: Author Won Yun was employed by the company Woosung Feed Co., Ltd., Daejeon, Korea. The remaining authors declare that the research was conducted in the absence of any commercial or financial relationships that could be construed as a potential conflict of interest.

References

1. Lebret, B.; Meunier-Salaun, M.C.; Foury, A.; Mormède, P.; Dransfield, E.; Dourmad, J.Y. Influence of rearing conditions on performance, behavioral, and physiological responses of pigs to preslaughter handling, carcass traits, and meat quality. *J. Anim. Sci.* **2006**, *84*, 2436–2447. [CrossRef] [PubMed]
2. Faucitano, L.; Conte, S.; Pomar, C.; Paiano, D.; Duan, Y.; Zhang, P.; Drouin, G.; Rina, S.; Guay, F.; Devillers, N. Application of extended feed withdrawal time preslaughter and its effects on animal welfare and carcass and meat quality of enriched-housed pigs. *Meat. Sci.* **2020**, *167*, 108163. [CrossRef] [PubMed]
3. Čobanović, N.; Stajković, S.; Blagojević, B.; Betić, N.; Dimitrijević, M.; Vasilev, D.; Karabasil, N. The effects of season on health, welfare, and carcass and meat quality of slaughter pigs. *Int. J. Biometeorol.* **2020**, *64*, 1899–1909. [CrossRef] [PubMed]
4. Cornish, A.; Raubenheimer, D.; McGreevy, P. What we know about the public's level of concern for farm animal welfare in food production in developed countries. *Animals* **2016**, *6*, 74. [CrossRef] [PubMed]
5. Cornale, P.; Macchi, E.; Miretti, S.; Renna, M.; Lussiana, C.; Perona, G.; Mimosi, A. Effects of stocking density and environmental enrichment on behavior and fecal corticosteroid levels of pigs under commercial farm conditions. *J. Vet. behavior.* **2015**, *10*, 569–576. [CrossRef]
6. Maes, D.G.; Dewulf, J.; Piñeiro, C.; Edwards, S.; Kyriazakis, I. A critical reflection on intensive pork production with an emphasis on animal health and welfare. *J. Anim. Sci.* **2020**, *98* (Suppl. S1), S15–S26. [CrossRef] [PubMed]
7. Pietrosemoli, S.; Tang, C. Animal welfare and production challenges associated with pasture pig systems: A review. *Agriculture* **2020**, *10*, 223. [CrossRef]
8. O'connor, E.A.; Parker, M.O.; McLeman, M.A.; Demmers, T.G.; Lowe, J.C.; Cui, L.; Davey, E.L.; Owen, R.C.; Wathes, C.M.; Abeyesinghe, S.M. The impact of chronic environmental stressors on growing pigs, Sus scrofa (Part 1): Stress physiology, production and play behaviour. *Animal* **2010**, *4*, 1899–1909. [CrossRef]
9. Morgan, K.N.; Tromborg, C.T. Sources of stress in captivity. *Appl. Anim. Behav. Sci.* **2007**, *102*, 262–302. [CrossRef]
10. Kuczynski, T.; Blanes-Vidal, V.; Li, B.; Gates, R.S.; de Alencar Naas, I.; Moura, D.J.; Berckmans, D.; Banhazi, T.M. Impact of global climate change on the health, welfare and productivity of intensively housed livestock. *Int. J. Agric. Biol. Eng.* **2011**, *4*, 1–22.
11. Hennig-Pauka, I.; von Altrock, A. Managing housing and stocking density to optimize health, welfare and production in pig herds. In *Optimizing Pig Herd Health and Production*; Maes, D., Segalés, J., Eds.; Burleigh Dodds Science Publishing: Cambridge, UK, 2023; pp. 333–376.
12. Vitali, M.; Bosi, P.; Santacroce, E.; Trevisi, P. The multivariate approach identifies relationships between pre-slaughter factors, body lesions, ham defects and carcass traits in pigs. *PLoS ONE.* **2021**, *16*, e0251855. [CrossRef]
13. Vitali, M.; Sardi, L.; Martelli, G.; Nannoni, E. Literature review on the pre-slaughter welfare of italian heavy pigs. *Animals* **2021**, *11*, 3352. [CrossRef] [PubMed]
14. AOAC [Association of Official Analytical Chemists]. *International Official Method of Analysis of AOAC International*, 16th ed.; AOAC International: Washington, DC, USA, 2000.
15. Oliveira, T.F.B.; Rivera, D.F.R.; Mesquita, F.R.; Braga, H.; Ramos, E.M.; Bertechini, A.G. Effect of different sources and levels of selenium on performance, meat quality, and tissue characteristics of broilers. *J. Appl. Poult. Res.* **2014**, 23, 15–22. [CrossRef]
16. Koćwin-Podsiadła, M.; Krzęcio, E.; Przybylski, W. Pork quality and methods of its evaluation- a review. *Pol. J. Food Nutr. Sci.* **2006**, *15*, 241–248.
17. Laves, J.; Herbrandt, S.; Van Meegen, C.; Kemper, N.; Fels, M. Effect of pens with an elevated platform on space utilization, skin lesions and growth performance in nursery pigs. *Animal* **2021**, *15*, 100002. [CrossRef] [PubMed]
18. Weeks, C.A. A review of welfare in cattle, sheep and pig lairages, with emphasis on stocking rates, ventilation and noise. *Anim. Welf.* **2008**, *17*, 275–284. [CrossRef]
19. Fraqueza, M.J.; Roseiro, L.C.; Almeida, J.; Matias, E.; Santos, C.; Randall, J.M. Effects of lairage temperature and holding time on pig behaviour and on carcass and meat quality. *Appl. Anim. Behav. Sci.* **1998**, *60*, 317–330. [CrossRef]
20. Endris, M.; Feki, E. Review on effect of stress on animal productivity and response of animal to stressors. *J. Anim. Vet. Adv.* **2021**, *20*, 1–14.
21. Goumon, S.; Brown, J.A.; Faucitano, L.; Bergeron, R.; Widowski, T.M.; Crowe, T.; Connor, M.L.; Gonyou, H.W. Effects of transport duration on maintenance behavior, heart rate and gastrointestinal tract temperature of market-weight pigs in 2 seasons. *J. Anim. Sci.* **2013**, *91*, 4925–4935. [CrossRef]
22. Marchant-Forde, J.N.; Lay, D.C., Jr.; McMunn, K.A.; Cheng, H.W.; Pajor, E.A.; Marchant-Forde, R.M. Postnatal piglet husbandry practices and well-being: The effects of alternative techniques delivered separately. *J. Anim. Sci.* **2009**, *87*, 1479–1492. [CrossRef] [PubMed]
23. Simitzis, P.E.; Kalogeraki, E.; Goliomytis, M.; Charismiadou, M.A.; Triantaphyllopoulos, K.; Ayoutanti, A.; Niforou, K.; Hager-Theodorides, A.L.; Deligeorgis, S.G. Impact of stocking density on broiler growthperformance, meat characteristics, behavioural components and indicators of physiological and oxidative stress. *Br. Poult. Sci.* **2012**, *53*, 721–730. [CrossRef] [PubMed]
24. Li, W.; Wei, F.; Xu, B.; Sun, Q.; Deng, W.; Ma, H.; Bai, J.; Li, S. Effect of stocking density and alpha-lipoic acid on the growth performance, physiological and oxidative stress and immune response of broilers. *Asian-Australas. J. Anim. Sci.* **2019**, *32*, 1914. [CrossRef] [PubMed]
25. Cheng, W.; Cheng, J.H.; Sun, D.W.; Pu, H. Marbling analysis for evaluating meat quality: Methods and techniques. *Compr. Rev. Food Sci. Food Saf.* **2015**, *14*, 523–535. [CrossRef]

26. Klehm, B.J.; King, D.A.; Dilger, A.C.; Shackelford, S.D.; Boler, D.D. Effect of packaging type during postmortem aging and degree of doneness on pork chop sensory traits of loins selected to vary in color and marbling. *J. Anim. Sci.* **2018**, *96*, 1736–1744. [CrossRef] [PubMed]
27. De Antonio, J.; Fernandez-Alarcon, M.F.; Lunedo, R.; Squassoni, G.H.; Ferraz, A.L.J.; Macari, M.; Furlan, L.R. Chronic heat stress and feed restriction affects carcass composition and the expression of genes involved in the control of fat deposition in broilers. *J. Agric. Sci.* **2017**, *155*, 1487–1496. [CrossRef]
28. Kim, Y.H.B.; Warner, R.D.; Rosenvold, K. Influence of high pre-rigor temperature and fast pH fall on muscle proteins and meat quality: A review. *Anim. Prod. Sci.* **2014**, *54*, 375–395. [CrossRef]
29. Xu, D.; Wang, Y.; Jiao, N.; Qiu, K.; Zhang, X.; Wang, L.; Wang, L.; Yin, J. The coordination of dietary valine and isoleucine on water holding capacity, pH value and protein solubility of fresh meat in finishing pigs. *Meat Sci.* **2020**, *163*, 108074. [CrossRef] [PubMed]
30. Mkwanazi, M.V.; Ncobela, C.N.; Kanengoni, A.T.; Chimonyo, M. Effects of environmental enrichment on behaviour, physiology and performance of pigs—A review. *Asian-Australas. J. Anim. Sci.* **2019**, *32*, 1. [CrossRef] [PubMed]
31. Leeb, C.; Troxler, J.; Schuh, M. Skin lesions and callosities of group housed pregnant sows—Animal related welfare indicators. *Acta Agric. Scand.* **2001**, *51*, 82–87. [CrossRef]
32. Smulders, D.; Hautekiet, V.; Verbeke, G.; Geers, R. Tail and ear biting lesions in pigs: An epidemiological study. *Anim. Welf.* **2008**, *17*, 61–69. [CrossRef]
33. Fraser, D.; Duncan, I.J.; Edwards, S.A.; Grandin, T.; Gregory, N.G.; Guyonnet, V.; Hemsworth, P.H.; Huertas, S.M.; Huzzey, N.G.; Mellor, D.J.; et al. General principles for the welfare of animals in production systems: The underlying science and its application. *Vet. J.* **2013**, *198*, 19–27. [CrossRef]

MDPI AG
Grosspeteranlage 5
4052 Basel
Switzerland
Tel.: +41 61 683 77 34

Animals Editorial Office
E-mail: animals@mdpi.com
www.mdpi.com/journal/animals

www.ingramcontent.com/pod-product-compliance
Lightning Source LLC
Chambersburg PA
CBHW042011260926
51850CB00020B/3115